LANDSCAPE ECOLOGY IN ACTION

Landscape Ecology in Action

by

Almo Farina

Lunigiana Museum of Natural History, Aulla, Italy
Venice University, Venice, Italy
Urbino University, Urbina, Italy

KLUWER ACADEMIC PUBLISHERS
DORDRECHT / BOSTON / LONDON

A C.I.P. Catalogue record for this book is available from the Library of Congress.

ISBN 0-7923-6165-2 (HB)
ISBN 0-7923-6166-0 (PB)

Published by Kluwer Academic Publishers,
P.O. Box 17, 3300 AA Dordrecht, The Netherlands.

Sold and distributed in North, Central and South America
by Kluwer Academic Publishers,
101 Philip Drive, Norwell, MA 02061, U.S.A.

In all other countries, sold and distributed
by Kluwer Academic Publishers,
P.O. Box 322, 3300 AH Dordrecht, The Netherlands.

Printed on acid-free paper

Printed in the Netherlands.

Contents

Foreword by Frank Golley VII

Preface IX

PART I : Principles, theory, and methods 1

1. Preface 5
1.1 The birth of landscape ecology: An important heritage 9
1.2 Towards an unified theory 11
1.3 Definition of landscape 11
1.4 The dimension of a landscape 28
1.5 Landscape typologies and functions 36
1.6 Emerging properties of landscape 43
1.7 Ecological concepts incorporated into the landscape paradigm 66
1.8 Processes in landscape 82

PART II: Landscape evaluation 107

2. Preface 111
2.1 Exploring evaluation procedures 112
2.2 Landscape evaluation and scale 114
2.3 Concepts and procedures of landscape evaluation 117
2.4 Tools for evaluating the structure and functions of land mosaics 164

PART III: Management and conservation of Landscapes 193

3. Preface 197
3.1 Some concepts in management and conservation 197
3.2 Landscape approach to management and conservation: Exploring new dimensions 203
3.3 Large scale landscape management 204
3.4 Multiple scale investigation and management 205
3.5 Constraints in managing 208
3.6 Types of management 210
3.7 Managing patterns 213
3.8 Managing processes 220

3.9 Managing functional areas 231
3.10 Conserving the landscape 241
3.11 Perspectives in management and conservation 260

Acting in landscape ecology: Concluding remarks 265

References 269

Glossary & Acronyms 293

Species Index 299

Author Index 301

Subject Index 307

Foreword

In my office I am encased in bookshelves which hold an accumulation of literature on ecology that represents the papers and books over the last 50 years. My students enjoy rummaging through this collection because it contains a record of the history of ecology and is full of surprises. Some of the most recent material pertains to landscape ecology, a subject that literally emerged fully active at the Veldhoven International Congress organized by the landscape ecologists of The Netherlands in 1981. The subject has developed quickly. It has one or more journals, which publish short works. It has a series of text books. And, it has just begun a series on monographs. One of the textbooks in landscape ecology is titled Principles and Methods in Landscape Ecology and was written by the Italian ecologist Almo Farina in 1998. My students like this text especially well because it is direct, to the point and comprehensive. "Farina" is on loan much of the time.

In the present volume Almo Farina again addresses the subject of Landscape Ecology but from a different perspective than he took in his textbook. Landscape in Action focuses on the application of the principles and concepts to problem solving. The two books make a pair, with the first technical and conceptual and the second applied to problems of land and water at large scale. Landscape Ecology in Action begins with a recitation of principles and theories reminding or educating us to the meanings of technical terms, to scaling landscape, to properties and processes of landscape and to the connections between landscape ecology in particular and ecology in general. Then Farina turns to landscape evaluation, which is not only viewing the landscape from photographs or from space but also interpreting what is seen. In other words, the analysis addresses what the images mean in space and time. This step is a major part of landscape ecology and Farina is correct in emphasizing it at the beginning. The remaining pages of the book focus on broad problems of management and conservation of landscapes. In this part Farina uses his theme, carried from the introductory material, of considering scale first, and then pattern and process.

I have enjoyed reading Landscape Ecology in Action. The text is full of illustrations and examples of the principles and cases mentioned. There is a nice balance between the approaches. It makes an excellent companion to the textbook.

In this volume Almo Farina provides us a service in bringing together information on a broad subject from an international perspective. He is well qualified to write this book because he has exceptionally broad international experience. For example, he is one of the few, or maybe the only member of the United States association of landscape ecologists, to have attended every annual meeting of the association. As Secretary General of the International Association for Ecology he organized and managed the VII Congress of Ecology held in 1998 in the city of Florence, Italy. These experiences have been used effectively in his writing. I predict that Landscape Ecology in Action will take its place in the list of important books in this subject and will attract the students, as well as the professional reader.

Frank B. Golley

Preface

Recent decades have been characterized by a globalization of many problems and information. Human society is increasingly aware of the finite dimension of natural resources, and new technologies applied to food production, health care and disease control, have improved the quality of human life in many countries. Despite these positive results many processes are markedly reducing the carrying capacity of our ecosphere, eroding resources, biodiversity , cultural diversity and ecodiversity (Naveh 1994).

The battle between an unsustainable development and a permanent unsustainable under-development is very uncertain (Lubchenco et al. 1991). The magnitude of the gap between how the world works, modern societal development and resource use and depletion, is increasing. It is clear that new forces function as drivers of the processes at work on our planet. It is human activity (development/under-development) that covers, masks and conditions many relevant and life-strategic natural processes.

Changes are moving faster and at a larger scale than in the past, producing unintended effects in the ecosphere. Solar energy input and its related cycles (water evaporation and condensation), and the general climatic events of terrestrial and marine ecosystems are still independent of human intrusion, but increasing fuel emissions and other chemical pollutants are likely to significantly disturb such cycles in the near future. Human activity is moving from a local scale to a global scale, impacting more and more on the dimension within which global processes operate. When the two scales become closer we can expect an increase of feedback between them.

Technology has been employed to force natural processes creating unsustainable situations and acting always in terms of resistance engineering systems. Dikes, marsh reclamation, water pumping stations, and consumption of fossil water are some examples.

Economy and information are robust drivers of human-related processes. Economy tests the possibilities of accumulating and manipulating resources and information stored and processed by new computer network (O'Neill 1997). The combined action of economy and information greatly accelerates the speed of changes in the real world which move natural, cultural, financial, and technological resources.

A new era of large transformations is expected and human society will run at a higher speed compared with natural processes. This time-lag process is destined to be a new challenge for our future. The human-related processes are moving progressively further from the natural rhythms and ecosystem scale. The distance between human society and the natural system will increase bringing consequences which will modify the entire planet.

Reaction to these global challenges requires new scientific knowledge in order to predict the future scenarios in which human society should live in an integrated and compatible way.

Ecology has played a fundamental role in describing patterns and processes acting at the ecosystem scale but, as recently warned during the VII International Congress in Florence (Farina 1999), ecology has to fill a deep gap between a paradigmatic, theoretical vision of the world and the real world. This passage is essential for the future of ecological systems and for the persistence and quality of human life.

Thus it seems urgent and realistic to insert into the ecological realm many other processes which have been previously confined into separate "cultural" compartments. A new technology, a real revolution of thinking, should use new approaches, to mitigate human intrusion, to mimic natural systems, to assure the resilient characteristics of systems and to use natural strategies in conserving gene diversity and biomass in crop and animal production. Although the need for an integrated approach is recognized by many scientists (Shea et al. 1998) , this new vision of the possible human technological world, compatible with the natural system, is not yet popular in the ecological sciences.

Remarkable progress in the direction of a more integrated real-world oriented ecology has been made by Landscape Ecology, a new branch of ecology (Naveh & Lieberman 1984, Forman & Godron 1986, Zonneveld 1995, Farina 1998). Landscape ecology provides the possibility of linking together patterns and processes acting in the real world as part of a unified theory of earth functioning. This ecological discipline can find the tools to relate and integrate the knowledge gained by separate scientific disciplines and explore the relevant processes across a set of spatial and temporal scales.
Landscape ecology, born from a new integration between the more advanced ecological theories, models and paradigms and the human dimension, appears a promising path in the jungle of sophisticated, hyper-specialized and separated cultural components of biological and ecological sciences.

The discipline of landscape ecology may be considered as the study of processes happening at large scale or at a specific spatial dimension in which humans live. This represents the scalar niche in which most of landscape ecology resides. Landscape ecology considers an important factor to be the space in which the phenomena appear and work, but has also extensively utilized and incorporated many theories, paradigms and models of "ecosystemic" ecology. The landscape dimension is now very popular in biological as well as in ecological conservation and is appearing also in educational curricula.

The aim of this book is to illustrate the utility of this new approach in which natural and human-related processes are investigated, described and interpreted across a dynamic scale. It is the intention also to orient decision-makers, planners and educators to a new vision of the environment as a multi-scale overlap of species-specific living systems.

I am convinced that the landscape paradigm, although in many aspects quite weak, is a great opportunity to address the real world. The novelty of this approach lies in the possibility of analyzing natural, wild ranges as well as disturbed and human modified areas, in order to restore quality and biodiversity, and to mitigate some human impacts. For this reason my aim in writing this book is not to speculate on the landscape as a distinct ecological paradigm, in contraposition with the ecosystem approach, just to gain a cultural space, but to explore the potentialities and limits of this new ecological tool and to open an honest confrontation between old ecological paradigms like niche, ecosystem, and island biogeography, and the new one, connected with the spatial distribution of processes and organisms (mosaic theory).

This represents a contribution toward conveying a fringe discipline into a more organic and compacted theoretical body, useful for many purposes and able to link the "traditional" ecological perspectives with more advanced theories based on the spatial arrangement and connection of physical and biological objects.

The style of the book is simple and the content organized in order to serve a broad audience and especially for use as a textbook for undergraduate and graduate students, as well as for planners and decision-makers. Due to the relative novelty of landscape ecology the first part will be devoted to explaining and exemplifying principles and methods, trying to clarify the relationships but also the gaps between ecosystem ecology and this new ecology.

Concepts like habitat and niche are not considered in landscape literature or are ignored because they are not as compelling as the landscape paradigm. Without a clarification on such topics, it is not possible to make progress with this new approach.

The second part will describe the evaluation procedures, the best synthetic indicators like integrity and fragility, and the most affordable metrics used in landscape analysis.

The third part will deal with management and conservation measures applied especially at the large scale (human) landscape level, focusing on the problems connected with maintaining process dynamics.

An extensive reference of the existing literature in the theoretical and applied field of nature conservation (resources, species and communities) is provided. Key studies illustrate the successes and limits of landscape ecology in different fields of investigation, management and conservation of natural and human modified resources.

Almo Farina

PART I

Principles, theory, and models

Contents

1. Preface 5

1.1 The birth of landscape ecology: An important heritage 9

1.2 Towards an unified theory 11

1.3 Definition of landscape 11
1.3.1 Different perceptions of the landscape 12
1.3.1.1 The "process" perception 13
1.3.1.2 The human perception of landscape 15
1.3.1.3 The animal perception of landscape 16
1.3.1.4 The plant (geo-botanical) perception of landscape 19
1.3.1.5 Natural versus human landscape 20
1.3.1.6 Landscape and eco-fields 21
1.3.1.7 A new model for combining ecosystem approach and the landscape approach 22
1.3.2 The landscape paradigm as a bridge between ecological disciplines: Confronting concepts 24
1.3.2.1 Ecosystem versus landscape 25
1.3.2.2 Habitat versus landscape patch 27

1.4 The dimension of a landscape 28
1.4.1 Scaling the landscape 29
1.4.1.1 Scaling communities within the landscape 29
1.4.1.2 Scaling populations within the landscape 31
1.4.1.3 Scaling complexity 34

1.5 Landscape typologies and functions 36
1.5.1 Sky-scapes 36
1.5.2 Water-scapes 37
1.5.3 Terrestrial-scape 38
1.5.3.1 Geobotanical landscape 38
1.5.3.2 Animal landscape 39
1.5.3.3 Human landscape 39
1.5.3.4 Vertical landscapes 39
1.5.3.5 Suspended landscapes 41
1.5.3.6 Other classifications of landscape 42

1.6 Emerging properties of landscape 43
1.6.1 Spatial autocorrelation, a measure of distributional patterns 44
1.6.2 A new view of matrix 45
1.6.3 The Patch paradigm: Definition 46
1.6.3.1 Population patches 48

1.6.3.2 Patch dynamics 48
1.6.3.3 Micropatch dynamics 49
1.6.3.5 Ecotope versus patch 50
1.6.4 Corridors 52
1.6.4.1 The importance of linear habitats 57
1.6.4.2 Corridors: an open issue 58
1.6.4.3 Perspectives 60
1.6.5 Ecotones, patterns and processes within discontinuities 61
1.6.5.1 Origin of ecotones 63
1.6.5.2 Structural and functional attributes of ecotones 64
1.6.5.3 The role of ecotones in the landscape 65

1.7 Ecological concepts incorporated into the landscape paradigm 66
1.7.1 Hierarchy theory 66
1.7.2 Diversity into landscape 71
1.7.2.1 Diversity in the soil 71
1.7.2.2 Diversity and information 72
1.7.2.3 Regional biodiversity 73
1.7.2.4.The role of biodiversity in assuring landscape functionality 73
1.7.2.5 Biodiversity hotspots 74
1.7.3 Source-Sink model 75
1.7.4 Metapopulation theory and landscape 81

1.8 Processes in landscape 82
1.8.1 Physical and biological processes 83
1.8.2 Centripetal and centrifugal processes 83
1.8.3 Local and regional processes, some examples and evidences 85
1.8.3.1 Decomposition processes across soil and river systems 85
1.8.3.2 Snow cover and vegetation 87
1.8.4 Regional landscape approach 87
1.8.5 Connectedness and connectivity 90
1.8.5.1 Patchiness and resource allocation 93
1.8.5.2 Meta-community concept 95
1.8.5.3 Patch selection 95
1.8.5.4 Organism density and heterogeneity 95
1.8.5.5 Foraging behaviour and patchiness 96
1.8.5.6 Predatory-prey interactions and habitat fragmentation 96
1.8.6 Fragmentation versus aggregation 98
1.8.7 Homogeneity and heterogeneity 99
1.8.8 Disturbance 99
1.8.8.1 Disturbance and diversity 103
1.8.8.2 Disturbance processes important for large scale landscapes 104

1. Preface

When we observe a detail of an object, it appears simple and neat, but when observed as part of the world, it appears complex. As recently argued by Goldenfeld & Kadanoff (1999), "the world contains many examples of complex "ecologies" at all levels". An object is complex when it has structures composed of different working parts. Each part can function in a different way according to the context. Many systems are both complex and chaotic (unpredictable), and the entire planet can be considered in such way.

The real world is complex, this complexity gives rise to a system which is self-maintaining, self-regulating and self-defensive with regards to external or internal perturbations. Complexity is more than the summations of the components, and represents the functioning of the components and the interactions between the parts.

Complexity has spatial and functional components and often the approach to this complexity discourages people and as a consequence we have disciplines which are oriented more towards investigating separated pieces of complexity than towards analyzing the entire block.

Complexity can be appreciated at all spatial and temporal scales and pertains to physical (Goldenfeld & Kadanoff 1999, Werner 1999), chemical (Whitesides & Ismagilov 1999), biological (Weng et al. 1999, Parrish & Eelstein-Keshet 1999), as well as economic processes (Arthur 1999).

A sand deposit along a river may appear as a homogeneous patch. But if we look with more attention using for instance, methods of remote sensing, emerging patterns appear which describe the erosive and sand transportation mechanisms across a river during flooding events. Turbulence, current distribution, velocity and amount of water are key elements that work at different spatial and temporal scales to produce the complex river deposit (Fig. 1.1). This example should clarify the way in which the world is complex. Of this complexity we perceive only an overview of the all processes in action. Many natural processes can create complex structures even in simple contexts and can obey simple laws even in complex situations. The natural evolution of ecology from the study of relationships between organisms and their environment is moving towards the study of ecological complexity "tout court". But when we address this complexity we find enormous difficulty in approaching such a target. We understand conceptually the complexity but at the same time we need to find practical tools to study such aspects of the ecological system. Separating complexity into different components is a very frustrating exercise that has often represented the major cause of knowledge failure. Nevertheless we can't escape this difficulty and landscape ecology paradigms seem very useful.

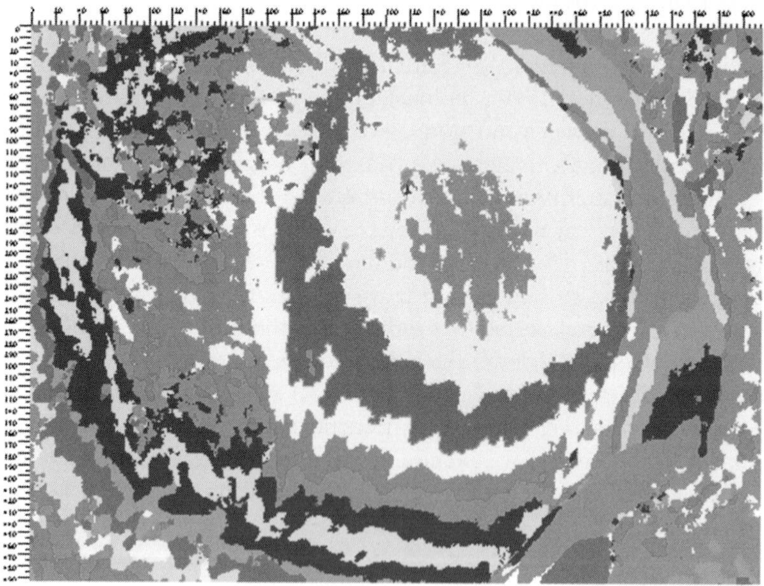

Fig. 1.1 - Example of a complex system. A recent gravel and sand deposit along the Magra river (Italy). A : aerial view of the deposit. The active bed is on the right part of the image. B: The same image after automatic classification and GIS elaboration. (The image clearly shows the stratification of at least seven events (energy waves) during the flood period. From this complex sedimentation will spring, in a short time, patches of different vegetation types.

Two approaches are actually available in landscape ecology: the organismic approach and the process approach, but the two approaches have rarely been considered in a common framework. Both approaches require a multiscalar attitude because the explicit characteristics of a function or of a species trait appear only when observed at the right spatio-temporal scale.

Ecological research is full of trivial questions and elegant explanations. One such trivial question could be : how important is the organismic approach compared with the process approach? The effects of a water drop are negligible when compared with beaver disturbance long a stream. But again the beaver activity is negligible when compared with the flooding regime of a stream. The importance of the different components of complexity is again a matter of scale. In particular the spatial scale contains relevant information to address complexity. The exercise of moving up or down the spatial scale has created modern landscape ecology.

The organismic approach has been considered deeply by scientists and there exists a powerful literature on the subject. By this approach plants, animals, bacteria and virus are considered like as customers of ecological systems. These organisms interfere in a species-specific way with the landscape and have a specific perception.

The process approach represents the more integrated and related approach to the concept of complexity. We can imagine that process intercepts much better with complexity than organisms. This is especially true if we consider the spatial and temporal scale at which an organism intercept it's surroundings. A deer or an ant relate to the environment (habitat) inside an ecological niche. The dimension of such a niche is species-specific and can change little during the life time of a species. But a process like water erosion can be observed at a very broad range of spatial and temporal scales. When a water drop falls, it produces the first small erosive event. Moving down-slope the erosive process changes it's scale and the effects on it's surroundings the process is completed when water reaches the sea or a lake. However, the erosive process can find a linkage with other processes, such as marine erosion/sedimentation, and if the flux of transportation by rivers is scaled to seasons the cycle of erosion/sedimentation is primarily linked to sea movements and secondarily to the cycle of river transportation.

From this example it emerges quite clearly that processes have a strong relationship with each other and that this relationship is maintained by a system of filters that transport and modify the information. The majority of ecological processes have a separate engine that activates the event, but it is in the context in which the event acts that the relationship with other processes and spatial patterns occurs.

Without doubt the complexity that we can appreciate is related to our capacity to track the physical and biological components, and by using a landscape approach our capacity is greatly expanded. We need to use flexible and more real-world related approaches, without the conceptual biases that "reductionistic" research imposes.

Landscape ecology seems, today to be one branch of ecology that can strongly contribute to the study of complexity (physical, biological and ecological). Landscape paradigms refer to hierarchical theory and to the effects of spatial arrangement of an object into a space. Landscape ecology adopts a multiscalar approach to investigate such complexity and uses sophisticated tools like remote sensing, geographic information systems and geo-statistic routines.

The landscape paradigm is actually one of the best synthetic approaches to describe and understand ecological complexity. It is an efficient way to approach ecological complexity because it is possible to distinguish the functional components of our living system without losing information. Landscape ecology is not a separate body of ecology, it is part of ecology, but more integrated and more related to the "real world".

Landscape ecology was born using the human perception of the landscape but this is not the only approach. In fact the capacity of modern landscape ecology should be to link together in a broad framework different ecological theories, models and paradigms. Most of the literature, especially from Europe deals with human perception of the landscape (f.i. Brandt & Agger 1984, Naveh 1992, Austad et al. 1993, Plachter & Rossler 1995, vanDroste et al. 1995, Zonneveld 1995).

A landscape has internal structures and functions but according to the attributed size, must be considered an open system in which the exchange of energy, material and information assure connection with other geographical regions.

It is extremely important to bear in mind that the geographical delimitation of a landscape may be based on key processes like the water drainage of a river basin, or by the religious boundary of a parish, or the living range of organisms.

The landscape ecology paradigm may find application in several earth contexts, moving from the urban system to the more remote and depopulated areas of the planet. From air to deep sea, the holistic, trans-disciplinary approach of landscape ecology finds several possibilities to describe processes, to find relationships between patterns and processes, to link different organism-oriented processes, and to encourage the human-oriented research to explore the more sophisticated relationships between human beings and all other natural processes (see also Pickett & Cadenasso 1995). There is one quality that is necessary in order to act as a landscape ecologist: the awareness that, when an environment is explored, it is part of a whole, and that the discerned patterns and processes are the result of the integration of complex systems.

1.1 The birth of landscape ecology: An important heritage

Although the origin of this ecological discipline is embedded in an uncertain patronage and it seems inadequate to remark that Troll, back in the 1930's, was one of the first biogeographers to discover the importance of considering ecological processes on a large scale, at least two fundamental schools of thought were created between 1930's and early 1980's.

The first has its roots in Europe, especially in the northern part, and the second in North America (Farina 1993c). The European school focuses on the managing, planning and evaluating procedures of a human transformed landscape. The second deals with the behavior of patterns and processes in space (from micro to mega landscape scale). These two approaches have been exported to many countries according to their differing cultural affinities (Fig. 1.2).

A rich literature describes the genesis, history and development of landscape ecology (Farina 1998). This discipline has been recognized by the international community for around 80 years, when it moved from an applied discipline in Europe, adopting a science basis in North America where basic philosophical statements (Naveh & Lieberman 1984, Naveh 1987), were integrated with solid theories and models (O'Neill et al. 1986, Turner & Gardner 1991, Turner 1998), evolving landscape ecology into a modern ecological science (Forman 1981, Risser et al. 1984, Forman & Godron 1986, Risser 1987, Zonneveld, 1995, Forman 1995, etc.)

Geographical ecology (MacArthur 1972), insular theory (MacArthur & Wilson 1967), percolation theory (Stauffer 1985, Gardner et al. 1992), hierarchical models (Allen & Starr 1982, Allen & Hoekstra 1992), metapopulation (Levins 1970, Gilpin & Hanski 1991, Hanski & Gilpin 1991, Hanski et al. 1994), source-sink models (Pulliam 1988, Watkinson & Suntherland 1995, Pulliam 1996), heterogeneity and disturbance regimes (Pickett & White 1985, Turner 1987b, Hastings 1990, Kolasa & Pickett 1991, Kolasa & Rollo 1991), and ecotones and buffer functions (DiCastri et al.

1988, Holland 1988, Holland & Hansen 1988, Harris 1988, Hansen & DiCastri 1992, Hansen et al. 1992a,b, Forman & Moore 1992) are some of the most relevant theoretical approaches that have contributed to create the disciplinary body of landscape ecology.

Powerful tools to analyze the ecological complexity of a space, have been arranged and adapted to the landscape arena, by using neutral models (Gardner et al. 1987), lacunarity indices (Plotnick et al. 1993), image analysis procedures (Haralick et al. 1973, Musick & Grover 1991), fractal geometry (Mandelbrot 1975, 1982, Feder 1988, Milne 1991, 1997), extensive remote sensing techniques (Haines-Young 1992, Goossens et al. 1991, Hall et al 1991), Geographic Information Systems (Burrough 1986, Hulse & Larsen 1989, Tomlin 1990, Maguire 1991, Maguire et al. 1991, Baker & Cai 1992, Burrough & McDonnell 1998), Global Positioning Systems (Hofmann-Wallenhof et al. 1993, Trimble Navigation 1994) and geostatistical analysis (Ripley 1981, Diggle 1983, Isaaks & Srivastava 1989, Hof & Bevers 1998).

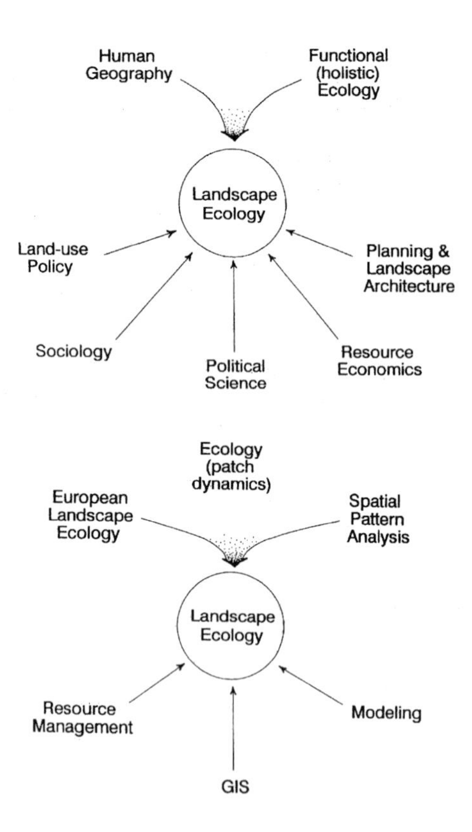

Fig. 1.2 - Exemplification of the historical development of landscape ecology in Europe (top) and in North America (bottom) (from Wiens 1997).

1.2 Towards an unified theory

The decisive contribution of many world class ecologists like Zev Naveh, Wolfgang Haber, Richard Forman, Michael Godron, Isaak Zonneveld, Frank Golley, John Wiens, Monica Turner, Bob O'Neill, regarding the definition of landscape and of landscape ecology is well known, although the different views have created some embarrassment among scientists. Is it possible to define a landscape? And is it possible to find a common paradigm to explain the complexity of the landscape? believe it is, on the condition that we have to consider the landscape not as a specific level of organization (see f.i. Lidicker 1995) but as a physical space in which some patterns, organisms, or processes interact fully with the "eco-field" sensu Laszlo (1996). In this way we can assume that a process-oriented, geo-botanical, animal and human perspective may be used to create the basis for the interpretation of the landscape dimension. This distinction has been a first attempt to create a bridge between the different schools and the different approaches (Farina 1993c, 1998).

The human perspective of the landscape is absolutely correct but it is not the only one. There are many landscapes according to the organisms or processes with which we are dealing (Wiens & Milne 1989, Farina 1995). Habitat and landscape assume in many cases a similar meaning, whilst in other cases they appear as distinct objects within different scales. The landscape must be considered not only as a system at the scale at which humans perceive it, but as a geographical entity within an organism or a process scale, thus we may be dealing with a microscopic or a kilometer wide geographical space. This view allows a unified vision of ecological complexity and certifies the approaches from the different schools of thought.

1.3 Definition of landscape

A landscape is a spatial configuration of patches of dimensions relevant for the phenomenon under consideration. Landscape, from a morphological view, is a spatial extent, followed by a spatial heterogeneous area. Landscape is more than a place, more than a geographical site or a mosaic of land covers, it is in reality the physical and functional context in which ecological processes and related organisms take place at different spatial and temporal scale (Swanson & Sparks 1990, Magnuson 1990). From several definitions of landscape presented, discussed and contested during recent decades (see for instance Naveh & Lieberman 1984, Zonneveld 1995), we could define landscape simply as "a piece of real world" in which we are interested in describing and interpreting processes and patterns.

Many definitions of landscape are available and, though at first glance they appear different, most of these definitions have a common cultural root, merely focusing on different aspects of the landscape (human oriented) approach as stated 200 years ago by German geographer Alexander von Humboldt who regarded landscapes as "the total character of a region":

"*The total spatial entity of human living space*" (Troll 1968).

"*Landscapes dealt with in their totality as physical, ecological and geographical entities, integrating all natural and human ("caused") patterns and processes*" (Naveh 1987).

"*Landscape as a heterogeneous land area composed of clusters of interacting ecosystems is repeated in similar form throughout*" (Forman & Godron 1986).

"*A particular configuration of topography, vegetation cover, land use and settlement patterns which delimits some coherence of natural and cultural processes and activities*" (Green et al 1996).

The real world is an overlap and integration of patterns and functions scaled by organisms and/or processes. If we view the environment as an organism world, the landscape is a mosaic of natural and artificial structures connected and perceived across a cultural filter. This filter is more active in close and isolated societies and becomes weak in modern open multiethnic societies.

1.3.1 Different perceptions of the landscape

Two categories of perceived landscapes can be recognized: a process and an organismic-centered perception. Processes like erosion and nutrient cycles "operate" in a specific landscape in the same manner as organisms interacting across a range of scales will adapt to other components of environmental complexity. This approach is not very popular and pertains often to geomorphological sciences. According to Farina (1998) three broad types of landscape can be considered when we use the perception of organisms: Geobotanical, animal and human landscape. All these types can exist contemporarily in the same place but will function in differently scaled space, time and process interactions. This differentiation is not a novelty in ecology, in fact niche theory (Hutchinson 1957) emphasizes the hyper-volume as a way of avoiding interspecific competition capturing most of the specific traits. Every organism builds a landscape or reacts to basic structures and functions adopting species-specific innate or learned strategies.

Despite a unique energy input of solar radiation, soil chemical processes, microbial activity, plant photosynthesis, animal energy balance and human adaptation are differently stored, manipulated and used. The strength of linkage between the different landscapes depends on the dominant process in each landscape type. So the microbial landscape of a cultivated field is strongly influenced by plant biomass but also by fertilization input and by plowing frequency. One type is not exclusive of the others. Often, the human landscape embraces the two other types, but in this case the driver is the human related processes (policy, economy, culture). In fact, these are the dominant processes that characterize and affect a landscape typology. This view allows us to analyze and better understand the ecological complexity, whilst maintaining contact with the real world.

The main difference between a geo-botanical-animal and a human approach is based on the more complex nature of humans. In fact, learning, experience and transmission of learning (culture) coupled to economy create a holistic view of the landscape that does not contradict the two other views.

According to these three broad categories, it is possible to relate the structures and functions that respond to a process, perhaps soil erosion for the geo-botanical landscape, migration for the animal landscape, or land use change for the human landscape.

But beyond these three different "organismic" approaches and the "process" approach, the human approach based on the large scale human perceived ecosystem, often dominates the scene of landscape ecology. Such dominance and popularity has often created confusion in the science of landscape ecology.

1.3.1.1 The "process" perception

Detailed exploration of the possibility of studying our neighboring living system and of considering the geobotanical, animal and human perception of the landscape, highlights a gap in this sequence. An upper level probably exists ranging from a few kilometer to many km and in which the mosaic is represented by entities that are not simply plant communities or agricultural land cover, but the full interactions between the different ecological components including edaphic factors, and climatic events. This level is easily patterned but at the same time is very difficult to study from a functional point of view.

Such a level often escapes our attention because the weakness of the cause/effect relationship is high when scaled against our small spatio/temporal scale.

This level could be defined as an upper-ecosystem and should close the series begun with the ecosystem, through the "organismic" landscape, and concluding with large scale patterning.

Under a deep analysis we discover that the level is determined by processes such as heat flux, water flux, erosion, disturbance, succession, connectivity, economy, digital information, etc. Thus, there is a world of processes in conjunction with a world of patterns and organisms. This idea needs deeper discussion in the future but represents a new deal in which ecology could find new challenges and new perspectives to interact with the real world. Processes do not have a perception of the landscape like an organism, but nevertheless react when crossing their mosaic in a very sensitive way. A water flow can modify it's speed depending whether the substrate is covered by vegetation or is composed of bare soil (Fig. 1.3). Wind movement is affected by the roughness of the morphology but also by the presence and type of vegetation.

Fig. 1..3 - Rounded sediments of large size being moved down stream by the Orinoco river (Venezuela), "perceive/interact" differently with the landscape according to water transportation capacity, presence of other free running debris, and shape and roughness of river bed.

Nutrients not absorbed by crops are removed by water flux towards the catchment area where their fate is influenced by the presence of riparian vegetation or by hedgerows. According to this approach a landscape can be patterned using a relevant process common to a selected area. For instance the landscape can be described using plant productivity (see f.i. Groffman & Turner 1995), or the movement of nutrients distinguishing input and output patches. Components of such processes as absorption, adsorption, transformation and immobilization can be used to balance the process in the landscape (Risser 1989). Particularly suitable for such an approach is the study of gravel river transportation (Powell & Ashworth 1995) and it's biological implications, especially in rivers that have a heterogeneous bed material allowing several sedimentological forms.

1.3.1.2 The human perception of landscape

It could seem that landscape, as perceived by human's, is an integration of mental processes with reality. According to this view we can assume that two types of landscapes (natural oriented and cultural oriented) occur; the first is the indirect product of our needs, like an agricultural landscape, while the second is the result of a cultural and life-style evolution and is the stochastic product of environmental-random oriented policies (Fig. 1.4). Often these two conditions are in contact and borders are characterized by a high unpredictability.

Fig. 1.4 - Human perception of the landscape is filtered by genetic and cultural attributes. Such a complex vision may be represented by these bronze statues of a Tucson plaza (Arizona) in which age classes and social posture of humans have been represented in a urban surrounding.

1.3.1.3 The animal perception of landscape

In animals, landscape "perception" is a result of genetic fixed cues and accumulated experiences (Wiens 1986, 1989, Wiens & Milne 1989).

A swift's flock flying around a breeding tower can't manipulate or change the neighboring landscape but will be strongly conditioned by the presence and physical geometry of buildings, by daily wind movement and also by the specific energy balance imposing circular flights to optimize the energy consumption (Farina 1998b) (Fig. 1.5).

The effect of a swift colony on the urban landscape is not as strong as the presence of termites in a savanna landscape. Nevertheless the impact of swifts on other birds, like starlings (*Sturnus vulgaris*) and house sparrows (*Passer domesticus*), or on aerial-plankton, is not secondary. By definition this species creates a functional temporary landscape based on competition and predation processes.

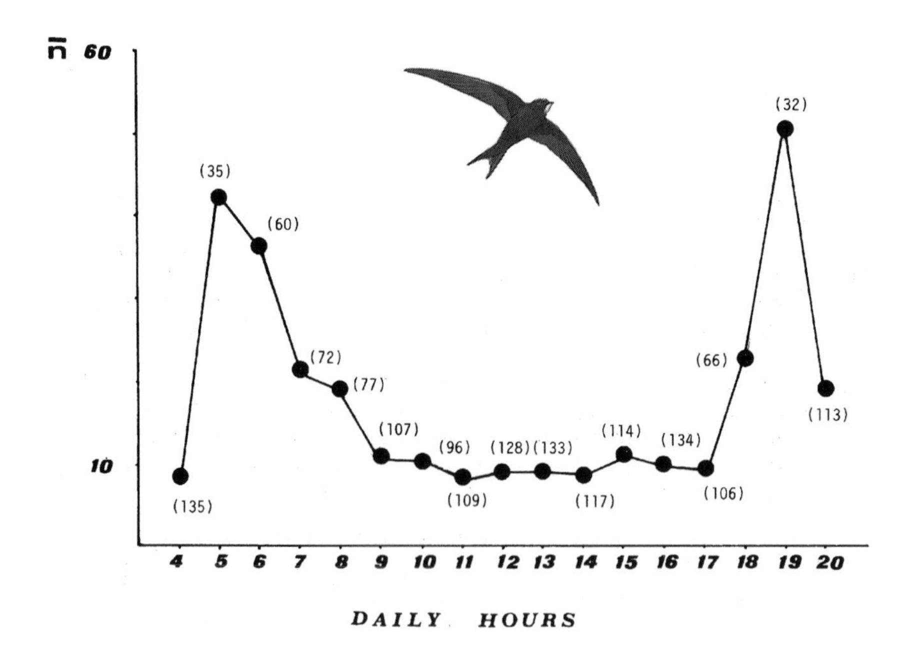

Fig. 1.5 - Average hourly number of birds (Swifts (*Apus apus*) counted at 5 min intervals around the nesting site, with the CV (Coefficient of variation) in brackets (from Farina 1988b).

Recognition of the surrounding space is vital for most animals. In fact, an organism uses the surrounding space extensively for many functions like breeding, mating, or roosting (Fig. 1.6).

Tab. 1.1 - Swifts fly around the nesting site creating a temporary landscape according to complex dynamics affected by climatic conditions and the physiological status of birds. Daily duration of observable flight (DDOF), daily duration of anti-clockwise (ACLW) and clockwise (CLW) flight, ratio ACLW/CLW, daily number of changes of flight rotation (DCRF) and duration of rotation flight (DDRF), number of cases (N) (from Farina 1988b).

	DDOF	ACLW	CLW	ACLW/ CLW	DCRF	DDRF	N
21 June	63.34± 47.47	3.77± 3.88	5.77± 5.62	0.65	2.71 ±3.05	19.49± 11.54	122
27 June	57.53 ±46.90	2.99± 3.28	5.70 ±5.66	0.52	2.30 ±2.62	18.47 ±12.62	174
4 July	58.04± 46.60	3.64± 3.54	5.82± 5.08	0.62	2.83± 2.89	16.10 ± 8.66	173
12 July	55.22± 43.05	3.10± 3.26	6.05± 5.85	0.51	2.48 ±2.90	18.14± 14.55	155
19 July	50.52± 40.37	3.95± 4.46	4.44 ±4.38	0.88	2.36 ±2.79	17.43± 11.21	100
25 July	64.23± 42.16	4.46± 4.45	7.35± 6.27	0.60	2.61 ±2.56	20.80± 14.06	60

There are at least three important behavioral processes involved in this spatiality: the capacity to return to home (homing), the capacity to disperse across hostile environments, and the ability to find the necessary resources to survive in a continuous balance between energy output and energy input.

Most animals have the capacity to create a spatial map which is sense oriented (visual, acoustic or olfactory) and consequently know where they are (Healy 1998).

There is nothing obscure in such mechanisms because we could use our experience to explain that a place may be well known or that we have a low level of personal knowledge. The level of geographical confidence is important for preventing predation or escaping predators, for optimizing food searches, for defending territory, etc. To accomplish this goal animals use landmark extensively.

Experiments on rats have demonstrated their capacity to appreciate geometric space more than other strong cues like smelling. This property has also been found in chicks and in human children and is called a geometric module (Cheng & Spetch 1998).

Fig. 1.6 - Animal perception is species specific. The landscape perceived by a badger (*Meles meles*) is quite different to the landscape explored by a dormouse (*Muscardinus avellanarius*) in terms of space and in terms of morphological and vegetational details.

Some animals have a clear perception of the distance based on the retina metric but this has not been demonstrated in many species. Vertebrates use landmarks extensively, especially the spatial configuration, but the importance of such configurations compared with separate landmarks varies between species.

When animals move long distances at a regular yearly time we call these migrations. The mechanisms for migration are different to the mechanisms of homing. During migration animals generally move along linear routes and

at many times the speed used during the sedentary phase. They use astronomic or magnetic cues, simplifying the landscape view in the same way an airplane computer traces the route from one continent to another (f.i. Berthold 1993).
Scaling the view of the landscape is important in order to optimize resources and is done in the same way that road indications on a highway are less detailed than those on a local road.

Many invertebrates have the capacity to internalize the path that they have covered moving in a relatively straight line on the return journey. Among insects (tested especially in ants and bees) the landmarks are important, but the context in which they operate is paramount. In fact, it seems that a memory of the mosaic in which landmarks are embedded is of major importance in site recognition (Plowright & Galen 1985, Southwick & Buchmann 1995).

Pivotal flyers are common in wasps and also honey bees when they fly for the first time or when they emerge from the nest in the morning. The maintenance of some positions in a space depends on the capacity of the insect to maintain the surroundings in the same retinal position.

Landscape and animal is a very popular topic in landscape ecology research and applications. In particular, landscape ecology and animal behavior represents an obbligatory passage towards the learning and planning of conservation measures (Lima & Zollner 1996).

Animal behavior has been used extensively to evaluate the effect of forest fragmentation, patch size (Schieck et al. 1997) and patch isolation (f.i. Root & Kareiva 1984, Sinclair 1984, Shorrocks & Swingland 1990, Knick & Rotenberry 1995, Redpath 1995, Scoones 1995).

1.3.1.4 The plant (geo-botanical) perception of landscape

Plants appear to have a different 'perception' to animals or human, but if we move from an extremely short temporal scale, at which animals react, to a more relaxed temporal scale we can find similar patterns in plant-landscape interactions (see f.i. Allen et al. 1989) (Fig. 1.7). The trial and error mechanisms used by animals and especially by humans, are utilized also by plants favoring the more adaptive forms. Plants react in a very short time to disturbance processes like fire, nutrient fluxes and water supply. When we move from plant species to plant communities we can find very close resemblance to animals.

Fig. 1.7 - Plants intercept physical, chemical, climatic and competition cues in a neighboring space (landscape). For instance a clump of *Sempervivum montanus* on a slope debris fan shares the limited local resources with only a few other species (in this case *Hieraceum* sp.).

1.3.1.5 Natural versus human landscape

Every piece of our ecosphere can be considered a landscape and in this new definition we have to include material, energy and information. In the case of the human landscape, information is also structured across a cultural mechanism. This mechanism mimics evolutionary long-term adaptive mechanisms. Information can be stored and matter manipulated with effects on the developing substructure.

It is quite common to differentiate landscape into a natural and a human landscape. Though this view partly contrasts with the above assumptions, it was entered into the common scientific practice. In the first case a portion of land is geographically delimited according to natural constraints like rocky fissures, water deficit or surplus, wind exposure, animal activity, plant dominance, etc. The second is the product of human interactions (cultural, economic, social, etc) with the neighboring landscape.

1.3.1.6 Landscape and eco-fields

Landscape ecology can be considered the study of the interferences between objects, particles, organisms, quanta and processes with a specific eco-field (sensu Laszlo 1996). Such a field, like an electromagnetic field reacts in a different way according to the typology of the particles. The field may be modified by the interference activity of the particles and/or can create new conditions for the eco-fields of other particles.

The relationships between particles (organisms, processes) are based mostly within two processes, one is a "consciousness" of the specific field, and the second is represented by the interference of elements that are modifying the field, as in external perturbations.

Interferences create new objects and are stored in each eco-field. They represent feedback mechanisms that are stored in the "memory" of each eco-field and transmitted to successive generations, in such a way that the new organisms have a better fitness if compared with the previous scenario. The differing quality of the eco-fields placed in different conditions are important for the survival of a species (Fig. 1.8).

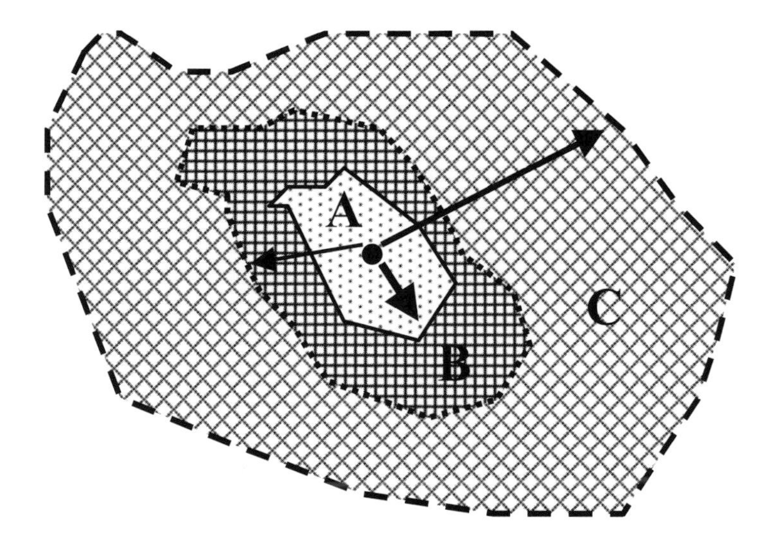

Fig. 1.8 - An interference field is created around each object (organism, particle) and represents the relationship between the object and the surrounding perceived space. Every organism has a specific eco-field, that can be associated with the spatial component of the niche. A to C the different processes/patterns intercepted by the organisms in the eco-field. Line width indicates the perceived constraint at the border.

Moving from individuals to aggregations (population, community) it is not reasonable and probably unrealistic to attribute new intrinsic qualities to an aggregation of individuals "tout court" maintaining the same scale. A level of aggregation is the result of an integration across hierarchical levels of individual information. The transfer of such information is perceived by the system at another scale and in this way the individual signals flow into another eco-field that is, for the hierarchical theory, considered to be pertaining to another layer.

1.3.1.7 A new model for combining ecosystem approach and the landscape approach.

One of the main difficulties in describing landscape ecology as a unified science, is represented by the heterogeneous contribution furnished by scientists belonging to different disciplines and by an intrinsic inclination to dominate. This dominant attitude is quite common in the sciences but seems extremely deleterious for landscape ecology and ecology too.

If the landscape is a spatial entity where are the top and bottom borders? To solve this intriguing puzzle I suggest a simple tri-dimensional model in which the topological characters linked to the organisms-ecosystem functions (Topological dimension) are displaced on the z-axis; the functions linked to geographical space (the chorological dimension) are indicated on the x-axis; and time is indicated on the y-axis (see also Zonneveld 1995).

Moving from the topological dimension to the chorological dimension (space) we meet different objects that are the hierarchical synthesis of lower level of organization. The inferior limit of the space may be moved as far as the sub-atomic particle families, and the superior limit as far as the infinity of the more distant galaxys, but such limits are beyond our present needs.

It is possible to place every object (organism, aggregation of organisms, scaled entities) in this model. Each object is sized according to an intrinsic ecological niche or hyper-space (Fig. 1.9, 1.10).

Again the function of a system may be maximized at a very small scale with all the functions expressed by an organism. Moving across the spatial scale we observe changes in particle organization. At a certain spatial scale distinct individuals are transformed in a single group assuming, at the largest scale, other characteristics. But if we move the same organism across the spatial scale it is possible to evaluate the level of 'degradation' in its intrinsic information.

Also, moving across the space, the scale changes from fine grain to coarse grain. At each scale it is possible to identify new scale-dependent processes or organisms.

If each object placed in the model is assigned a finite volume that represents the ecological niche, by moving the object across the model it is possible to change the shape of the volume but not the volume itself.

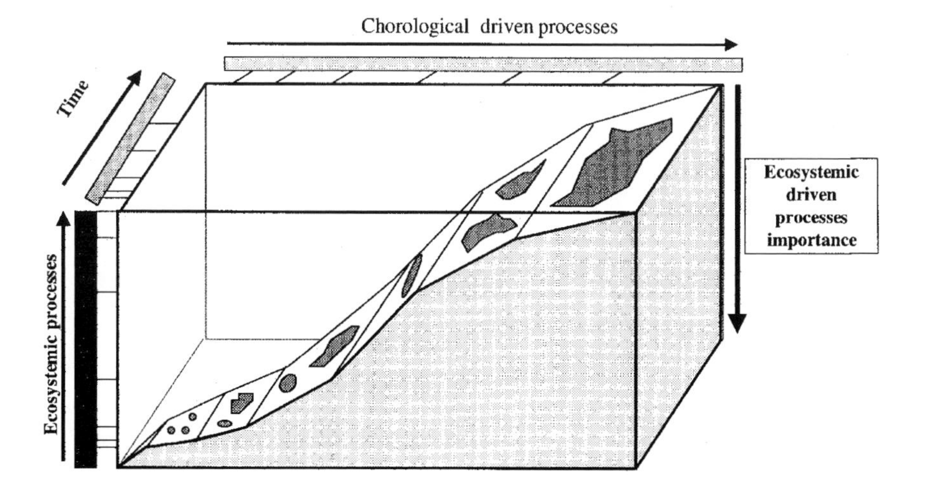

Fig. 1.9 – The Ecosystem-landscape conceptual model: z-axis represents ecosystem processes, the x-axis represents by chorological driven processes (spatial relationships and their effects on processes) and the y-axis, time.

We don't know the quantitative relationship between the topological and the chorological dimension, but we know that the two dimensions are scale related. The change from the chorological to the topological dimension represents the ecological context of the eco-field in which we are operating. The components of an un-scaled landscape react to an object only at the spatial scale of that object and only at that time do they represent a field.

The entire theory of ecotones is based on this assumption. Ecotones are tension zones that exist if they have an interference with an organism. This is true for the objects posed in a landscape matrix.

An "ecological" landscape can be "created" by a modification of a "pseudo-ecological" landscape like the one appreciated by satellite sensors, or by another organism-scaled organism like a man. By using this paradigm it is also possible to select the right scale for describing the landscape extension by processes like river dynamics, desert sand dune dynamics, water and wind erosion, glacial dynamics, and sea and atmosphere dynamics. For instance the topological dimension of water evaporation can be found into the chorological dimension of cloud formation. The extension of this last dimension pertains to cloud system dynamics.

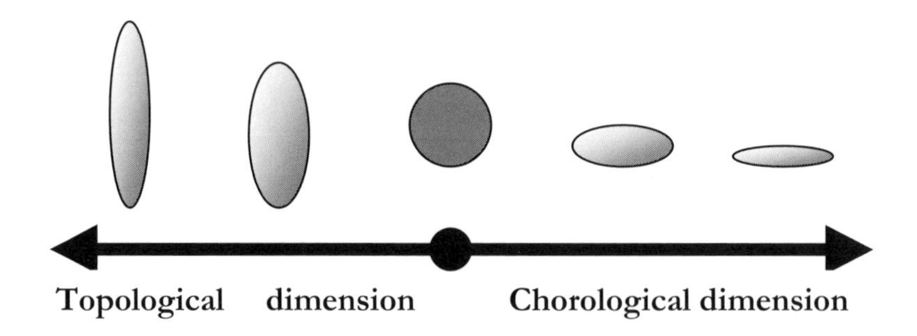

Fig. 1.10 - Every event (process/pattern) can be perceived at a maximum level of information at a restricted range of the spatio/temporal scale. Moving from this optimum position toward the topological dimension or to the chorological dimension, the object change the typology and the quality of information.

1.3.2 The landscape paradigm as a bridge between ecological disciplines: Confronting concepts

It is the primary goal of this book to clarify the relationship between current ecology concepts and the landscape ecology paradigm, and to demonstrate the necessity to create an integration between functional and spatial ecology, contributing to the formulation of a "mosaic theory". Ecosystem versus landscape and habitat versus patch seem two promising topics to introduce such debate.

1.3.2.1 Ecosystem versus landscape

In recent literature the term ecosystem has often shared the pool position with landscape in a sort of "semantic competition", especially in the applied fields of environmental management. It is time to create a bridge between the "Ecosystem approach" and the "Landscape approach". Ecosystem is considered as a functional unit in which biological, physical and chemical components interact (Tansley 1935, Golley 1993). This unit has a prevailing vertical component due mainly to sun light and gravity gradients. Sun light shows a gradient moving towards the earth filtered by atmospheric gases and by upper vegetation layers. Gravity creates a developing gradient.

The ecosystem approach often does not consider the entire organism (f.i. root system), or populations, or communities, but only functional parts (energy flow, productivity) and related processes.

There is a real possibility to translate such an approach into a more chorological approach adopting a proper spatial/temporal scale to investigate the trends and functions, for instance of the root system or the effect of the higher part of the crown cover on phytophagus insects.

Landscape represents the geographical projection of the ecosystem. Landscape and ecosystem are both important approaches in describing the complexity of the real world. In some cases ecosystem assumes a more relevant role compared with the horizontal component, in other cases the horizontal component plays a major importance. The vertical (ecosystem) component may be considered nearer to a unit (homogeneous status), the horizontal component is linked more to the heterogeneity of the medium.

The fundamental functions of a species can be described according to these two components. For instance, an herbaceous species like *Taraxacum officinale* spends most of his life cycle in the vertical component: from sprouting from a seed to the complete flowering. The horizontal component assumes a relevant importance during the seeding when the plant disperses seeds far from their origin. We can assume as a general rule that the vertical "ecosystemic" component is mainly related to growing functions and that the spatial component is linked to the mechanisms of dispersion and reduction of intra-specific competition (Fig. 1.11).

The extension in the space of the ecosystem creates the landscape dimension in which the ecosystem factors are conditioned and are actors in the spatial realm. This vision of ecosystem-landscape allows the connection of ecosystem oriented paradigms, like the fluxes of nutrients or inter-specific competition, to the spatial dimension. Moving across a spatial scale we can see different objects that receive the best of the information at a specific level of spatio-temporal resolution.

Apparently it would be a nonsense to study heterogeneity only at one large scale when it also occurs at other (smaller) scales. In conclusion this is the main reason by which I consider the landscape approach not restricted only to the large human perceived scale. In fact the self-similarity of most of the processes is a good indicator that environmental complexity spreads along a huge range of spatio/temporal scales.

A **B**

Fig. 1.11 - The topological dimension is dominant in *Taraxacum officinale* plant (A), the chorological dimension appears important during seed dispersion and soil implanting (B).

If we look at the right scale of resolution every place, every position of our planet (from atmosphere to land and sea) are heterogeneous. This heterogeneity is composed of spatially distributed 'homogeneous' units. This may be defined as "landscape". Old and new definition of landscape, the processes that have created such a heterogeneous pattern, the processes that maintain, and the effects on organisms is the main interest of landscape ecology. Landscape from a few square millimeters to square kilometer areas is appreciated only by moving the spatial scale through some order of magnitude.

1.3.2.2 Habitat versus landscape patch

Habitat refers to a place where a species normally lives in a heterogeneous landscape, and may be considered a combination of physical and biological factors important for a species or for an assemblages of species. Habitat can be characterized in terms of properties in space and time. In space, a habitat can be continuous, patchy or isolated. In time, a habitat may be constant, seasonal, unpredictable or ephemeral.

The easiest way to incorporate the habitat concept into the landscape paradigm is to consider the habitat as a function of the geographical scale: Large geographical areas have patchily distributed habitats.

The landscape is perceived by a species in terms of suitable, unsuitable habitats, thus habitats can be seen as a combination of suitable patches of different quality or supporting different life trait functions.

Each habitat can be divided into separate functional and structural microhabitats that refer to a precise physical location for a species (Calow 1998). Breeding habitat, foraging habitat, migratory habitat are fine distinctions of preferential use by a species during its life cycle, and can be used as synonymous of breeding, foraging, migratory patches.

Some species require a specific habitat in which to spend all their life-cycle, such as sessile organisms, but it can occupy the habitat at different concentrations according to the quality of the composing patches. "Habitat patch" can be used to describe the inter-dispersion of a specific habitat in a matrix of non-habitats. Habitat means suitability but does not measure the level of such suitability. Each habitat may be more or less heterogeneous, and such heterogeneity can be expressed in term of source-sink patch/habitat. Mobile organisms that have the capacity to survive in different types of environment show the same patterns as sessile organisms but at a different spatial scale, not based on a few centimeters, as for a marine sponge, but at some kilometers wide. In this case the habitat is more difficult to describe and often we refer to more restricted microhabitats like breeding microhabitat because our scale of resolution is not at the right focus. But if we use the right scale, again a habitat is one part of a landscape, and patches are components of this habitat.

Landscape structure in turn greatly affects the habitat quality for an organism. The geographical extension of habitats and their external shapes, produce a quality appreciation of these habitats by organisms. Also in this case the habitat concept is linked to the landscape structure.

Some American warblers that need forest sites for reproduction are good examples of such interrelationships (Probst & Weinrich 1993). In this case the habitat requirement is not only fixed by the presence of the forest alone, but the location (spatial arrangement) of the forest patch is also important. Habitat, patches and landscape are strictly linked and not in contradiction.

In conclusion at least two levels of relationships between habitat and landscape exist. The first is a coarse level in which habitats are considered functional patches in a large scale landscape. The second finer level is represented by the subdivision of each habitat in functional sub-units or micro-patches.

1.4 The dimension of a landscape

Landscape is the geographic entity at which processes and organisms (people included) perceive the environment and react to physical and biological constraints. The size of a landscape is not fixed "a priori"; we can consider a landscape as a few meters of soil for an ant colony, or a hundred square kilometers for an elephant flock.

The landscape scale must not be limited to "a few kilometers in diameter" as this is a view of the landscape paradigm centered on the human perception of complexity; the scale can range according to the process or the organisms on which we are focusing.

For this, as discussed by King (1997), the term landscape level and also landscape scale must be avoided because landscape is not a level of organization but a geographical extent in which some processes and patterns have the maximum of information.

An organism (process)-centered view seems a very promising approach in landscape definition and data interpretation to fill the gap between different disciplinary contributions that have created the theoretical basis of modern landscape ecology.

1.4.1 Scaling the landscape

Dimension, size and location of a landscape are determined by processes, patterns and organisms in which we hare interested in investigating, describing, managing, and manipulating. For this reason the scale assumes a central role in landscape ecology because the changes of spatio-temporal scale allow us to track the processes that create the complexity of our living systems (see f.i., Delcourt et al. 1983, Delcourt & Delcourt 1987, 1988, Meentemeyer & Box 1987, Carlile et al. 1989, Dayton & Tegner 1984, Holling 1992, Levin 1992). In fact there are many landscape scales depending on whether the landscape is perceived by an organism or by a process centered point of view.

We have stated before that a landscape exists according to a process or organism perception and that landscape is not a level of spatial resolution nor a level of organization, in fact these views must be abandoned although still popular (Odum, 1959, Forman & Godron 1986, etc.) substituting the structural anthropogenic approach with a functional process/organism approach.

Often the scale is defined by an observer, and disturbance, succession, evolution, community, ecosystem, habitat, niche, population, symbiosis and competition are explored using the human scale.

Comparing plants and animals from a significant sample of the ecological literature, Hoekstra et al. (1991) have confirmed that often the scale of human perception is used extensively to investigate the different aspects of ecology.

The organism oriented landscapes are often perceived at different scales of resolution according to the phenology and the function of the organisms. Animals are more vagile and occupy a larger niche than plants.

In conclusion, individuals, populations and communities are entities that have scalar relationships with the landscape.

1.4.1.1 Scaling communities within the landscape

It is quite different to approach populations and communities in a scaled environment. Communities are often quite intangible entities composed of different species that in turn have different ecological requirements.

Community landscape is a real generalization because we know that species composing a community are changing with time. In fact a species reacts first to the biogeography and species-oriented habitat.

Nevertheless some communities are quite stable and can be classified according to semi-permanent taxa attributes (Fig. 1.12).

In general plant (tropical forests, mountain prairies, arid grasslands, etc.) and animal communities may be associated with climatic conditions. Often the landscapes of communities are more related with the unique characteristics of the site. The size of the site can span from the few meters of a thermal spring to the hundreds or thousands of kilometers of the boreal forest.

Some species, like humans, are the first actors in this process. Placing a community in a landscape is not an easy task because community boundaries are difficult to detect. A community changes according to the quality of species and not the quantity of individuals. So, often communities show a continuum that does not fit the visible patterns (structures like patchy mosaic). In a community there are the interspecific relationships that compose the "cement" of the community matrix, and these relationships are often active on a broad range of spatial and temporal scales. Of course we can define a community as a specific assemblage of species within geographical and temporal boundaries.

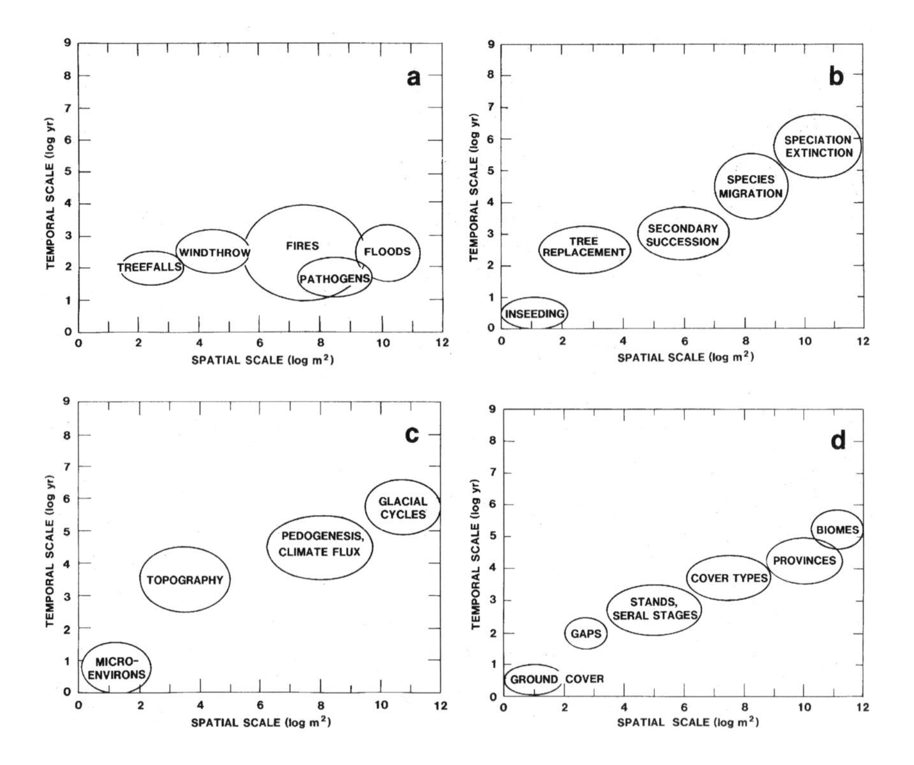

Fig. 1.12 - The spatio/temporal domains in which disturbance regimes (a), Forest processes (b), Environmental constraints (c), and Vegetation patterns (d) can be observed (from Urban et al. 1987 and from Delcourt et al. 1983).

Sessile organisms are the preferred subjects of the community approach because they often fit the human scale and, in particular, small plant communities with a short life span are favoured. More difficult is the study of large plant communities to the wider geographical area and the long time scale which surpasses the human temporal scale.

According to the new perspectives of landscape ecology, the use of remote sensing techniques coupled with spatially explicit models trace new possibilities for the investigation of communities of large organisms. In addition, this approach allows the exploration of a more complete range of inter-specific interactions that are probably only visible at different spatial and temporal scales.

1.4.1.2 Scaling populations within the landscape

It is more simple to scale populations within a landscape because we can measure the density of organisms in a geographical area and, following such an abundance, understand the processes by which a species intercepts and uses resources in space and time. Source-sink models and the metapopulation theory are efficient tools to solve the relationship between a species and the environment. Again Hoekstra et al. (1991) have found that animals are the preferred subject of such investigations due to the difficulties of individual plants counting in the field.

For instance Turner et al. (1997) reported the different spatial scales at which ungulates forage in winter when the soil is covered by snow and food must be discovered by escavating craters. A hierarchy of decisions allows the selection of the area in which to forage and additional decisions at a finer scale allow individuals to select patches of different palatable plants. The landscape heterogeneity created by a disturbance regime constraints these animals to multiple scale decisions. For instance, investigations on foraging habits in Yellowstone National Park have found that a feeding station is apparently randomly selected without distinguished patches of high or low biomass. But studies of the distributive patterns of elk and bison, indicated that burn intensity and slopes were important predictors of feeding location.

At a smaller scale the grassland habitat type appears important. From a spatial scale of 81 to 255 ha, elk and bison were sensitive of spatial heterogeneity. But at fine scale (900 m^2) response patterns to quality and food availability were not observed. It is not surprising that, considering the heterogeneity present at every spatial and temporal scale, organisms should have the capacity to recognize scale across this pattern. We have a rich collection of evidence that many animals have the capacity to distinguish heterogeneity, thus changing their scale of perception.

The robin (*Erithacus rubecula*) one of the commonest birds of Europe, selects woodland at a broad scale during the breeding season but outside the breeding season, it selects at a finer scale, e.g. small fields and gardens showing a high sensitivity to fine landmarks (Nardelli 1996).

The capacity to change the scale of perception may also be induced by climatic stress such as sudden fall in temperature in winter or a decrease in atmospheric pressure. It is well known, for instance, that in winter before a heavy snow fall many species of birds move from woodland to gardens and fields around the farms. In this case birds are searching for food at a finer resolution than in forest and woodland, selecting with accuracy dunghills, vegetable gardens, small burned stands, etc. There is increasing evidence that perception of the landscape is species specific but the scale of resolution is difficult to find. The relationship between species and the mosaic habitat in which a species lives assumes central importance in exploring the habitat requirements of a species and in detecting the consequences of environmental change (Fig. 1.13).

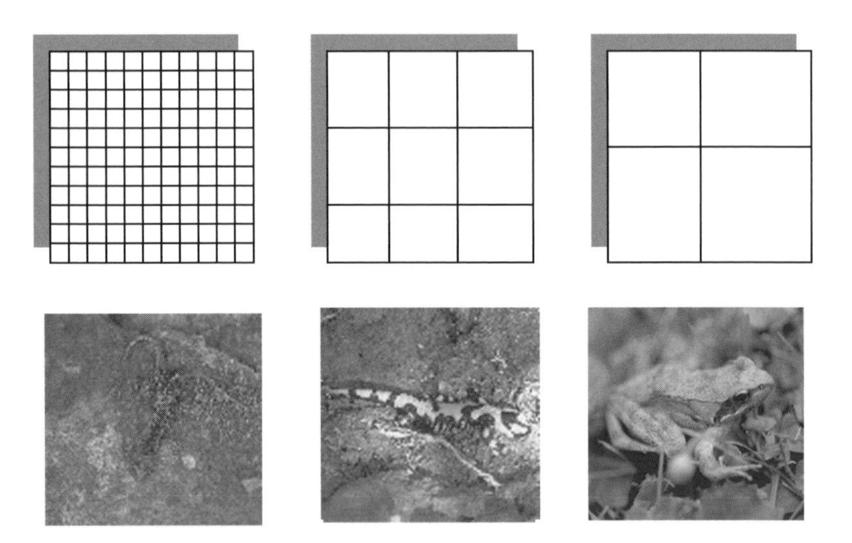

Fig. 1.13 - Depending on their mobility, animals have a differently scaled landscape. The more vagile a species is, the coarse will its perception of the landscape be. In this case, Italian cave salamanders (*Hydromantes italicus*), fire salamanders (*Salamandra salamandra*) and agil frogs (*Rana dalmatina*) are sorted according to their degree of mobility.

The difficulty of finding the right scale at which a species reacts with the mosaic is a real obstacle to collecting genuine data. To reduce this common bias Kurki et al. (1998) have correlated landscape data from satellite remote sensing with the position of red fox and pine marten in two regions of Finland subjected to a different land tenure (Fig. 1.14).

A circular landscape with a radius of 3000, 5000 and 10,000 m around selected wildlife triangles was analyzed according to the different land uses. The radius of 5000 m seemed most related with patch perception by both species. Agricultural land with a cover of 20-30 % explained 26% and 11% of the spatial variation in fox abundance in two selected areas (northern and southern respectively). This relationship is weaker in pine marten. However the two species were not negatively related, showing a capacity to share common resources. In fragmented boreal forest's, the fox appears to be a species with the capacity to exercise a strong predator control, but this is not so evident in pine martens.

Landscape patterns are often not enough to explain species abundance in time and space, but nevertheless it is the most efficient approach for investigating the unpredictable effect of large scale patterns.

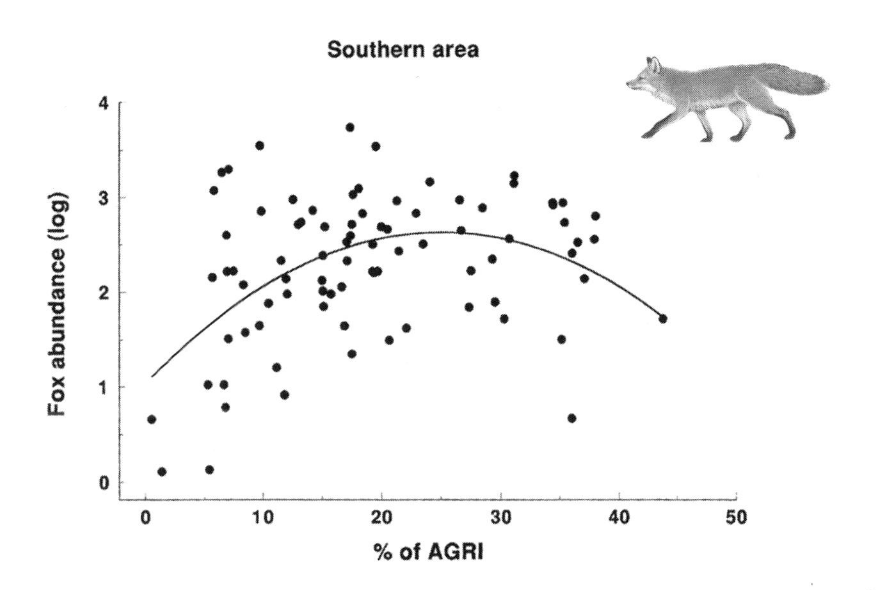

Fig. 1.14 - The abundance (log) of red fox compared with % of agricultural land (AGRI) in Finland, shows a max relationship for an agricultural cover of 20-30% (from Kurki et al. 1998).

1.4.1.3 Scaling complexity

Recently, it has been recognized that most physical and ecological processes act at different scales and that, to better understand phenomena it is necessary to incorporate spatial scales into an explicit experimental and sampling project of field studies. With this approach, called by Hughes et al. (1999) "a broader, landscape view of ecology" it is possible to better understand the local, and the regional effects of processes on biotic realms.

These authors have four sampling scales (250-500 km (sector), 10-15 km (reef), 0.5-3 km (site), 1-5 m (replicates)) along Australia's Great Barrier Reef. At each scalar point, coral adult abundance and juvenile recruitment were measured based on the two main coral types: spawners (those that release sperm and eggs simultaneously), and brooders (release fertilized larvae).

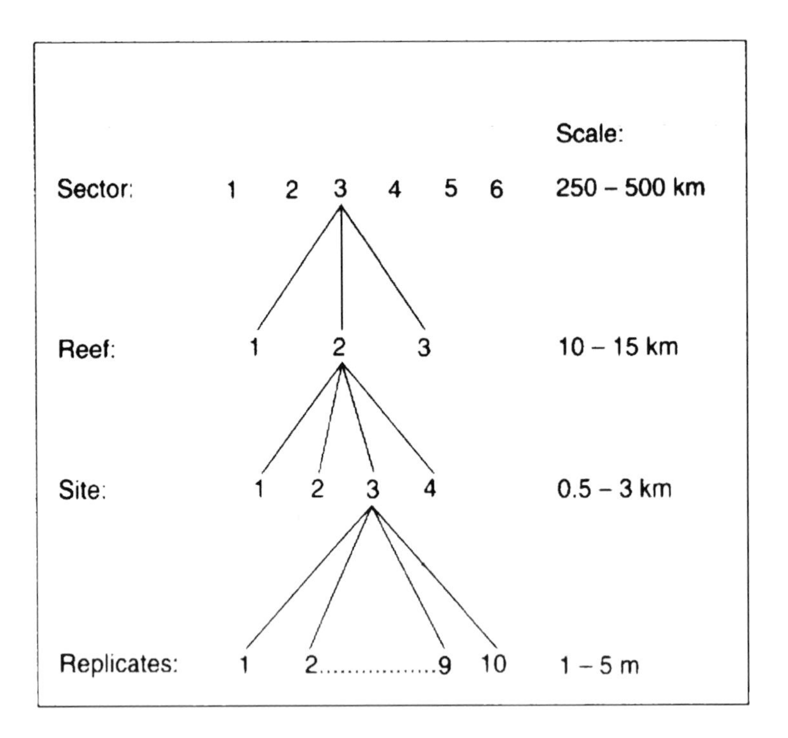

Fig. 1.15 - The sampling design organized according to a hierarchical order used to evaluate recruitment and abundance of corals along the Great Barrier Reef of Australia (from Hughes et al. 1999).

According to this powerful approach it has been possible to evaluate the variability at each scale, demonstrating the necessity of investigating across a broad range of scales to intercept the variability that maintains the presence of organisms across their distributional range. Fig. 1.15 depicts the sampling design and fig. 1.16 the components of variation at the four scales (sector, reef, site, within site), for juvenile recruitment of spawners, brooders, and spawning and brooding adults. Although a full explanation of the differing behaviuor in recruitment and adult abundance is not available, the results of this investigation highlight the methodology and its use of a multiscale approach to interpret the variability observed (or not observed) where a unique scale of observation is utilized.

This approach moves from local conditions (mosaic of nutrients and substrate conditions) to large geographical conditions (mosaic of climatic conditions), capturing the multiscale variability that seems implicit in most demographic events.

The implication of factors acting from local to large scale populations allows investigation of the ecological constraints that affect populations and their fundamental traits.

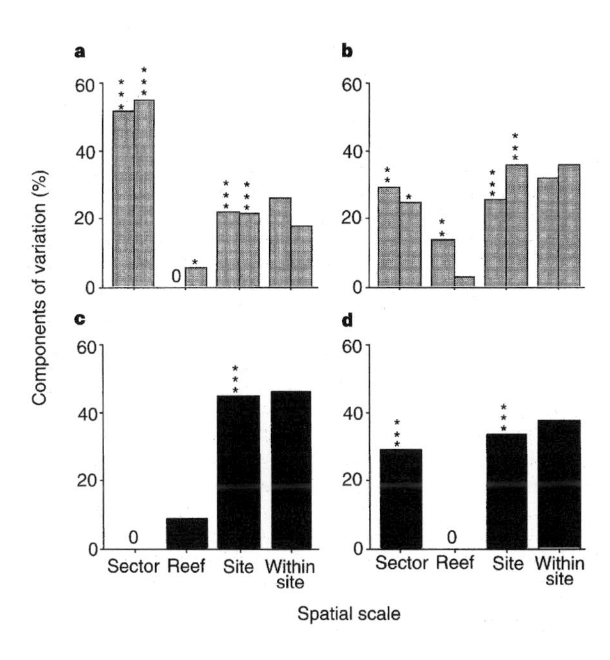

Fig. 1.16 - % of variation in numbers of coral a) spawning recruits, b) brooding recruits, c) spawning adults, d) brooding adults. The variation is divided according to the scale of investigation. Bars with asterisks indicate a significant difference in a specific spatial scale (Hughes et al. 1999).

1.5 Landscapes typologies and functions

Many forces concur to create and model landscapes; these forces may be physical forces like wind, atmospheric pressure, and water, acting in the different states of plants, animals and humans. Size and functioning of landscapes are strongly affected according the different drivers. This classification based on the dominant processes responsible for landscape shaping is really elementary but for the purpose of this book it should serve to steer the reader towards the new paradigm of landscape ecology.

We present some intuitive examples of such landscape types, all are characterized by a high dynamism.

1.5.1 Sky-scapes

The most dynamic "landscape" is the "sky-scape" composed of macropatches of air at different level of humidity, temperature and turbulence.

A change in one of these attributes generates winds and clouds. The sky-scapes play a foundamental role for many organisms because their transparency drives photosynthesis and phototropisms.

The sky-scape is becoming more and more important for airplane navigation but also for pollution levels in the low strata of the atmosphere. Of course meteorology is a well developed science without the new paradigms of landscape ecology, nevertheless landscape ecology could generate new interesting statements and encourage discoveries.

In a sky-scape, an appreciation of the dynamic model moving from function to spatial extension is quite clear. When we select a narrow column of atmosphere the main phenomenon is represented by the transformation of liquid water into water vapor. Enlarging the spatial scale we can observe the formation of different types of clouds and at a very large scale we can appreciate the movement and evolution of cloud systems and their interactions (processes) with the land and sea-scape.

The analysis of a sky-scape is a good exercise for understanding the potentialities of landscape ecology paradigms. The sky-scape has three-dimensions and ground observation differs greatly from high altitude airplane vision.

A cloudy sky-scape filters solar energy through patterns created by the cloud masses. Pollens, spores, arthropods and birds have strict relationships with the sky-scape for moving, feeding and resting, although cloud patchiness changes so quickly that many processes and organisms hare not able to handle such information properly.

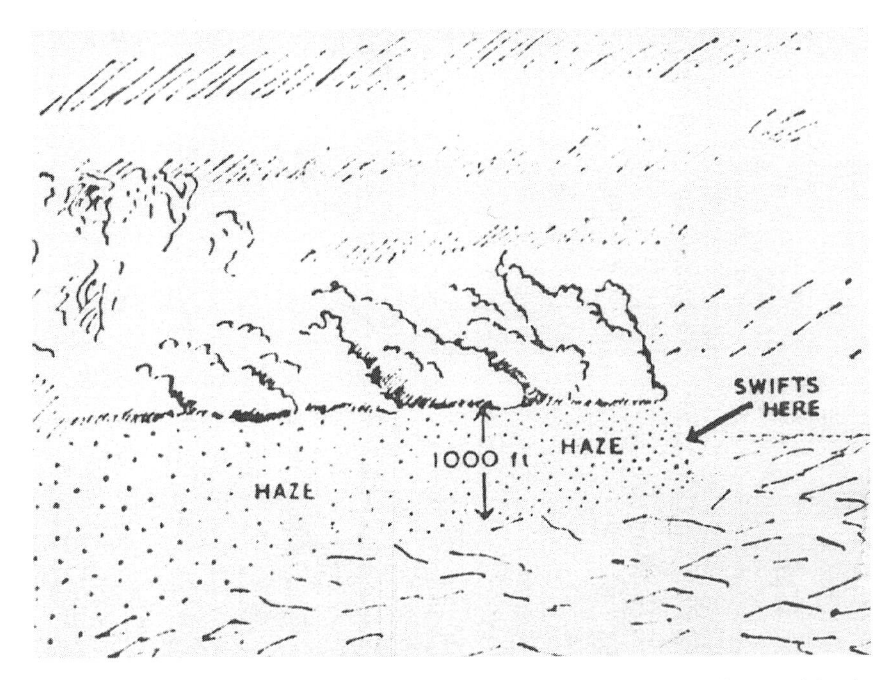

Fig. 1.17 - Cloud typology, dynamics and spatial arrangement are important factors of the sky-scape for air-plankton concentration, and consequently for swift position and their foraging behavior (redraw from Simpson (1967)).

The relationship between sea-breeze fronts and the concentration of arthropods and feeding swallows and swifts is well known (Simpson 1967) (Fig. 1.17).Often the relationship between the sky-scape and terrestrial or marine landscapes is too weak at the same spatial and temporal scale and therefore neglected by research, but it still exists and probably has a larger effect than expected. For instance there is a strong relationship between airplane navigation, flying safety, passenger comfort and sky turbulence.

In the future the sky-scape will be more and more important for humanity and not only a cause of sentiments and mental reactions!

1.5.2 Water-scapes

On the other hand, large water bodies create sea-scapes and freshwater-scapes. Sea water is characterized by deepness, salinity, temperature and light transparency. All these factors concur to characterize the sea-scape, again, in a three-dimensional space composed of great bodies of water masses with relatively homogeneous characters moved by internal thermal energy coupled with Coriolis forces. Sea currents are well known patterns created by these processes.

Changes in the sea-scape, as in El Nino, have dramatic effects on plankton, fishes, seabird starvation and the economy of southern hemisphere countries.

If the "liquid" sea-scape may be appreciated only by using bio-indicators like krill, fishes or seabirds, the superficial frost sea-scape can be easily appreciated directly by our cues. The shape of frozen sea has direct effects on vessel navigation and also on migration and movement of large animals like otters, penguins and white bear.

The freshwater –scape has strong similarities with the sea-scape and differs only by the reduced scale of the processes.

1.5.3 Terrestrial – scape

The terrestrial (land)-scape is better studied by geographers, geologists and ecologists. We can distinguish broadly between two categories: one dominated by abiotic processes (abiotic landscapes) and one by living organisms (biotic landscapes).

The abiotic landscape is mainly represented by deserts, high altitude mountains and saline depressions, of which the desert landscape is most common and can be defined as a portion of land in which most of the processes are dominated by wind erosion and transportation and in which water-related processes and biological processes are negligible.

In this case wind is the main actor again and the physical constraint is represented by the geological nature of rocks. Water and living organism processes are so diluted in time and space that have only secondary effects on the entire system. Terrestrial biotic landscapes may be distinguished in three different groups of organisms: plants, animals and humans.

1.5.3.1 Geobotanical landscape

The distribution of plants in space represents the most obvious patterning of a landscape. I have called such a landscape a geo-botanical landscape (Farina 1998) because plants are the main forces, but the geological substrate is important in determining the plant assemblage types. This landscape is very common and dominates from tropical to artic regions. The scale of this landscape spans from a few centimeters to thousand of kilometers. Plant typology and structure modify soil chemistry (humus forms) and soil micromorphology (depressions where trees fall), acting as a direct control of landslides. Plants have the capacity to move geographically at a large temporal scale, this capacity is impossible to detect at a reduced temporal scale, but assumes particular importance across a landscape because the spatial arrangement of vegetation types is controlled by multi-scaled processes like seed dispersion, competition for light, etc. (see also Bennett 1998).

1.5.3.2 Animal landscape

The animal landscape is more common than believed, for instance, the coral reefs created by a group of cnidarians secreting an external skeleton of calcium carbonate and spreading for thousands of kilometers across tropical seas. Less visible but still very important in tropical zones, is the action of termite detritivore habits and the capacity of many of the 2600 species to degrade cellulose. Some species build a mound nest that can be 30 m in diameter as in the *Odontotermes* spp., a group of subterraneous termites of Eastern Africa. These organisms are considered keystone species. They can represent 10% of all animal biomass in the tropics.

1.5.3.3 Human landscape

The human landscape is becoming the dominant terrestrial landscape in most continents. It is produced by human activity and can be differentiated into rural, urban or metropolitan landscapes with an incredible differentiation of types according to the cultural diversity of populations and the importance of economic forces.

Landscape ecology was born from the study of the human-related landscape and human perception of the landscape. Often the human landscape overlaps all the other landscapes, and may enter into conflict with the shaping factors or mimics these factors. Human landscapes are dominated by economic processes if we exclude relicts of ancestral landscapes such as pluvial forests or warm deserts where primitive human societies (hunters and harvesters) shape the land with the only goal being to survive. We will see that often the change from a natural to a human landscape is not a linear process, that some processes are reversible and that many conservation policies are based on the maintenance of remnants of other landscapes.

1.5.3.4 Vertical landscapes

At first sight it would appear that we are forced to distinguish vertical from plane landscapes. In effect there are many intermediate categories between the two possibilities.

If, for simplicity, we consider only a prevailing flat landscape and a prevailing vertical landscape important figures appear between the two types. Firstly gravitational force moves objects (soil particles, spores, bacteria, seeds and animals) in a downward direction but rarely, and only by passive transport (wind and animal dispersion), are organisms moved upwards. Nevertheless, if we look closely at a horizontal and a vertical landscape we have difficulty in distinguishing between the two types. Organisms are spatially distributed in both cases. One example of a vertical landscape is represented by cliffs where the matrix conserves very little soil, but plants can select micro-sites in which to develop. Lichens, mosses and ferns are the dominant plants.

Fig. 1.18 - Vertical landscapes are important for conserving rare and endangered species, like *Primula apennina*, an endemic plant of the Northern Apennines.

We have very little information on vertical landscapes. We know that there are sites in which diversity is high (Farina 1971), and rare species find refuges. It would be of great interest to investigate the dynamics responsible for the biological organization of such extreme environments.

Slope aspect and thermal currents are the main drivers of vertical mosaics. In many cases rain can't intercept vertical walls, hence they can be considered as dryer microclimatic islands (Fg. 1.18). Many plants and animals are perfectly adapted to such vertical landscapes. Crag martin (*Hirundo rupestris*) is a small pale brown West Paleartic Hirundinidae that spends most of the breeding season strictly linked to a specific cliff (distributed from sea level to uplands). The behaviour of this solitary bird is quite characteristic (Farina 1978). Despite other swallows, the crag martin spends a long time roosting on the cliffs and searches for insects directly on the cliff surface or by flying in proximity to the cliffs using the upwelling air currents. For this species the living space (landscape) has a prevailing vertical component and it defends such territory by intra and interspecific mobbing behavior.

Vertical landscapes are not very common but can be easily found because they emerge from the horizontal view. The investigation of these landscapes is important because they often serve as a refuge for endangered and rare species, and contribute greatly to the local and regional biodiversity as recently discussed by Krajick (1999). An increasing pressure on these landscapes is represented by climbing activity that causes the disturbance of vegetation (generally it is removed to assure a better path) and causes the escape of breeding birds like raptors.

1.5.3.5 Suspended landscapes

Most of our attention in studying the land mosaic is focused on patches placed at ground level. When working in an old-growth forested mosaic such an approach is often not enough to describe the complexity. Field ornithologists (MacArthur & MacArthur 1961, MacArthur et al. 1962, Blondel & Cuvillier 1977) have recognized the importance of vertical structures of vegetation for birds. I propose to extend the approach of such pionering ecologists to describe, using surface maps, the distribution of vegetation in different layers opening new interesting perspectives for understanding the complexity of the systems. It is well known that in tropical forests organisms have a different altitudinal distribution on the trees, based on humidity and solar radiation, but the chorological effects are not considered.

An organism moving in a suspended mosaic has more information to consider: a vertical optimum belt with upper and lower limits, and a horizontal patchiness. The hostile mosaic is distributed below and beyond a belt that may be very narrow or quite broad. The hostile mosaic may also be perceived across the layer. Many animals move exclusively on the soil, for instance mice, and others like sloths (*Choloepus didactylus*) avoid the ground considering it too hostile. The three-dimensional perception of the landscape is common to many organisms, but it is mobile animals that have a higher perception of this three-dimensional heterogeneity.

At present, there are too few studies on this subject to dedicate a chapter but in my opinion the study of a three-dimensional perception of the landscape and its possibility with new electronic devices, will open a new unexplored field in which the spatial scale should be the unifying framework. The separation of a mosaic at a specific layer will improve the knowledge of complexity, especially in the water-scape, in which vertical separation is well perceived by organisms as differences in light, temperature and pressure.

1.5.3.6 Other classifications of landscape

According to the role of processes and organisms we can have passive landscapes and active landscapes. We define passive landscape areas as those in which a process or an organism finds suitable conditions. A small soil depression can produce the accumulation of water so the flowing water processes can find a step, and the water retention process acts properly.

The passive phase may be temporary and restricted; for instance, at the first stage of a process or at the establishment of an organism. The successive phase is more active and produces modifications in the landscape with a "personalization" of the landscape.

An individual based landscape is a level of organization that is intercepted by an individual. Often the relationships, competition, or simply multiple use of the same "piece of land" by assemblages of processes or organisms, produce a new assemblage functioning landscape.

In conclusion, applying a well known hierarchy which spans from individual, population to community, we have an individual based landscape. In vagile animals the behavior of a dominant organism differs significantly from the other components of the flock. For instance, in deer or buffalo the dominant male has a landscape greater than the sub-dominant individuals..

At a population level we can find different attributes so heterogeneity across a population landscape is composed of the summation of individual based landscapes. Population landscape can be easily checked using the density of individuals and is one of the more promising scale approach for studying the relationship between species and land mosaic types. By the introduction of the meta-population model, population landscape is progressively moving to a new promising way.

There are a number of theories and models which have been prepared for land-scapes; these theories and models can also be applied to the "sky-scape" and the "sea-scape" but we have to consider the different behavior of land, gas and water. The spatial and temporal scales of the processes occurring in these three different dominions are quite different. Air movements are characterized by high speeds of gas turbulence and the spatio/temporal ratio is completely different from the ratio found in the land-scape context. The situation in the sea-scape is intermediate. Unfortunately, little information is yet available about the behaviour of sky and sea-scapes.

1.6 Emerging properties of the landscape

Any landscape can be described as a mosaic of patches of different sizes, shape and composition in which aggregation is a function of a process or of an organism (Fig. 1.19).

These patches may have a double composition, which can be physically appreciated by the senses like a wood-lot or a lake, or they may be appreciated as functional patches which are species-specific related.

An agricultural landscape may be composed of fields of different crop varieties separated by edges, roads, channels, natural forests.

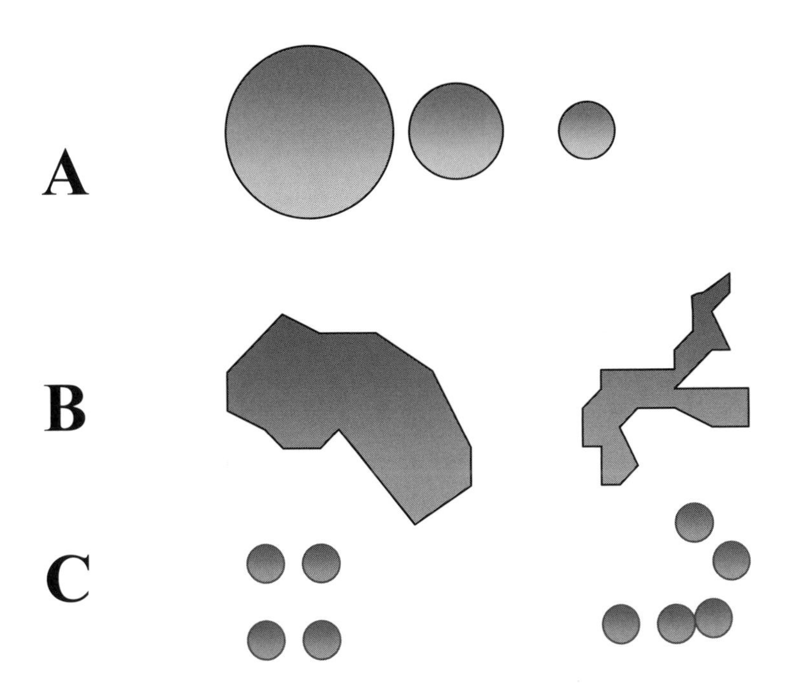

Fig. 1.19 - Size (A), shape (B), and spatial arrangement (C) are important attributes of patches in a landscape, influencing many processes like: preferences, reproduction, movement, predation, migration, extinction, colonization.

A clearing in a forest may be considered structurally homogeneous but in reality it is perceived as a patchily composed predatory-risk. So, there are parts of the clearing with a high risk of predation and other patches with a lower or null predatory level.

The spatial arrangement of the patches and their isolation have a strong influence on the cycle of water and nutrients and create different habitat requirements for organisms, which also affect animal dispersion and migration (Blake 1986, Blake & Hoppes 1986, Blake & Karr 1987).

The shape and size of patches, and their spatial arrangement have a great influence on many processes including climate (Miller 1981, Pielke & Avissar 1990, Klaassen & Claussen 1994), nutrient movement (see f.i. Peterjohn & Correll 1984, Schimel et al. 1985, McDowell & Likens 1988, Correll et al. 1992, Hantush & Marino 1995, Schlesinger et al. 1995, Schwarz et al 1996), biogeochemical cycles (Woodmansee 1990), soil genesis and evolution(Huggett 1975, McAuliffe 1994, Huggett 1995), animal behavior (Gross et al. 1995) and plant dispersal.

More then 2/3 of the earth's has been modified by human activity and this trend is growing very fast. More and more natural areas have been transformed in agricultural or in urban landscapes. Natural heterogeneity is replaced by human induced heterogeneity and natural patterns like vegetation mosaics, sand deposits, marshes and lagoons, river meanders, have been transformed, destroyed or permanently disturbed.

This story is not new, the whole 'saga' of humanity has been characterized by these types of modification, but the rate of transformation has been tremendously accelerated by new technological tools such as airplane transportation which facilitate an easy and relatively rapid connection with remote regions, and by computer networks. The natural world is increasingly fragmented into "land" islands and the human world is increasingly connected by new roads, faster land vehicles and airplains.

1.6.1 Spatial autocorrelation, a measure of distributional patterns

Moving towards large scales and the counting of abundance of species, highlights two possibilities: a species may show an autocorrelation in its abundance distribution or a species has no evident spatial autocorrelation.

These two possibilities indicate two different processes. In the first case, a species comprises populations that show a synchronization in some traits of their life and such species are more prone to episodes of disease or other processes that can affect the entire population. These species are more sensitive to land fragmentation.

On the other hand, an asynchronous pattern allows a species to be divided into sub-populations with distinct traits of local extinction and recolonization. This pattern has been observed in Californian birds by Koenig (1997). Using 100, 250, 500, 1200 km distance categories, the author found a lower autocorrelation in resident birds when compared with migratory birds, and also a lower autocorrelation during breeding time than in winter time (Fig. 1.20). The asynchronous behavior of species assures a higher resistance to episodes of disease contagion and of environmental crisis.

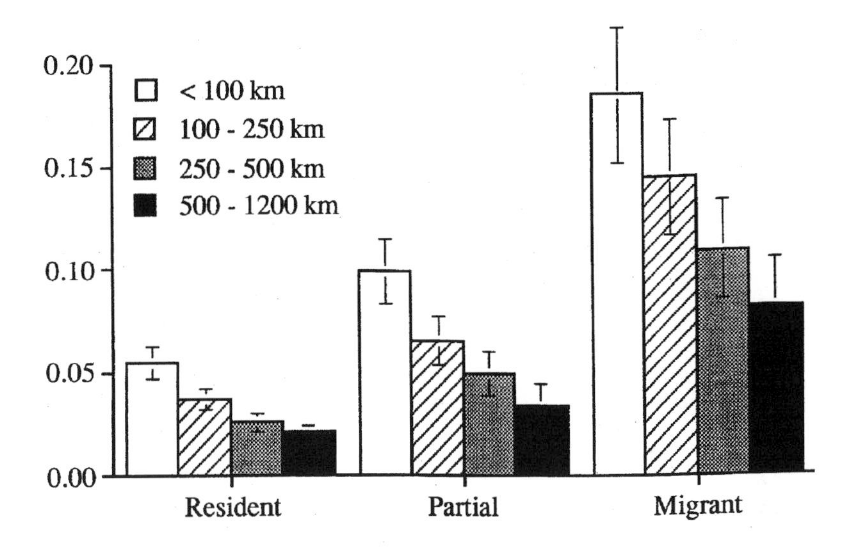

Fig. 1.20 - Differences in autocorrelation value between resident, partial migrant and migrant birds of Californian land birds. Resident species have the lowest value of autocorrelation interpreted as a mechanism of defense against disease contagion and environmental crisis (from Koenig 1997).

1.6.2 A new view of matrix

The formulation of a new mosaic theory seems a realistic objective in landscape ecology development, supported also by a core of new spatially explicit paradigms like metapopulation and source-sink systems. This theory should integrate the chorological and topological aspects of the real world.

The matrix is considered the medium in which different types of patches are embedded. This is true, for instance, during fragmentation processes of the forested landscape. When deforested areas dominate, these open spaces are considered as a matrix of a generally hostile surroundings, the ocean in which tree islands are emerging.

This vision has accompanied the growth and the popularity of many landscape principles. To day we have more reasons to critically reconsider this simplified vision of the mosaic.

Too much emphasis has been devoted to a precise separation of matrix and patches and often this difference is more apparent than real. Only in highly disturbed landscapes is it possible to find such sharp divisions of the roles between matrix and patches.

In particular, it seems very difficult to separate the matrix from the patches along the Mediterranean basin. We have much evidence that the matrix in such a complex system plays a fundamental role in the maintenance of the diversity of ecological processes and the diversity of life (see later Rosenberg et al. 1997).

In this sense, what is the role of patches? It seems reasonable to consider patches as the "phenology" of a matrix in which the flux of energy, nutrients and organisms can find a higher constraint, where the local "micro-gradient" is higher, than in the surrounding matrix. It is the concomitant presence of local constraints that creates the conditions for the "organization" of a distinct patch emerging from the "matrix bed". In this respect the ecotone paradigm appears extremely useful, an ecotone acts as an "active front" for a matrix in which abiotic and biotic components are mixed especially by animals, favoring aeration of sediments and increasing the rate of recycling of macro-nutrients (see also Covich et al. 1999).

1.6.3 The Patch paradigm: Definition

Heterogeneity is a key attribute structuring natural and human induced processes. The discrete element of a landscape is called a patch, considered also as a definable area on the Earth'surface whose structure or composition differs from adjacent areas (Pickett & Rogers 1997). The surroundings of patches creates a mosaic, that must be considered as a descriptive attribute of a landscape.

We have to remember that any mosaic we can observe from a the scale of a few millimeters to kilometers, has a meaning only if we can attribute a function to the described elements.

We can easily appreciate the mosaic composition of a human landscape by finding the different colors, compositions and shapes of the composing patches. The 'patch" word is used in many ecologically related fields. This simplified view is essential to define a patch as a piece of the sky-sea-land-bio mosaic that differs in some properties from its surroundings.

A patch can be considered as the elementary units of a mosaic, composed of a few millimeters of rocky soil covered by bryophytes, or of thousands of hectares of pine forest.

However, patches cannot be defined only by patterns (structural patch) but also by different processes or by different levels of processes (functional patch). Inside a patch, according to a specific organism perception, it is possible to distinguish a core area (interior part) from a buffer zone positioned between the core and external border (edge).

Different spatial abundances can be considered in a patchy way by discrete classes. A functional patch can be formed by a flock of foraging mammals or flying birds, or by an area in which the predatory risk is high. A functional patch may be a foraging patch in which resources can be tracked by a species. Vagil animals can find suitable patches in which the probability of finding food is higher and the duration of patch exploitation is determined by several factors like the balance between food harvested and energy spent (Stephen & Krebs 1986). Behavioral ecology defines a patch as the site where animals feed. The ephemeral attributes of such a patch depend on the physiological status of the species. For instance, many species spend more time in a patch when hungry than when satiated (Fig. 1.21) (Krebs & McCleery 1984).

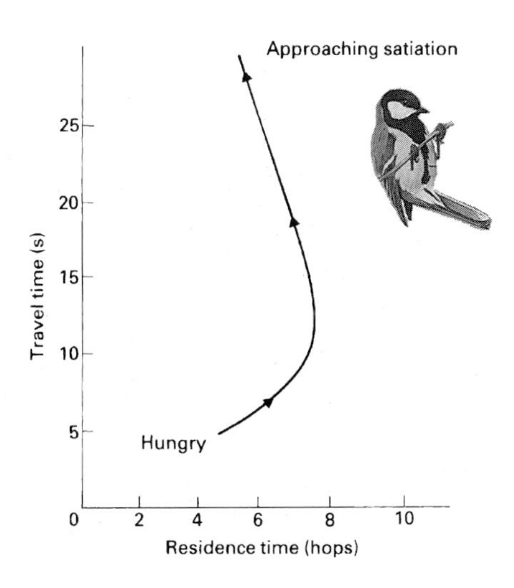

Fig. 1.21 - A male great tit (*Parus major*) observed in an aviary shows more resident time when it is hungry but increases it's travel time when satiated (Krebs & McCleery 1984).

Patchiness exists at all scales but every organism or process interacts specifically within a range of scales; the minimum element perceived is known as the grain (Kotliar & Wiens 1990). Behavioral ecology and landscape ecology are strictly correlated and so a common benefit can be found from developing an integrated framework.

There are several criteria for selecting a territory but food availability is one of the proximate factors increasing the probability of establishment. For instance, in pied wagtails it has been found that a territory is maintained after creation even when resources are depleted. The owner moves away for supplementary food but may return to the old territory. Patch selection and the time spent in a patch are some attributes of foraging (Ward & Saltz 1994). Other factors can influence foraging patches such as temperature, season, density of animals, etc.

1.6.3.1 Population patches

The patch paradigm can be used to evaluate the distribution of organisms in space. Populations are aggregations of individuals and when they are small and isolated can be compared to a fragmented landscape. If these populations are connected to each other by a flux of individuals (immigration/extinction and emigration) these populations can be considered as metapopulations.

The patch approach of interpreting the complexity of the landscape started with Island Biogeography Theory (MacArthur & Wilson 1967). According to this theory the isolation and the size of the island are the driving factors for colonization and extinction. This theory opened new views which consider an ecological system as an open system in which extinction and colonization are two extremes of the same process.

1.6.3.2 Patch dynamics

Patch dynamics is a useful approach to study the changes and the evolution of landscapes. This is especially useful when we are working at the anthropogenic scale. Patches are created by different agents like wind, water stress, plant development, or animal movements. Patchiness may be synonimous with heterogeneity. Before the development of a clear landscape ecology paradigm, patch dynamics theory has played a fundamental role in developing spatial ecology (Kolasa & Pickett 1991).

The mosaic created by the aggregation of patches of different types and shapes does not have a stable configuration and changes during times of internal and external disturbance regimes. This creates the shifting mosaic, a mosaic of patches that change size, shape and position. The shifting mosaic model considers the stable presence of habitat patch types in time but their position changes in space. This approach is quite interesting in terms of maintaining diversity in ecosystems, communities and in landscape too.

There are mosaics like the managed grasslands created in the Mediterranean in which, according to the season, patches are composed of temporary aggregations of annual plants, especially geophytes that have a limited epigean biomass. Fires, tornado or logging activity can produce a change in the tree mosaic of a forest in a few years. A special mention must be made of ephemeral patches that can create in turn an ephemeral landscape. Temporary ponds created by heavy rains can develop plant and animal communities that develop survival strategies during long periods of inactivity.

1.6.3.3 Micropatch dynamics

That the medium (e.g. soil, vegetation cover, etc.) is heterogeneous at many scales may appear today a boring refrain, but new evidence of such heterogeneity and the role played in nutrient exchanges, also emerges from the micro-spaces, for instance from the bacteria world.

The pelagic medium apparently is an inhomogeneous nutrient environment for micro-organisms. Chemotactic bacteria have been observed to concentrate around nutrient patches released by zooplankton excreting plumes of nutrients or by the dicharge of undigested organic matter and inorganic nutrients from food vacuoles (Blackburn et al. 19989)(Fig. 1.22).

The presence of micro-patches produced by such processes is exploited by chemotactic bacteria that can use these resources. The nutrients are more easily intercepted when they are patchily distributed than when they are homogeneously spread around.

Patches of 1 mm in diameter may have bacteria clusters composed of over 1000 motile organisms, with a concentration up to 10^7 ml-1, that act for a time span of 10 min (Blackburn et al. 1998). The time of a temporary mosaic is no different in it's functioning from the mosaic perceived by large animals in a seasonal environment.

Patches allow a mechanism to support high bacterial concentration and at the same time assure high growth rates if compared with the background concentration.

The bacteria food web in this way was the available resources at a higher rate of efficiency when the resources are offered clumped in a heterogeneous medium.

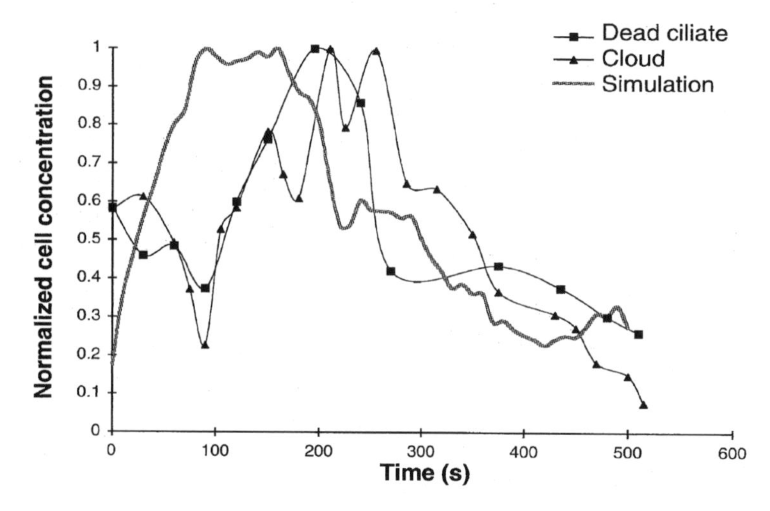

Fig. 1.22 - Normalized bacteria concentration in nearby lysed ciliates, compared with simulation (from Blackburn et al. 1998).

1.6.3.4 Ecotope versus patch

Whittaker (1975) defines an ecotope as the "species' response to the full range of environmental factors affecting it, including both the intensive or intra-community factors that define its niche, and the extensive or inter-community factors that define its habitat.

In recent European literature ecotope refers to the smallest unit of the landscape (Naveh & Lieberman 1984, Vos & Stortelder 1992, Zonneveld 1995) and this often creates confusion when compared with American literature in which "patch" is a more popular term.

Apparently the two terms may be a cause of confusion but let me explain the difference which is mainly due to the approach used to analyse landscape patterns.

Ecotope is used especially to aggregate physical characteristics of the land (Physiotope) related to a biotic function; it has a more integrated meaning and always needs a detailed description.

Fig. 1.23 - Meadows in Logarghena sub-montane Apennines range (Northern Italy) can be considered an ecotope for migratory and wintering birds. In this case ring ouzel (*Turdus torquatus*).They are maintained by human stewardship on calcareous soil on south-western exposed gentle slopes.

Ecotope is considered a functional unit delimited by a prevailing factor (process) in which we are interested in investigating. Like other functional units, the overlap does not always form a distinct pattern. For instance an olive orchard may be considered an ecotope in the Mediterranean region for wintering birds. In fact olive orchards are situated on south-western slopes and are maintained by an "ad hoc" stewardship (grasses and shrubs are periodically removed, and olive trees annually pruned). A warmer microclimate is also an important characteristic of this ecotope. Another example of an ecotope may be represented by upland prairies for bird foraging. These prairies are a "functional" ecotope (Fig. 1.23).

The "patch" approach is, in general, more patterned and not functional, and represents mainly the effects of other processes like the disturbance of a forest. A fragmented forest is composed of remnant patches. For instance, the clearings in a beech forest along the Apennines (Italy) represent patches.

When are patches coincident with ecotopes? When patterns and functions are coincident. So, the Apennines clearings are situated on southern slopes, and they may be considered ecotopes for some species of butterflies, but not the clearings which have a northern exposure, although both have the same vegetation physiognomy. As you can see the distance between the two definitions is not great and patch and ecotope are often used as synonyms. But both terms are quite inadequate to describe integrated processes like "the risk of predation", the preferred path of animal movement (corridors), territorial behaviour in nomadic animals, etc. In this case, we can consider the functional mosaic for that organism as a functional path.

During the initial stages of development of landscape ecology as a distinctive discipline, great importance was devoted to classification procedures. This attitude has dominated the landscape scenario for at least two decades creating conflicts and false expectations among landscape ecologists and practitioners. Classification is a descriptive procedure, and it is important to have in mind what reference system or process is being described. Ecotope is defined as the land unit, the "pixel" of a landscape (see also Zonneveld 1995). It represents the topological dimension of the landscape unit. Microchore is considered the horizontal arrangement of ecotopes. Mesochore is a pattern of microchores, macrochore is a pattern of mesochores, and megachore is the larger landscape.

Although I recognize the importance and the validity of such a classification, I think that a unit operates only if it has an individual functionality. So "a priori" classification may be a good "geographical" exercise but nothing else. A classification has a meaning only if each class is linked to explicit processes. This point can be considered controversial but most of the more recent literature recognizes the importance of the scaled context of a process or living trait of an organism more than theoretical classifications of units that have no functional significance. It is not possible to utilize this as a standard classification because each process or matter in which we are interested to investigate and describe acts at a specific scale. So an ecotope for a crop land classification has a smaller size of some order of magnitude than a forested ecotope in a pristine region. But nevertheless the hierarchical paradigm of the different classifications responds to a recognized pattern across a landscape.

1.6.4 Corridors

When we are dealing with a mosaic of patches of different quality we have to consider the hostile matrix in which these patches are embedded. In many cases a corridor is the special configuration of patches where this configuration appears narrow like a road. The connection between the different patches is species specific according to the ecological relationships of each species. The perceived configuration of the mosaic can be considered suitable or unsuitable to movement. The main function of a corridor is to provide suitable patches for movement. Corridors are not simply linear habitats, but are areas that function for connectivity. Corridors may be considered a source of connectivity for plants and animals, providing connectivity between isolated remnants of a land cover (forest, prairie). But the application of the corridor concept to land conservation and restoration becomes difficult if the target species or group of species under concern is not distinguished.

The corridor concept is utilized ambiguously in too many situations and some confusion on this point is present. In fact we have to distinguish between linear habitats and corridors *s.s.* Linear habitats like rivers seem to be corridors only due to the shape of such a mosaic. For example, species like Cetti's warbler (*Cettia cetti*), a small passerine living along rivers and streams, has linear habitats and the concept of a core and edge component of the habitat is not easily distinguished. Recently ecological corridors have received a lot of attention (Harrison 1992, Beier & Noss 1998, Beier 1993, Simberloff et al 1992, Andreassen et al. 1996, Morreale et al. 1996, Naiman & Rogers 1997). This is largely due to the necessity to recreate the minimum conditions for survival for some relevant species of animals (Lindemayer & Nix 1993).

Due to different rates of movements and different sizes, animals vary in their utilization of corridors created by human activity, such as shelterbelts. For this reason it is important to carefully consider corridors. Corridors are important for migration, dispersive movements reducing genetic imbreeding and local prey over-exploitation. But there are species that migrate for long distances without using terrestrial corridors. Corridors are not cure-alls against the fragmentation of habitat and must not be an alibi for increasing fragmentation by compensating for the loss of habitat with an increase of strip corridors. For wildlife conservation, corridors assure the movements of individuals between the different sub-populations reducing the rate of local extinction and assuring the permanence of a gene flow. The balance about the demographic oscillation can assure a more stable set of (sub)-populations in a region. Genetic and demographic mechanisms are both considered important when applying the corridor paradigm. Two main types of corridors exist: a corridor created by physical constraints like deep valleys or mountain passes (Fig. 1.24). Such corridors force organisms to move within certain regions during migratory or dispersal movements. These corridors pertain to the habitat of a species. The European wolf (*Canis lupus*) move along ranges using mountain ridges saving time and energy and assuring a good connection between the different sub-populations.

Fig. 1.24 - Two examples of corridors based on physical (natural) and artificial constraints. Mountain (Rocky Mountains) allows the movement of large mammals across the divide . In this case mountain lion (*Felix concolor*). A secondary road along the Western plains increases movements of mule deer (*Odocoileus hemionus*) and weeds.

The second type of corridor is represented by movements inside a home range for a species living in a patchy environment. For instance, a fox (*Vulpes vulpes*) explores it's hunting territory moving with high fidelity along preferred pathways within it's home range. This is common to most terrestrial animals from ants to mice.

The European hare (*Lepus timidus*) living in southern European mountains moves from a grazing area to it's roosting area by using livestock trails. Nomadic animals like African ungulates move in a predictable way along fixed routes from one grazing area to another, but inside these areas the movement is apparently casual (Sinclair 1984). This nomadic movement is conditioned by seasonally clumped resources (Fig. 1.25).

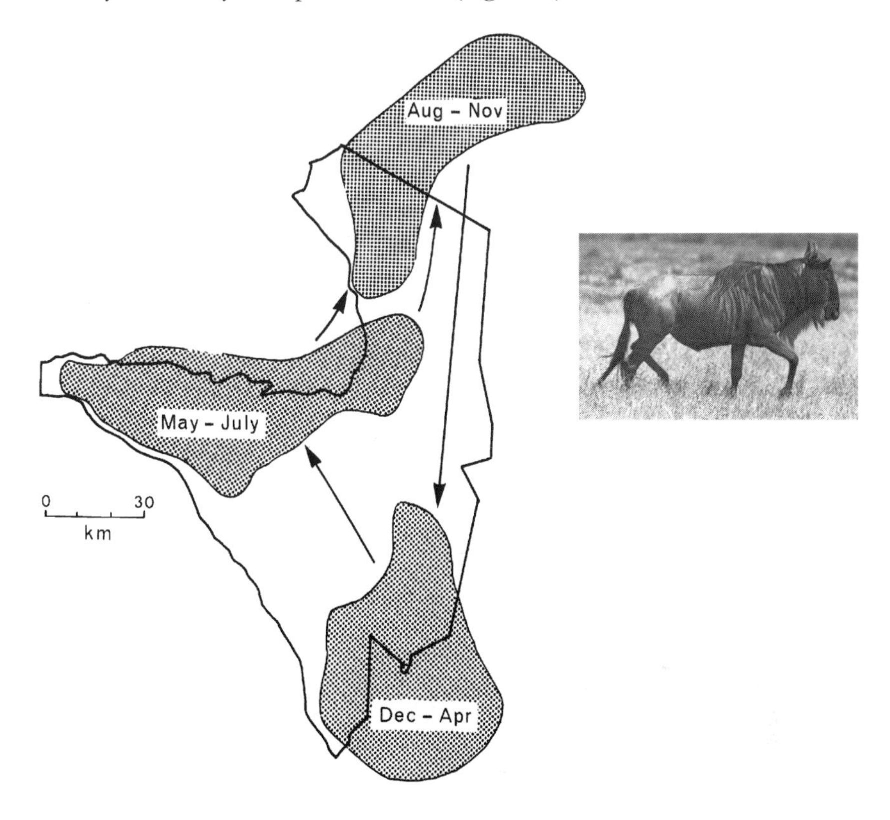

Fig. 1.25 - The movement of wildbeest (*Connochaetes* sp.) in the Serengeti Park, Tanzania follow fixed trails through the seasons. This trail can be considered a functional corridor (from Sinclair 1984).

For example, many species of amphibians require corridors to move from non breeding habitats to ponds. The corridor paradigm is probably widely utilized as a strategy to move from one habitat to another, or to explore territory inside a habitat patch. In fact random walking is quite rare in nature and, from bacteria to elephants, organisms optimize their movements assuming preferred directions or maintaining a quite regular path across the habitat. The species-specific strategy of searching for food of insects, birds and mammals, has been well studied at the scale of individual movements.

Such movement, and the associated strategies for finding food is fixed genetically in each species. The classical studies on European tits by ethologists have confirmed that this instinctive pattern allows each foraging guild to reduce internal competition. More information on the ecology of animal movements can be found in Swingland & Greenwood (1984).

It is important to recognize that corridors can be observed at two scales. One scale is centered on the individual inside the home range, the second is centered on the population living in a patchy environment. Their function is quite clear but the way in which it is perceived requires more investigation.

Despite a wide literature it has not been demonstrated that corridors definitely increase the rate of animal movements, but it is clear that more individuals can be found in linear patch habitats.

The difficult of replicating the empirical observation seems to be the major bias in the corridor dispute. Often, studies on animal movements have concentrated on corridors as patterns and not in the hostile matrix, and as a consequence we don't have a controlled experiment protocol. Study of small mammals have demonstrated that linear corridors are used more for dispersal than other surrounding habitat patches like forests and open cropland (Wegner & Merriam 1979).

Often in the manipulation of the environment, as in an experiment conducted by Rosenberg et al. (1997) on the Oregon Ensatina salamander (*Ensatina eschscholtzii*), climatic conditions (wet, drought period) played an important role in the selection of linear corridors. Non-corridor pathways were used less than corridor pathways, but non-corridor pathways were crossed faster (Fig. 1.26). This compensatory mechanism reduces the difference between corridor and non-corridor strategy. In another study on butterflies, Haddad (1997) demonstrated that the absence of corridors did not isolate the sub-populations.

It is difficult to understand in which way an organism perceives it's surroundings and for this reason the corridor paradigm appears weak and quite difficult to test under a fully controlled situation.

In many cases the so called hostile matrix is not a real barrier to dispersion but, using a compensatory mechanism, organisms often cross the hostile matrix faster, reducing the survival cost. A realistic hypothesis has been presented by Rosenberg et al. (1997). Corridor perception largely depends on the matrix quality.

In a matrix of low quality the probability of finding a quality corridor is higher than in a high quality matrix. The speed at which corridors are crossed is again a function of the matrix quality; the lower the matrix quality, the highest the speed.

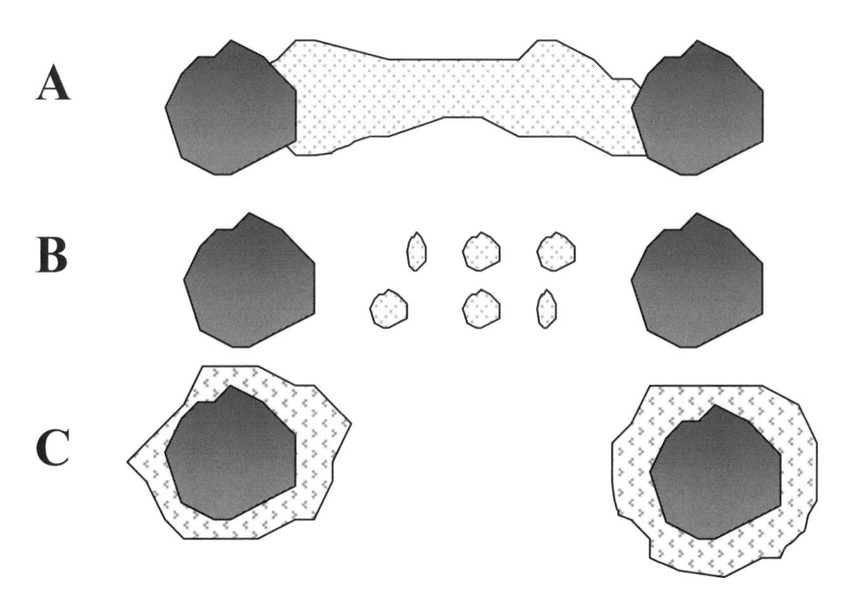

Fig. 1.26 - Three models to improve connectivity in a fragmented landscape. A) to create a corridor between two patches, B) to improve the quality of the matrix, including moderate quality patches, C) increasing the patch size (modified from Rosenberg et al. 1997).

Similarly, the speed within high quality corridors is lower than the speed in low quality corridors (Fig. 1.27). If animals do not select a corridor but cross a hostile matrix, they compensate for this "mistake" by increasing the speed with which they cross. Velocity appears a very important factor in reducing the risk of predation across corridor and non corridor patches.

1.6.4.1 The importance of linear habitats

The fact that many linear habitats are not true corridors does not authorize us to reduce the importance of such habitats for conservation . For example, the linear vegetation along rivers plays a fundamental role in conserving biodiviersity and buffering nutrient cycles. The creation of green ways, the green lines in urban and suburban areas, must be considered in a positive way both from an educational, scenic and ecological point of view. It is important to consider such structures as mitigation structures in conservation. In any landscape design the linear elements should be considered as additional and not substantial components. It would be a great mistake to accomplish such a design based only on linear elements. But of course in the restoration of past cultural landscapes these linear elements play a fundamental role (Fig. 1.28).

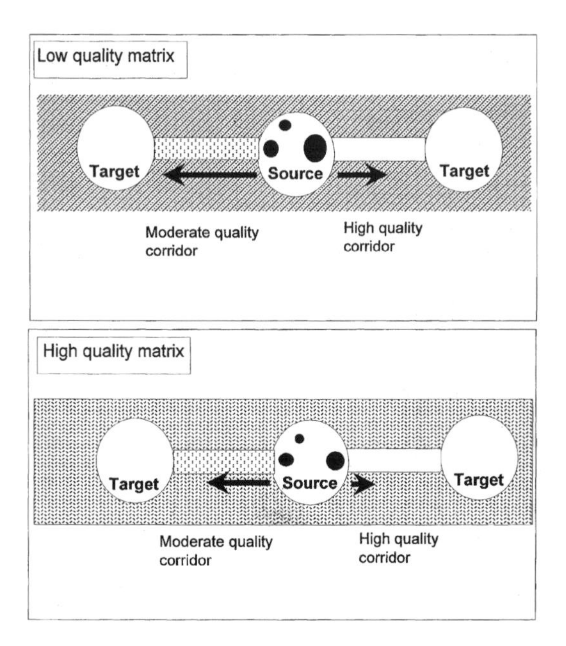

Fig. 1.27 - Animals behave differently according to the quality of the matrix and the quality of corridors. The solid circles represent the probability of dispersion from the source patch. In a low quality matrix the speed of movement of dispersing organisms is higher than in a high quality matrix. In both cases the speed of crossing moderate quality corridors is higher compared with high quality corridors (from Rosenberg et al. 1997).

1.6.4.2 Corridors: an open issue

Corridors may be created by remnants of native vegetation after deforestastion, by substitution of native vegetation with other plants in intensive treeless crop-lands or by high dynamic events like waterflood, snowfall and wind storms in natural environments.

The big expectation from corridors as guardians of wildlife conservation has created a great deal of interest for such structures, but there are both positive and negative effects to be carefully considered. For instance, in metapopulations corridors can reduce genetic variability.

As recently discussed by Beier & Noss (1998) many open questions remain about the biological value of corridors, but without doubt the presence of a "corridor" represents a useful tool for conservation. It is the same situation for small remnants of natural vegetation. The conservation of such remnants is important, especially if the matrix role in landscape dynamic is re-evaluated. Corridor functions must be measured in a more precise way comparing movements across corridors as well as across the embedded matrix.

In some cases, as pointed out by Beier & Noss (1998), the dimension of the study area was smaller than the home range of the animal considered, for instance birds. To study corridor functions it is necessary to use organisms with low vagility and with a home range smaller than the considered landscape.

Fig. 1.28 - A man made stone wall in an agricultural landscape represents a linear habitat for many species of vertebrates (especially micromammiferes) and their predators. In this case the weasel (*Mustela nivalis*). But often they act as ecological traps or simply as a sink. It is easier to cope with the human disturbance system by using such linear habitats, but we must be aware of the risk of reducing interior habitats that often are the source habitat for the species found along edges. For instance weed invasion is favored by such linear elements of a mosaic.

There are many gaps in the corridor matter, as in the effects of disturbance risk and spread, exotic species invasions and predation rates. The natural landscape is more connected than a human disturbed landscape. This could be an important reason for preserving and encouraging the presence of corridors in disturbed mosaics.

The unsolved problems connected with corridors require some actions at least:

Monitoring the movement of dispersing organisms in continuos way and not intermittently

Investigating the factors that determine the direction of dispersion

Investigating the mortality rates during movement across hostile habitats

Identifying the potential corridors for species that have a high degree of dispersion

Identifying the minimum area required for a species according to it's specific home range

Investigating human impact on corridors to find mitigation actions to survive the corridor function

1.6.4.3 Perspectives

The importance of corridors decreases when there are large core areas; also, a low quality matrix can be enough to ensure the connectivity between the populations of such core areas. So the necessity for high connecting between patches depends on the size of such patches. Moving from the strategy to adopt a highly "geometrized" landscape with "potential" corridors linking small isolated patches, we could adopt a strategy of improving the quality of the matrix. For instance the release of trees in a logged stand, favoring large core areas without the necessity of engineering a hypothetical solution and, finally, improving the quality of the matrix to mitigate the effect of habitat isolation. Landscape configuration and biodiversity conservation are very important issues to develop in the future. Most future conservation in more disturbed environments will be debated from this perspective. The difficulty of distinguishing whether a linear patch functions as a habitat *per se* or as a corridor, depends on many factors connected to the life history of each species and on the landscape configuration. Again, it appears important to distinguish between patterned landscapes and functional landscapes. All ecological studies deal with the functional landscape of course, which represents the way in which an organism or a process perceives its surroundings. The constraints in patterned landscapes survive also in a functional landscape. An island is perceived as an island by most species, and terrestrial organisms consider the sea a very hostile matrix when they are travelling close to the coast, whilst the offshore zone is considered a true barrier.

1.6.5 Ecotones, patterns and processes within the discontinuities

Every landscape is by definition heterogeneous, composed of discrete scaled patches. The size and level of heterogeneity are process or organism oriented. Patch size, shape and spatial arrangement are common attributes in all landscapes. These patches create a mosaic in which edges are the borders. Edges assume a central role in the mosaic (landscape) theory and from the first description of ecotones by Clements (1905), a tremendous amount of literature has been produced on edges or ecotones (Shelford 1913, Weaver & Clements 1928, Leopold 1933, Odum 1959, Daubenmire 1968, Ricklefs 1976, Wiens et al. 1985, Odum 1990, Naiman & Decamps 1990, Risser 1995, Farina 1995b), which are interpreted as true habitats, zones of tension between two different ecosystems (patches), areas of transition between two vegetation stages, meeting points of two biomes, etc.

Fragmentation and human disturbance regimes have created favorable conditions for edge formation and development. Edges are becoming a very common scientific theme in many natural landscapes (Wiens et al. 1985), and especially in human-dominated landscapes, thus expanding the argument to the socio-economic context (Desaigues 1990).

Ecotones exist at all scales (Delcourt & Delcourt 1992, Rusek 1992, Gosz 1993, Risser 1993) and can be considered as at least two main types - process ecotones and pattern ecotones. The first is not distinguishable by explicit patterns but only by the function of organisms. For instance, the distance from a shelter is evaluated by a prey, not as a pattern but by integrated information processed by the organisms senses. Ecotones intercept a great quantity of matrix information and represent the place in which different levels of information meet. Between the different views by which an ecotone can be interpreted, the ecotone *per se* can certainly be considered as the verification of the importance of the matrix. In fact the "activity" of the matrix is visible only by the spatially explicit organization of it's complexity. Within ecotone, where edaphic, climatic, vegetation cover, and successional stages change, the structural constraint produces a different aggregation of organisms.

There are several reasons for studying and considering ecotones as a valuable component of any land mosaic (f.i. Risser 1995, Naiman et al. 1988, Holland et al. 1991, Ranney et al. 1981). First of all, within ecotones diversity is higher compared with neighbouring zones. Another important characteristic of ecotones is their capacity to play the role of an active or passive filter (buffer) for organisms and nutrients (Baudry 1984, Burel & Baudry 1990a,b, Stamp et al. 1987, Burel 1996) (Fig. 1.29, 1.30). Buffer zones are generally areas with a very rich biodiversity. The interest of ecologists towards this component of the landscape has grown rapidly in recent decades.

Fig. 1.29 - Two examples of terrestrial ecotones: A) the edge between a beech forest and alpine prairies along the Northern Apennines range (Northern Italy) and a marshland bordering a boreal forest of Canada (Quebec province).

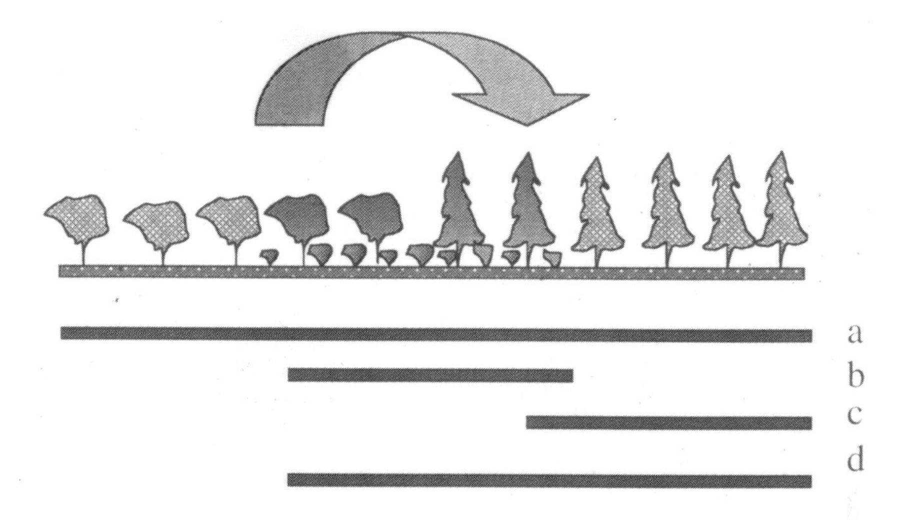

Fig. 1.30 - Animals can intercept ecotones in different ways according to their specific "eco-field". For some species ecotones are invisible (a), for others ecotones represent a true habitat (b), ecotones can be perceived as a hostile environment (c), or can be utilized as part of a habitat (d).

The autopoietic capacity of ecotones is important for creating new diversity that will be absorbed by the embedded matrix. The importance of the matrix is well explained by the ecotone paradigm, although distinct objects are considered more in landscape ecology than diffuse indistinct objects. Temporal events like flooding, drought and vegetation dominance, can create temporarly ecotones, that concur greatly with the diversity of a mosaic affecting soil formation and composition, which in turn will favor new processes.

1.6.5.1 Origin of ecotones

According to their origin it is possible to classify ecotones as "inherent" and "induced" (Thomas et al. 1979). In the first case ecotones are created by internal processes like succession, and in the second case by fire, logging, trampling and grazing. Holland (1988) distinguishes four cases: ecotones created and maintained by human activity (hedges, shelterbelts); ecotones created and maintained by natural processes (beaver dams, marshlands, seashores (Fig. 1.31); ecotones created by human activity and maintained by natural processes (marshlands after dam construction), and finally ecotones created by natural events and maintained by human activity (riparian forest conserved by human stewardship).

Fig. 1.31 – Ecotones between terrestrial and sea-scapes can be utilized as a roosting habitat patch by harbor seals (*Phoca vitulina*) (South from Monterey California).

1.6.5.2 Structural and functional attributes of ecotones

Dimension, shape, biological composition, structural constraint, internal heterogeneity, density, fractal dimension and patch diversity are some of the patterned attributes that can be observed within ecotones. These variables affect organism permeability of edges.

Stability, resilience, productivity, and porosity are some of the functional attributes. Both structural and functional attributes are species-specific and can be measured properly using an "organism" caliper.

1.6.5.3 The role of ecotones in the landscape

Ecotones are structures that operate as filters across the air-sea-land-bio mosaic modifying flux, behavior and quantity of material and nutrients, organisms and information (Forman & Moore 1992, Wiens 1992). The role of ecotones in assuring the stability of a system is not clear, but, for instance, it is clear that the role of buffer is assured by riparian ecotones between crop fields and ground waters. Ecotones control the water flux and in agro-ecosystems reduce the movement of nutrients favoring the complexity and the efficiency of bio-chemical cycles. The riparian zone can be considered as a buffer zone and represents an important component of the land mosaic, especially for it's role as a trap of sediments, nutrients and pollutants.

In disturbed landscapes, like agro-ecosystems, ecotones contribute to the maintenance of biological diversity and at the edges such diversity is often higher than in the surrounding patches (see f.i. Hunter 1990, Risser 1995) (Fig. 1.32).

Fig. 1.32 - The man-made ecotone between fields and woodland sustains a rich fauna, attracted by a dense and fruit-producing vegetation (Central Tuscany, Italy). Fruit of *Rubus* spp., a nest of blackcap (*Sylvia atricapilla*) and adult of grass snake (*Natrix natrix*).

Ecotones often represent ecological traps due to a higher predatory risk of many species that are attracted by the structure of vegetation and by an abundance of resources (Pasitschniak & Messier 1995, Farina, 1997a). In fact, the presence of ecotones can increase the number and the pressure of predators and parasites (Gates & Gysel 1978, Brittingham & Temple 1983).

Despite the global distribution of ecotones and the broad range of scales at which ecotones can be observed, their study appears difficult. This largely depends on the vanishing of the patterned characteristics of ecotones when observed at a short distance. The net separation visible from large distances, disappears progressively when moving closer to the ecotone.

1.7 Ecological concepts incorporated into the landscape paradigm

1.7.1 Hierarchy theory

A hierarchical system is a system of systems within systems organized into levels (King 1997) (Fig. 1.33). What are the relationships between the hierachy theory and the landscape perspective?

According to the general assumption that a landscape is a level of process/organism perception of complexity, the hierarchy theory can be incorporated into landscape ecology. King (1997) has presented the example of hypothetical wildlife species in which the female home range is incorporated into a larger polygamous male territory, and in which families are aggregated in geographically distant colonies. So, in order to understand the biology of such species we have to move across different levels of organization with implications for the spatial and temporal extent.

My example relates more to the human landscape. If we consider the "coltura mista", a typical north Apennines landscape, this is organized by the parish at the kilometer scale. In this case several properties are aggregated in a religious entity. Assuming this model is still functioning, though in reality it has disappeared during recent decades, the activity of farms is centered around the "church" of the village (parish). Thus social and religious activity was developed in this geographical context.

Moving down the scale each farm is composed of a mosaic of fields and woodlots. Farm extension responds to a precise criteria of energy invested and harvested at the crop scale at a familiar level. Each field in the farm has an internal structure composed of rows or randomly situated plants. For instance, a vegetable garden has a very complicated structure with vegetable rows (e.g. cauliflower) or sparse carrot squares. This is a nice example of the application of the hierarchical theory to the landscape paradigm.

Further, according to this theory the lower the level, the higher is the frequency of processes. At the lower level the change in composition of the vegetable garden occurs three or four times a season. At field level the turnover may be annual or biannual. At farm level the mosaic changes only at a centennial scale, whereas the re-aggregation of the parish may be a multi-century event.

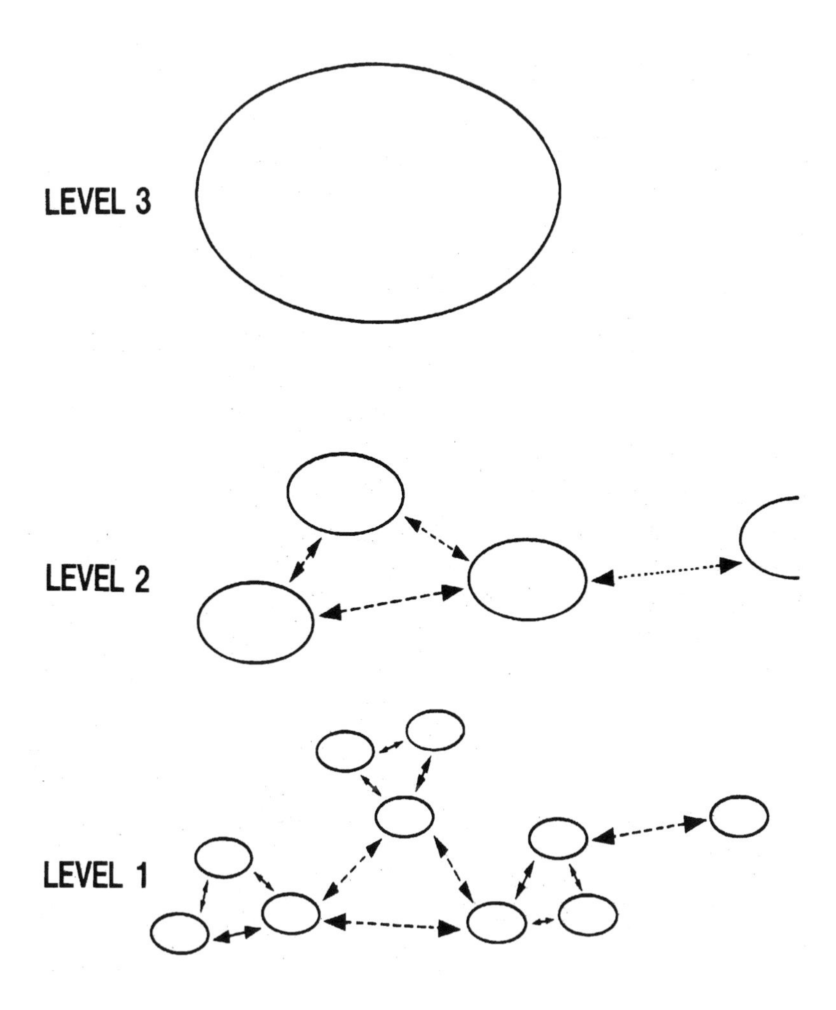

Fig. 1.33 - Representation of a generalized hierarchical system (from Urban et al. 1987).

We can deal with the carrot landscape, the vegetable garden landscape, the "coltura mista" farm, and the parish landscape. The "cauliflower landscape" is composed of rows of spaced plants generally with irrigation furrows, and has a linear geometry. Fertilizer, water and a repeated hoeing are the disturbances necessary for the maintenance of such a landscape. Tractors, marketing and social-political changes are the main forces acting as modifications from farm to parish.

The hierarchy theory seems very important for understanding the behavior of landscapes because it locates the context of a landscape at a particular scale level. The choice of the level depends on the quantity and quality of information achieved, so by definition every landscape is data oriented.

Another example of hierarchy can be proposed using the tree/forest *methaphor* of a leaf, a branch, a tree, a stand, a forest. We can assume that every leaf has a landscape composed of surrounding leaves that act as competitive elements or protecting elements. The position of a leaf is determined by branch development that in turn is influenced by other branches. The development of a tree is determined by the presence of other trees competing for light, water and nutrients. The development of stand depends on soil aspect and microclimate, and by disturbance events like landslides, fire and flooding. The distribution of forests can depend on land use and climatic changes.

Hierarchy theory can find a very interesting application in the explanation of the functioning of a landscape. We present the case of a mountain moraine gully from the Northern Apennines (Fig. 1.34a,b,c,d).

A 500 m long gully (Ospedalaccio pass, Northern Apennines) has been excavated by running water during rain peaks in Autumn and spring time. The gully is excavated in moraine deposits and, due to the high slope of the soil, has an increased depth at it's mouth.

At least three main processes are active: erosion, slope aspect and distance from the origin. The first is represented by water erosion which increases as it moves down along the gully. This process occurs at the gully scale (500 m). A second process, subordinated to the first is represented by the aspect of the two slopes. One is north-facing, whilst the second steeper slope is south facing. In terms of vegetation we have two different vegetation mosaics. A third level can be observed according to the distance from the gully surface – the deeper the gully, the wider the eroded slope. These processes have a strong influence on vegetation dynamics and their migration towards the bottom of the gully by gravitational movements. Every level has an influence on the two others moving up-down and viceversa.

In conclusion, a hierarchy of processes produces different process-oriented landscapes.

Fig. 1.34a - Large scale vision of the gully system at Ospedalaccio Pass (Northern Apennines, North Italy). Hatched line indicates the level of erosion origin.

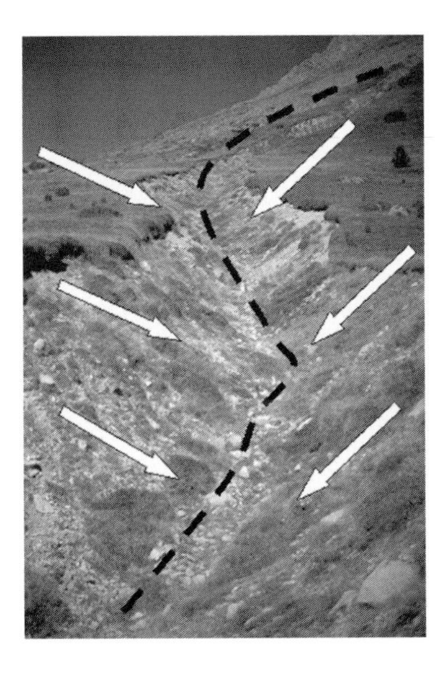

Fig. 1.34 b – Vision of different slopes along a gully. On the left the north facing slope is, on average, less steep than the southern aspect (right).

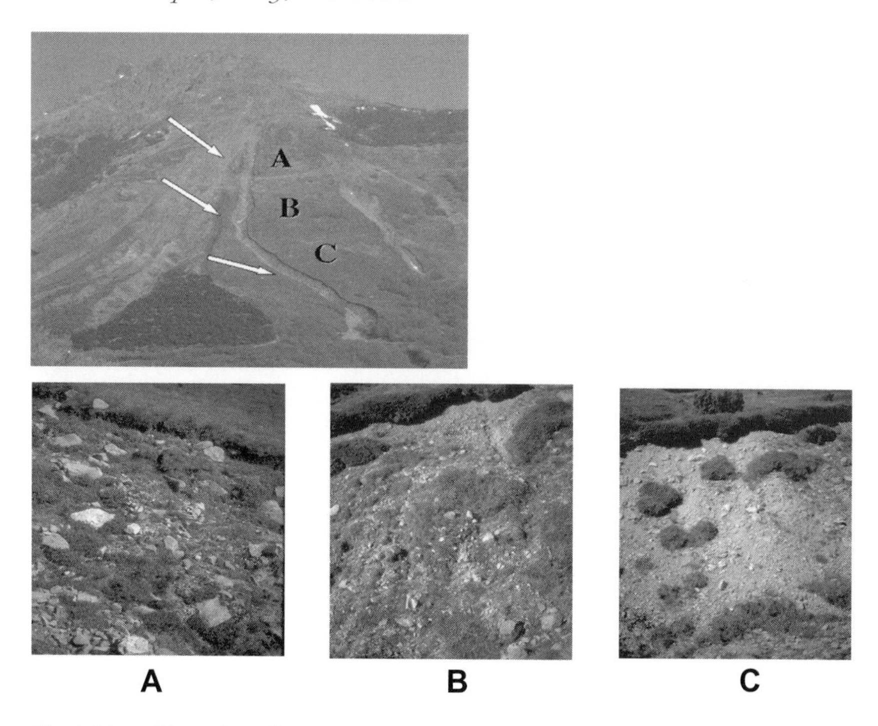

Fig. 1.34c – Along the gully different vegetation covers can be observed. Moving from the head to the mouth, vegetation (dominated by *Vaccinium myrtillus* and *Genista pilosa*) appears scarser on a more mobile soil.

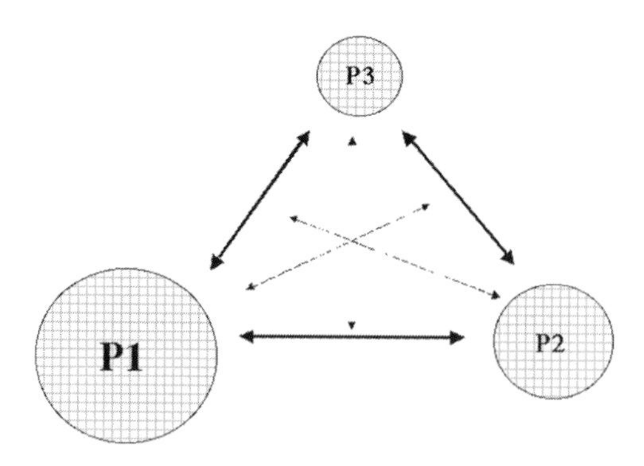

Fig. 1.34d – Schematic representation of the hierarchy of different dominant factors in gully dynamics. P1 erosion (increases as it moves down along the gully). P2 aspect (the southern slope is more exposed to micro-climatic constraints, and erosion is faster). P3 width of eroded surfaces (the larger the eroded slope and the more mobile is the soil, the greater the fragmentation of vegetation patches). Dashed lines represent the secondary feedback across this hierarchy.

1.7.2 Diversity into landscape

Diversity is considered an important index that measures the number of species and their abundance. Dominance and diversity are inversely related. A high nutrient content in the soil or favorable conditions for a species increases the dominance and decreases diversity. The richest landscapes are not the most productive, and disturbance often enhances the number of species (Fig. 1.35). The maintenance of diversity is one goal of conservation and to achieve this is necessary that some disturbance regimes persist.

Alfa, Beta, Gamma and Delta diversity have been used to evaluate the number of organisms at different spatial scales (local, regional, biogeographical). Alfa diversity measures the number of species within a specific habitat and at unitarial time. It describes how species co-occur in the same site. Generally this co-occurrence is possible because species have different body sizes. Beta diversity, or cross habitat diversity, is the diversity accumulated when new habitats are added, thus it measures the gradient of change in diversity along different sites or communities. Beta diversity can measure the amount of species uniqueness between different landscape patches defined according to a specific criterion (like the amount of wood cover, the amount of edges, etc.). Gamma diversity measures the total diversity of a whole geographical or ecologically defined region. In this way we use the diversity index at different extension.

1.7.2.1 Diversity in the soil

The "Soil-scape" appears an important and quite unexplored field of landscape ecology. The importance of soil nutrients is hampered by the action of mycorrhizal fungi diversity, as recently investigated by van der Heijden et al. (1998).

The complexity of the fungal system favors the diversity and the resilience of the above-soil system. Soil conservation is fundamental to assure the functions of the above biomass. In stressed agricultural soils the poverty in mycorrhizal fungi has negative effects on vegetation and on a large set of processes.

Diversity seems to help in maintaining stability in the system because the presence of many species means there is a redundancy in the processes; if different species develop similar strategies and if some become extinct, they may be replaced by species with similar functions.

Fig. 1.35 - The rich composition of this meadow is a combination of poor soil, a human disturbance regime (seasonal cut), livestock grazing and interspecific competition. For instance this system is less prone to weed invasion than an alfa-alfa field.

1.7.2.2 Diversity and information

In a complex system the diversity of flow facilitates landscape homeostasis such as at ecosystem level. But homeostasis, like information (indeterminacy), is difficult to measure (Ulanowicz 1997) and has been substituted by stability and information on diversity. The complexity of a system is generated by the number of combinations of possible encounters. In this case combinations increase in geometric proportion.

The logarithmic transformation, with a binary basis, of the possible combinations produces the possible number of events that have generated the observed complexity. Boltzman, and later Shannon, have applied this elementary logarithmic transformation to frequencies and/or probabilities postulating that the overall complexity is proportional to the negative algorithm of the probability that the configuration will occur.

$$S=-k \log p$$

Where p is the probability that the configuration will occur, S is the potential contribution to the overall complexity and k is a constant of proportionality. Considering that $\log(1)=0$ is a common configuration and has scant importance for explaining system complexity, this represents a very simple system.

1.7.2.3 Regional biodiversity

Regional biodiversity can be considered as a collection of relevant organisms that can be detected moving across a human scaled landscape. It is important to recognize that often when we are planning a reserve or protecting a landscape we use cues that are relevant for us but when the human perception is excluded, we generally ignore the real value of the landscape. When we consider large areas, biodiversity may often be represented by the summation of organisms living in small isolated remnants. These small (bits) of nature may play an important role in preserving regional biodiversity. Often rare habitats exist in small isolated patches (relict vegetation, vernal pool (ephemeral biotopes)). Such remnants may represent stepping stones for species movements.

1.7.2.4 The role of biodiversity in assuring landscape functionality

General empirical assumptions can be presented in favor of biodiversity in ensuring landscape functionality :

"*High species richness maximizes resource acquisition at each trophic level and the retention of resources in the ecosystem*" (Chapin III et al. 1998).
When more species are present in an area there is a higher probability that each species will capture energy and nutrients at different rates with full utilization of available resources avoiding loss from the ecosystem. If this assumption is correct, monocultures like croplands and forestry plantations should have a very low resource-use efficiency in some parts of the year and in some regions.

"*High species diversity reduces the risk of large changes in ecosystem processes in response to directional or stochastic variation in the environment*" (Chapin III et al. 1998).

It is extremely important to better understand the role of similar species considered as "redundant" in a rich environment. Such species could utilize resources better in a longer time lag thus improving the stability of the system. Diversity could be considered as an assurance policy against dramatic changes in ecosystems and in landscapes. This has tremendous implications for land management and conservation policies. In fact, the maintenance of a high degree of biodiversity should no longer considered an option only for developed countries but a global necessity to reduce the risk of unpredictable changes in ecosystems.

"*High species diversity reduces the probability of large changes in ecosystem processes in response to invasions of pathogens and other species*"(Chapin III et al. 1998).
If a system is rich in species the probability that a new invasive species will find a suitable niche is low. This assumption, yet to be proved, is important for future research especially in disturbed simplified croplands. The cost of reducing pest invasion could higher than the cost of managing ranges, croplands and forests using a more ecosystem perspective. The changes produced by ecological invasions could be reduced if high diversity were to act as resilient tool. This concept could be expanded to the genetic variability of cultivated plants.

"*Landscape heterogeneity most strongly influences those processes or organisms that depend on multiple patch types and are controlled by a flow of organisms, water, air, or disturbance among patches*" (Chapin III et al. 1998).
Many processes are influenced by landscape heterogeneity because they interface with patch borders.

1.7.2.5 Biodiversity hotspots

The term "biodiversity hotspot" was coined by Norman Meyers (1989, 1990). A biodiversity hot spot is considered to be an area in which the number of species is very high compared with the surroundings, though actually it is mainly applied to a geographical region in which biodiversity is particularly high. But this concept has been used to find conservation priorities regarding rarity and endemism. For instance, GAP analysis (Scott et al. 1993) has been extensively used to locate biodiversity hotspots.

There are megadiversity countries (for instance, Indonesia, Brazil, Colombia and Costarica), and regions with relatively few species but rich in endemism (Madagascar, Australia, New Zealand, atc.). Finally, there are hotspots for threatened species independent of local or regional richness. There is a weak geographical relationship between hot spots of different taxa and different spatial scales. A criterion to protect at least a small part of each eco-region appears a sufficient approach to intercept the diversity of many species.

1.7.3 Source-Sink model

The heterogeneity of the landscape is responsible for the different patch quality as appreciated by living organisms; there are more suitable and less suitable habitat patches (Fig. 1.36). Organisms share different habitat patches according to this "suitability" and their permanence, survival, and reproductive success largely depend on these patch attributes.

Recently Pulliam (1988) presented a demographic (source-sink) model to better understand this behaviour. The model born, which was developed to evaluate suitability by monitoring birth/death ratios of animal populations, has been recently expanded into a more general model of habitat patch suitability.

This model represents a more detailed descriptor of a complex reality of quality mosaic and is particularly useful for quality assessment because it is independent of the local abundance of organisms.

This model can consider mortality but also the maintenance of body weight of one set of organisms or population, and can be used in a comparative study (Blondel et al. 1992, Donovan et al. 1995a,b, Dias 1996).

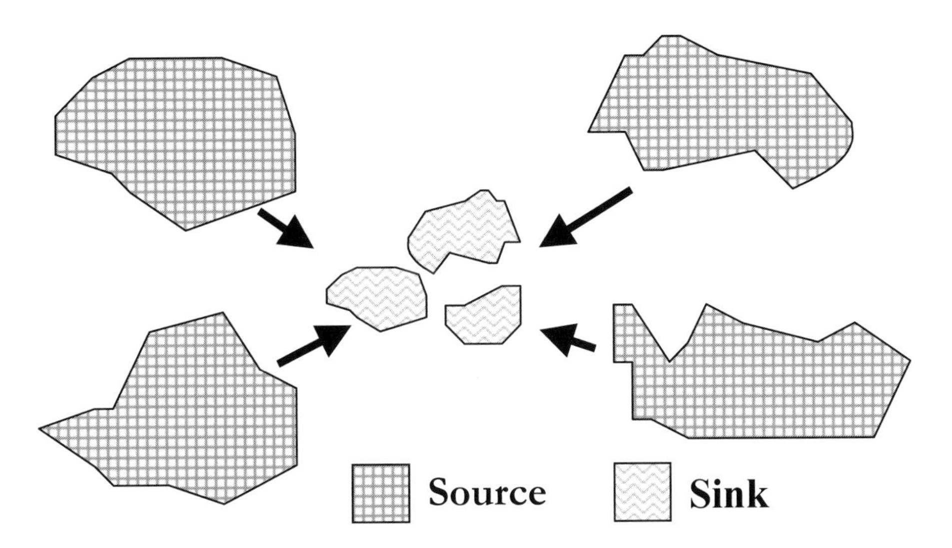

Fig. 1.36 - The source-sink model considers the source population (patch) as exporting individuals and the sink population (patch) as absorbing this surplus.

The demographic rates of organisms of the same species living in habitats of differing quality can be expected to vary at the level of sub-populations. In the source-sink model, a source is a population in which births exceed deaths, and emigration exceeds immigration (Pulliam 1988). Conversely, a sink is a population in which deaths exceed births and immigration is higher than emigration. Without new immigrants a sink population is prone to extinction. This model can be used to evaluate the patches that harbor source or sink populations. In this way we can call source patches as those that support a source population and the opposite for a sink patch.

The demographic model assumes that nT is the number of individuals at the end of winter time. Assuming that no individuals die during the breeding season, and each adult produces a β offspring, then at the end of the breeding season the population will be $nT+\beta nT$. If adults survive with probability Pa and juveniles with probability Pj, at the beginning of the next spring the population size will be:

$$nT+1=PanT+Pj\beta\, nT=nT(Pa+\beta Pj)=\lambda nT$$

$\lambda=(Pa+\beta Pj)$ where λ is the finite rate of increase for a population. This measures the growth rate of a population in a specific habitat.

Assuming that $n1^*$ is the maximum number of individuals that can stay in a source habitat, the number will be $\lambda 1 n1^*$ at the end of a non breeding season. Of these, only $n1^*$ can stay and the remaining $n1^*(\lambda 1\text{-}1)$ must emigrate in a marginal patch habitat (sink). In the sink habitat the population at an equilibrium will be composed of $n2^*=n1^*(\lambda 1\text{-}1)/(1\text{-}\lambda 2)$. $\lambda 1\text{-}1$ is the per capita reproductive surplus in a source habitat and $1\text{-}\lambda 2$ is the per capita deficit in a sink habitat. When the reproductive surplus of a source habitat is larger than the sink habitat, the density of individuals in the sink may be higher than in the source. Without a continuous immigration, sink populations would soon become extinct. The attribute of source or sink may be more significant if a longer time window is used than the variability of patch quality from one year to another. With regards to long time periods, a patch may be considered a source or sink after the calculation of the geometric mean of growth rates.

$$\lambda=(\lambda 1\lambda 2\lambda 3 \ldots)1/t$$

When mean λ exceeds 1 the patch habitat must be considered a source, if $\lambda < 1$, it is a a sink. Another way is to separate the periods in which that patch is a source from the periods in which it is a sink.

There are conditions in which a patch behaves as a pseudo-sink (a term coined by Watkinson & Sutherland (1995)). A pseudo-sink is a patch in which, in the absence of new immigration, the population declines but persists at a low level of density.

The source-sink model has been proved in many investigations of plants but generally plant behavior changes according to location, climatic regime and competition. Pulliam (1996) has recently revised some examples of source-sink models applied to plant demography. The attention of botanists has tended to concentrate on small plants with a short life cycle like *Plantago* or *Stipa*. Many plants have ambiguous behaviour according to the substrate. Some species have the demography of a source type in humid sites and a sink type in arid conditions. Others change their demography according to the seasons.

Fig. 1.37 – Wormwood (*Artemisia* sp.) is a common weed migrating from abandoned fields to river bed. Along the river, wormwood forms a sink population that disappears when flooding events occur.

Wormwood, a very common weed in the sub-humid Mediterranean, is a true pest in abandoned fields where it adopts a source strategy of exporting a great number of seeds, including along river beds. In these habitats that can be considered sinks, the growing season is favorable but the summer scarcity of water and the highly permeable sandy soil generally produces a severe drought, thus reducing the abundance of these plants before the floods in the Autumn (Fig. 1.37).

Pulliam (1996) reported many examples of vertebrates in which demography is one focal theme. Clear examples of the source-sink effect are very frequent especially in highly mobile species. The so called marginal areas are sink areas in which individuals moving due to forced dispersion, as in many small rodents, have to adapt to live in sub-optimal habitats. Caribou (*Rangifer tarandus*) live in large areas of forest and tundra from Eurasia to North America. In tundra the predator pressure by wolves during the breeding period is less than in forest habitats. Tundra may be considered a source habitat and forest a sink habitat.

Some habitat patches may be very productive for intrinsic reasons and in this case may support a population of source type, but with changing environmental conditions the same habitat patch may represent a sink habitat for a species.

The source-sink model is also very useful for assessing the quality of the land mosaic outside the breeding season. Wintering birds in the Mediterranean region select open fields and orchards. The quality of the composing patches can be measured based on the net intake of resources by individuals. Increase or decrease of body weight may be a good indicator of patch quality although other mechanisms related to specific physiology must be considered (Joalé & Benvenuti 1982). The selection of a patch is often determined by structural attributes more than by real foraging resources. Such types of conditions are often the result of man-made manipulation and such patches may be considered ecological traps.

Most agricultural areas, if rich in edges, attract breeding birds because edges often have a denser foliage cover than undisturbed woodlands, and food resources in nearby crop lands are very abundant. However human disturbance and predators interact with the edges increasing the probability of nesting failure (Fig. 1.38). For instance, red backed shrike (*Lanius collurio*) breeds regularly in Mediterranean wine-yards but disturbances by spray stewardship in late June often results in the abandonment of nests by laying females. A replacement brood, when successful, has less offspring and juveniles have only a short time to complete their development before the late summer migration.

Fig. 1.38 - Hedgerows (*Fraxinus ornus* and wine grape) in cultural landscapes can be considered ecological traps because they attracts many organisms and also predators like and Jays (*Garrulus glandarius*) (A) and stone martens (*Martes foina*) (B).

The concept of niche finds a good interpretation in the source-sink model. According to Hutchinson's model of niche it is possible to distinguish a fundamental niche as a set of environmental conditions within which a population exists (Hutchinson 1957). The realized niche is the set of conditions occupied by a species in the presence of another species. In general the realized niche is considered smaller than the fundamental niche because some habitat resources are used by the other species. The realized niche without the presence of competitors may be traced using $\lambda=1$. In the second example the realized niche in the presence of a competitor is smaller and the $\lambda=1$ delimits a more restricted space. In the last example we can consider into the realized niche areas in which $\lambda < 1$ is part of this "expanded" niche. The expanded niche is realized by the immigration of individuals in sub-optimal conditions which occupy more than the fundamental niche. These three models are very useful for understanding the possibilities of a species utilizing the space. The dispersal capacity and the probability of finding a suitable habitat patch are an important component of the demographic process.

It emerges from this model and the empirical evidence that not all suitable habitats are occupied and that the presence of high density and persistent populations is not always an indicator of source habitat but exactly the opposite; in many species most individuals are found in "unsuitable" sink habitats (Fig. 1.39).

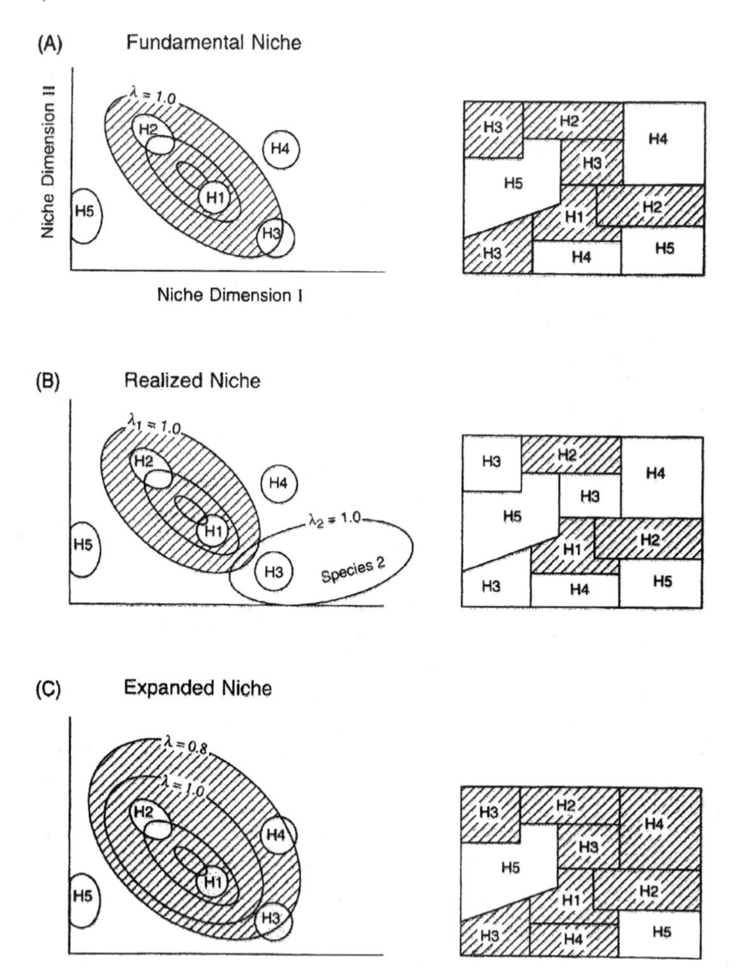

Fig. 1.39 - Schematic representation of niche dimension in relation to the source/sink value of a habitat patch. In A the condition of a fundamental niche in the absence of competitors is represented. In B the realized niche in the presence of a competitor is represented, and in C the expanded niche that has different λ values (from Pulliam 1996).

1.7.4 Metapopulation theory and landscape

Although the distribution of organisms in a space according to a patch model was already quite clear in the first part of this century, it is only recently that ecologists have shown an interest in the unequal distribution of organisms. Natural and human induced fragmentation of natural habitats has created conditions for a partial isolation of organisms in suitable "islands". Richard Levins coined the term "meta-population" in 1969 and 1970 (Levins 1969, 1970), defining a metapopulation as a population of local populations linked by exchanges of individuals. A metapopulation structure is based on dynamism, colonization and extinction rates and on genetic transferring. According to Hanski (1997) some basic assumptions or conditions are preliminary to the classic Levins model: that suitable habitat occurs in discrete patches; that patches can be occupied by breeding populations; that large local populations are at risk of extinction; that habitat patches must not be too isolated to prevent re-colonization after extinction; and that local populations do not have synchronous dynamics.

The classic Levins model is based on the following equation:

$$\frac{dP}{dt} = cP(1 - P) - eP$$

where c and **e** are the colonization and the extinction parameters. P is the fraction of occupied patches, and 1-P the fraction of empty patches.
At the equilbrium

$$P = 1 - \frac{e}{c}$$

and a metapopulation is predicted to persist when P is positive

$$\frac{e}{c} < 1$$

and progress towards extinction when this ratio is >1.
Today, many models of metapopulation have been created to describe different demographic and spatially explicit conditions (Hanski & Gilpin 1997). Such models can also be related to the source-sink model covering a broad range of geographical uniquenesses.

Natural metapopulations have several structures that affect persistence and coexistence, and often local (within-population) processes can strongly affect metapopulation dynamics (see for more details Harrison & Taylor (1997)).

It is not our intention to enter into further details on this subject, which has already been extensively reported in a rich literature, but Wiens (1997) has recently discussed the relationships between the metapopulation theory and landscape and it seems quite clear that a double benefit can be found from coupling the metapopulation theory with the landscape ecology paradigm. In reality, landscape ecology is quite poor in theory and it is important to create a more robust theory to develop this discipline for the future. That a landscape is not simply an heterogeneous medium composed of patterned patches embedded in a hostile/neutral matrix, but that the land mosaic is regulated by complex processes covering physical and biological entities, appears today a realistic vision.

From this vision it is important to develop new models to interpret such complexity. For instance, the linkage between metapopulation and percolation theory can be useful to explore these new fields.

1.8 Processes in landscape

There is a multitude of natural processes in our world but some are better known than others, for instance: erosion (water and wind (Zhang et al. 1995)), flux of nutrients (Kesner & Meentemeyer 1989), disturbance (secondary succession, biomass reduction (grazing, burning, storms)), animal and plant movements (dispersion, migration, extinction), toxic spreading (Fahrig & Freemark 1995), and the economy. The variety of processes that can be intercepted using the landscape paradigm is very high.

Most of the processes are active at different scales but are particularly evident at "mesoscale", the scale at which most organisms and their aggregations lose their distinctive characters and create a mosaic of "homogeneous" patches. No process is "created" rather it is the consequence of a continuum of events that are active by a cosmological input of energy. Energy accumulation, high potential energy at the topological level and a threshold of stability of a system, are the ingredients for the starting of a process. The accumulation of the effects creates the final process which we can describe using a multiscalar approach.

For example, economy is a process that can be usefully employed to evaluate the richness of a region. *Per se* this indicator is not enough to form a complete scoring of the quality of regions, it should be coupled with processes linked to physical and biological phenomena.

1.8.1 Physical and biological processes

According to their origin it is possible to divide processes into two broad categories - biological and physical - but often their effects are indistinct. In fact some processes of biological origin or causes, as in erosion by trampling, produce the same effects when fully developed as the physical ones. In this case the soil is easily eroded because it is exposed to erosive physical processes. The soil can experience the same type of erosion and transportation when exposed by a landslide or by a fire that has burned vegetation cover.

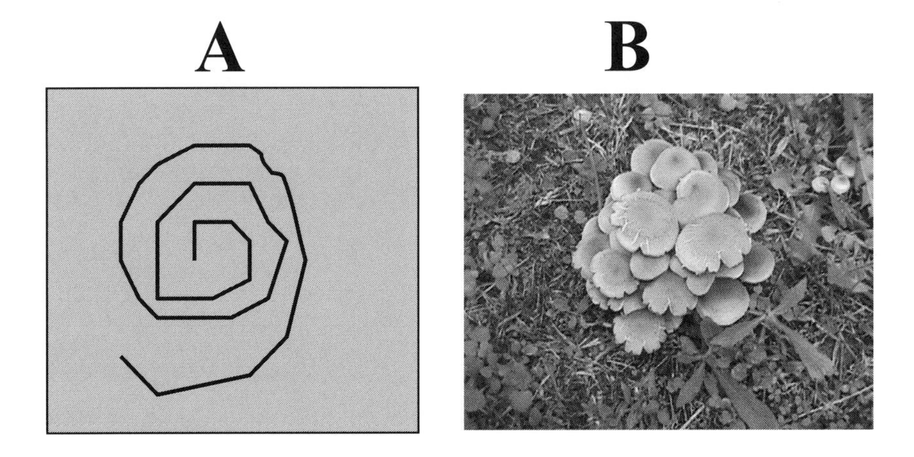

Fig. 1.40 - A centripetal process creates an ecological constraint and produces distinct patterned objects (A), like a mushroom clone (B).

1.8.2 Centripetal and centrifugal processes

According to their behavior across the landscape and their effects processes can almost be divided into two groups. One group creates "local" patterning, increasing local specificity, uniqueness or local complexity with a centripetal direction like succession or fragmentation (Fig. 1.40).

A second group, considered a centrifugal process creates a connection between different parts by acting across the space (Fig. 1.41). Pertaining to this group are heating, water and wind fluxes, organism dispersion and migration. Using well known objects the processes of the first type are housed in "patches" and the second type in "corridors".

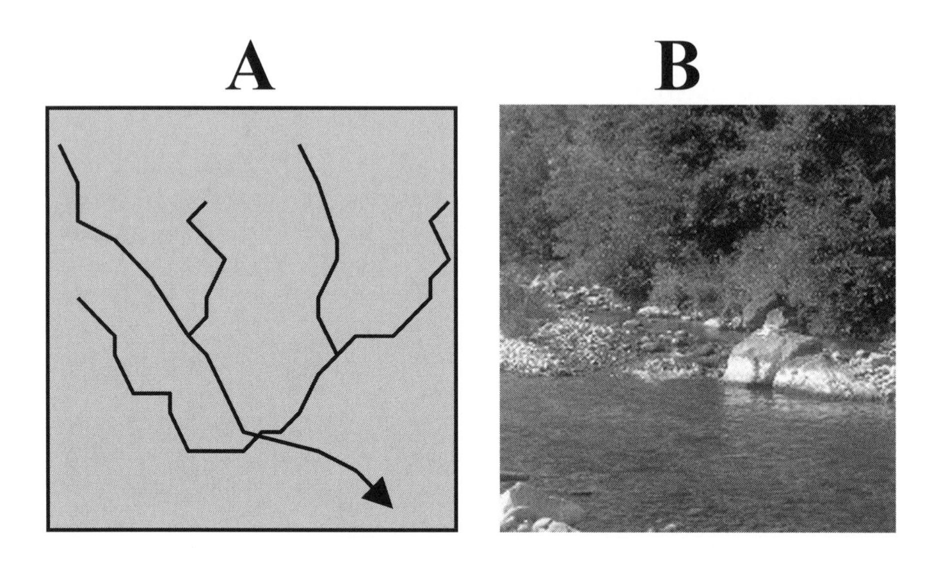

Fig. 1.41 - A centrifugal process creates a flux of material, energy or information between different objects spatially arranged in a landscape matrix (A), for instance a mountain stream (B).

Both groups increase their visibility thus enlarging the spatial scale. In large patches, succession or gap replacement acts in a less disturbed way, and in large areas erosion by wind fluxes accumulates energy from the surrounding components.

Interactions and alternations between centrifugal and centripetal processes create the complexity of a landscape. Often when a process has less energy the other force prevails. For instance, when the water flux along a river bed is reduced by water scarcity the centripetal process creates patches of vegetation but when the centrifugal process (initiated by water flooding) prevails, the centripetal process looses importance and vegetation is removed, dispersed or seriously damaged.

Organism dispersion (centrifugal process) occurs after a period of centripetal process like reproduction. Structuring (centripetal) and removing (centrifugal) are two important types of dynamics in ecological systems. This model can be easily transferred to human processes like money investment or viral contagion. Moments of "organization" are interrupted by moments of fluxes.

1.8.3 Local and regional processes, some examples and evidences

The limit between the summation of effects of a local process and the presence of a new genuine regional process is very confused. One of the more classic examples is represented by local (alfa) diversity, mosaic diversity (beta diversity) and regional (gamma) diversity (Whittaker 1977). The same process (biological diversity), when appreciated at different scales, changes the relationship with the ecological substrate. Diversity is a real process because when we explore the effects on the ecological system we can also observe severe consequences. For instance, a high bird diversity affects many invertebrate populations and also fruits and seeds. Climate appears to be the result of many overlapping and competitive processes that characterize a region. The hydrological regime of a river is the combination of area, climate, morphology, land cover, land uses, and disturbance regime.

By enlarging the geographical scale, processes will appear that have an *in situ* origin but at the local level it is external events that dominate. For example the heterogeneity of a woodlot of 10 ha can't influence regional climate, but a vegetation mosaic at the scale of 10,000 ha, has strong influences on the regional and local climate.

The difference between patterns (effects) and processes (causes) is not always easy to distinguish (Wiens 1989) but when we observe a pattern, it is evident that certain processes must have produced such a pattern. We expect to have a very short time lag between the two elements whilst in other cases the time lag may be so elapsed that is difficult to identify the process responsible for the observed pattern. In this case the patterns represent a type of landscape memory. This is the case of land abandoned in many regions in which the land cover mosaic (woodland and open spaces) has been conserved although cultivation has long since vanished (Farina 1994, 1995a, 1996b).

1.8.3.1 Decomposition processes across soil and river systems

Leaching is the main process that removes the soluble component of a substance. Litter comminution is the conversion of large particles into small ones, undertaken especially by invertebrates and microbial catabolism.

Decomposition may be considered a cascade process in which leaching and organismal activity is subjected to further self-similarity processes (Fig. 1.42).

The parallelism between carbon and nutrient processes of rivers and soil seems extremely important (Wagener et al. 1998). Particles reduce their size as they move down through the soil and downstream in rivers. The carbon-to-nitrogen ratio decreases for the immobilization of carbon and microbial respiration.

Invertebrates and microbes reduce the size of litter and in turn, through digestive processes, offer new food sources for further microbe or invertebrate assimilation.

Soils and rivers have a similar process at the beginning when the main source of biomass is represented by leaves. This material is decomposed mainly by fungi. The organic material produced by this first process is transported in the middle part, the sublitter region of the forest floor, where it is available for primary producers. In streams, macrophytes and periphyton are the main input. The complexity of litter decomposition in soils seems higher than decomposition along a river, because the spatial scale is different. In soil decomposition most of the processes take place within a few centimeters thick and the different phases are difficult to separate.

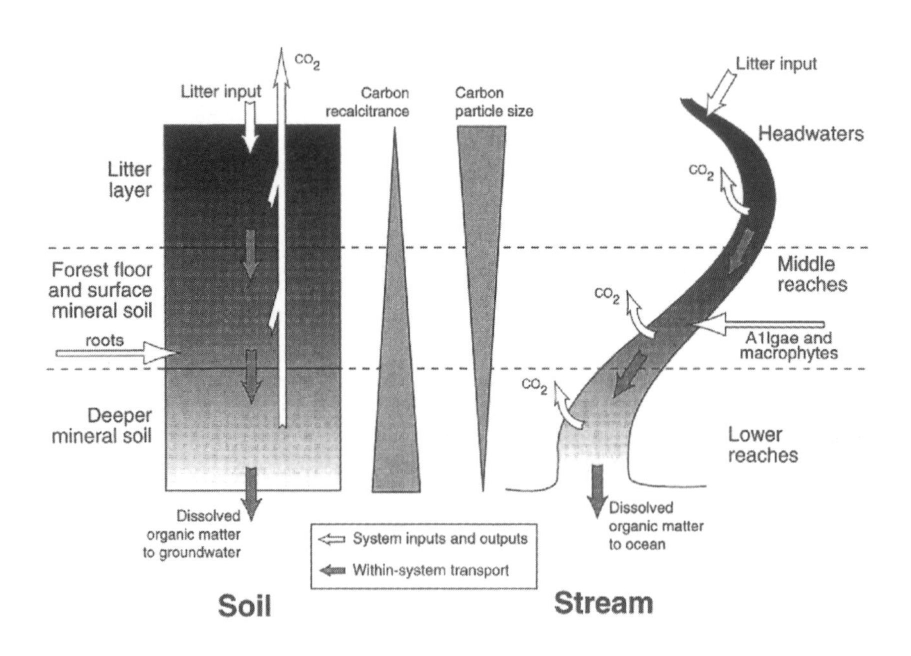

Fig. 1.42 - A comparative scheme of litter flux into soil and along a river. In both cases, the movement of carbon is accompained by a decrease of particle size of organic matter and microbial activity (from Wagener et al. 1998).

A river system is a good example of the complex integration between different components of a land-water mosaic. Often, river functioning has been studied as a separate component of the landscape. However, recent evidence demonstrates the complex relationships between aquatic and neighboring terrestrial systems (Gregory et al. 1991). Although the main ecotone is located between open water and river/stream banks, there are more complex interactions along a river. At the headwater, there is a large input of material into the active bed but, downstream, material is exchanged from the active bed to the adjacent terrestrial environment. Hence, the behavior of a river changes according to the distance from the source.

The effect of land use on the river basin is well known especially in terms of organic material, erosive particles, nutrients and pollutants moving downwater into the river sink. It seems evident that the simplification of land use, from natural habitats to croplands and urbanization, reduces the physical and biological complexity of rivers (see Schlosser 1991).

1.8.3.2 Snow cover and vegetation

Snow cover affects the distribution and type of vegetation in Alpine-type biomes (Fig. 1.43). This process is well documented by the great diversity and patchiness of Alpine vegetation that follows closely the process of snow cover, snow persistence and snow melting as observed by Walker et al. (1993). The moisture gradient in the soil is another important component in this process. In many alpine conditions, water represents a limiting factor for the high permeability of topsoils in summer time.

Snow accumulation is again linked to the wind pattern. Small scale effects are constrained by regional effects such as the varying behavior of the vegetation normalized index (NDVI) across the Great Divide in the Rocky Mountains. A significant regression has been found between NDVI and the elevation gradient. The West Divide has a distinct pattern compared with the eastern but in the East Divide on the west-facing slopes which are more exposed to western winds, there is no significant trend between NDVI and altitude.

1.8.4 Regional landscape approach

This terminology refers to the study of the land mosaic at a broad scale assuming that the adopted scale allows the comprehension of processes pertaining to the persistence of populations and communities. This approach is very popular between planners and landscape architects, but despite this popularity this approach is far from being considered fully integrated with the ecological processes that are at the basis of the expressed patterns. We recognize the importance of this approach but at the same time we must be alerted about the imprecision by which, following this approach, the reactions of a system are understood. Often the best result from this approach is the recognition of the complexity of a set of systems and the incapacity to interact with the systems that are present in the defined area. Sub-systems appear unbounded and some boundaries are an ecological non sense.

Fig. 1.43 - Snow accumulation along mountains creates vegetation patterns and also affects ground water circulation. Gran Sasso d'Italia Mt. (Central Apennines) in Spring. Snow accumulated in depressions will condition yearly summer plant diversity and vegetation dynamics.

This approach assumes a great validity when processes that have a geographical scale overlapping with the described patterns are considered.
It is possible to appraise trends in land cover quality and temporal changes of selected areas – both of which are useful exercises. The regional scale is particularly suitable for monitoring regional changes and for developing conservation strategies.

Human expansion in pristine environments and the changes which have occurred in long-term developed areas, like the Mediterranean basin, are interacting with a growing number of ecological processes.

The maintenance of large scale processes is vital for every small scale "ecosystem" and, considering the broad time scale at which most large scale landscapes change, long term monitoring actions are necessary. The inadequacy of the ecosystem approach is quite evident, because small conserved areas have a higher probability of biodiversity erosion. The hostile matrix in which the conserved areas are embedded assumes an increasingly unsustainable characteristic for the "imprisoned" populations. Conservation based on boundaries is often a human artifact and not an ecological function.

Research towards a homogeneous area as a prevalent criterion in classic ecology is, today, far from ecological functioning. It seems heterogeneity is more important for sustaining biodiversity and, again, heterogeneity is a property of the scale of landscapes.

The contemporary use of a coarse and fine filter approach to conserve biodiversity may have some meaning. In fact, homogeneity and heterogeneity are both implicit attributes of an ecological system. Homogeneity is a matter of scale as is heterogeneity. When we classify a large scale landscape of rural areas and woodlands we assume both land uses are homogeneous at that scale because we maintain the same resolution by reducing the viewing window. However, this reduction is a mistification of reality because the resolution at which we have distinguished the "homogeneous" pattern is typical of a large area; when we observe a pattern at an appropriate finer resolution, the heterogeneity reappears. Heterogeneity has played a fundamental role in assuring the diversity of life at all levels, and changes in this pattern will have effects over both short and long periods. Conservation strategies, today, relate to short term effects; attempts to predict the long term effects of the modification of heterogeneity are scarce. Heterogeneity is important for many species of organisms such as amphibians and fish. Changes in habitat preferences from larval to adult life require a heterogeneous environment in which the aggregation of some habitat patches must have a precise dislocation. Changes in habitat are evident in most Paleartic birds. Many species require different habitats according to the season. The robin (*Erithacus rubecula*) spends the breeding season in dense woodlands, but outside the breeding season it may be found in rural areas or scattered trees and shrubs in grasslands.

1.8.5 Connectedness and connectivity

The heterogeneous character of a landscape has a great influence on plant dissemination and animal movement. In the human landscape, heterogeneity is considered a mosaic of suitable and hostile environments.

Patches placed in a matrix in different positions assume a different level of geographical connectedness. The distance between different patches (connectedness) is an easy pattern to measure using simple Euclidean geometry. Basically, most of the mapping output is carried out using a simple distance measure between the different patches. The distance may be measured border-to-border or assuming a central point internal to each patch.

Connectivity is a species specific process by which a species perceives the neighboring areas as fragmented, less connected or full connected. So the same landscape is perceived in different ways by different species that interact with processes in a specific way.Connectivity is a very important attribute of a landscape. As discussed by Baudry & Merriam (1987), connectivity is species specific and must be distinguished from connectedness, which is a unspecified distance from selected patches. Connectivity allows animals that require different types of habitat patches to move freely, selecting the optimum place in which to spend a segment of their life.

Connectivity is one of the first attributes of a land mosaic to be reduced by fragmentation and by human intrusion with infrastructures like roads, intensive fields, urban and industrial settlements (Fig. 1.44). Dispersal is a fundamental trait of most vagile animals and the suitability of a landscape must meet such a requisite. Often animal dispersion is more related to the spatial configuration of habitat patches than to the distance between patches.

There is much empirical and theoretical evidence of this process. Animals are influenced by the spatial arrangement of resource patches and this may be of variable importance depending on the severity of climatic conditions occurred among a critical period.

For example, a severe winter with a long period of snow cover can affect the survival of species like elk and bison in Yellowstone National Park when burned patches are fragmented, but when burns are clumped survival is higher according Turner et al. (1994). In mild winters, the burn pattern becomes unimportant.

Landscape processes, ecosystem processes and biodiversity are linked to each other across scales. We have to abandon the paradigm by which the effects occurring at the small scale, move only toward the top and not vice-versa. The gully erosion on the Apennines slope represents a good example (see Fig. 1.34a,b,c).

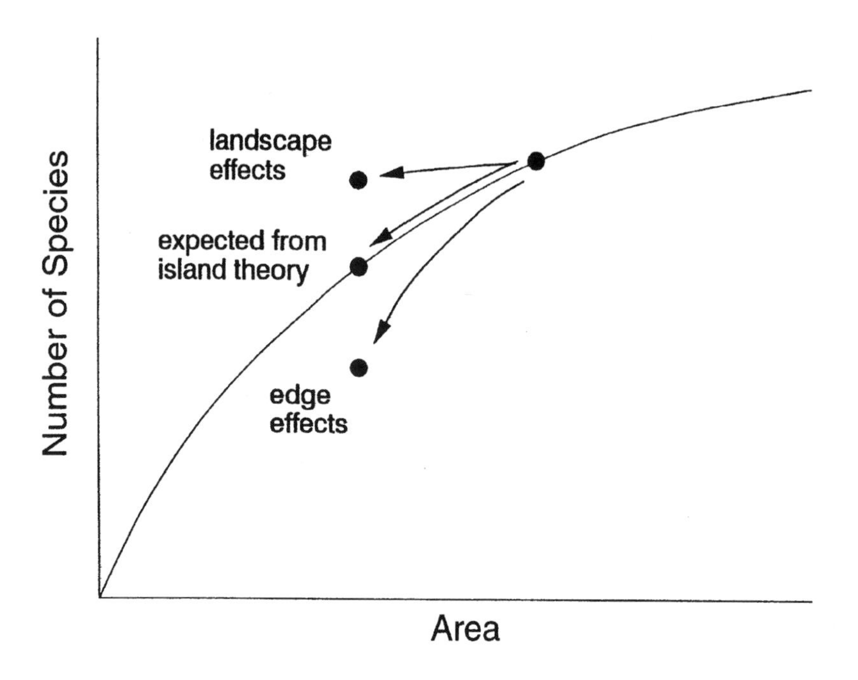

Fig. 1.44 – Landscape heterogeneity can produce differences in the relationship between number of species and available area. Connectivity, shape and patch size are some of the landscape components that can alter island biogeography prediction theory. In addition, edge effects can modifiy this prediction (from Wiens 1997).

Although erosion is produced mainly by the movement of surface water which finds it's initial channel for the transport of water in a depression, exposure of the gully and vegetation patterns are both responsible for the prosecution and arrest of the erosive process. In fact, if we plant pioneering vegetation and regulate gully with woody dykes, the erosive process along the gully is dramatically interrupted.

Connectivity is a measure of distance between favourable or hostile habitat patches for plants as well animals. The distance is measured assuming a different caliper according to the organism. The caliper may be of a few centimeters for a small terrestrial arthropod or several kilometers for a raptor bird (Fig. 1.45).

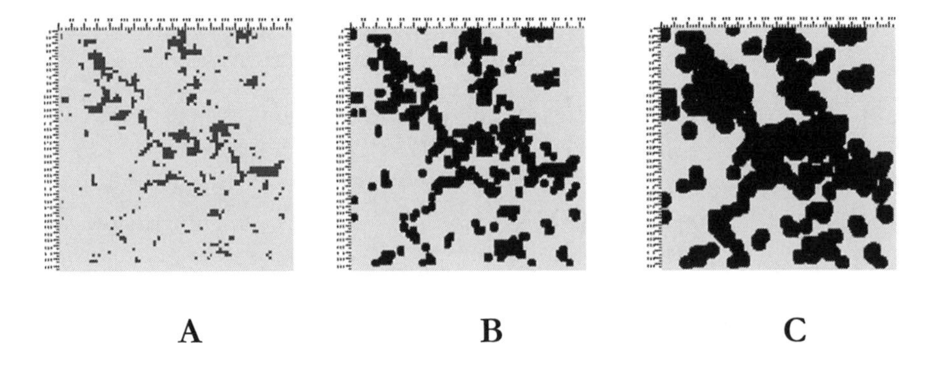

Fig. 1.45 - Two levels of connectivity based on a random landscape in which the target area (in black) (A) appears increasingly clumped if a larger neighborhood is assumed. For instance, a more vagile organism perceives the mosaic as being more connected than a sedentary species. In this case a neighboring of 10 (B) and 20 pixels (C) has been assumed to simulate the mobility of an organism.

But, due to the different behaviour and needs of a species throughout different periods of the year or it's life-cycle, the ruler size changes depending on the different processes with which we are dealing. Again we have to scale the environment in which a species is living.

For highly vagile animals and small seeds dispersed by the wind, the connectivity of a landscape may be high even in conditions of apparently high fragmentation.

The concept of connectivity is not isolated from the model of the metapopulation. In fact, if a species is structured as weak sub-populations that have a genetic interchange we must assume that a certain level of connectivity occurs. Also, it is strictly linked to the concept of corridors. The presence of a corridor, as already discussed, increases the value of connectivity. For plants, connectivity is strongly influenced by the availability and capacity of vectors for displacing seeds. In windy areas the circulation of seeds, pollens and spores is favored and the value of connectivity is high. An opposing case is represented by natural and man-made barriers like water bodies, desert areas, etc. (see also Henein & Merriam 1990, Hansson 1991) (Fig. 1.46).

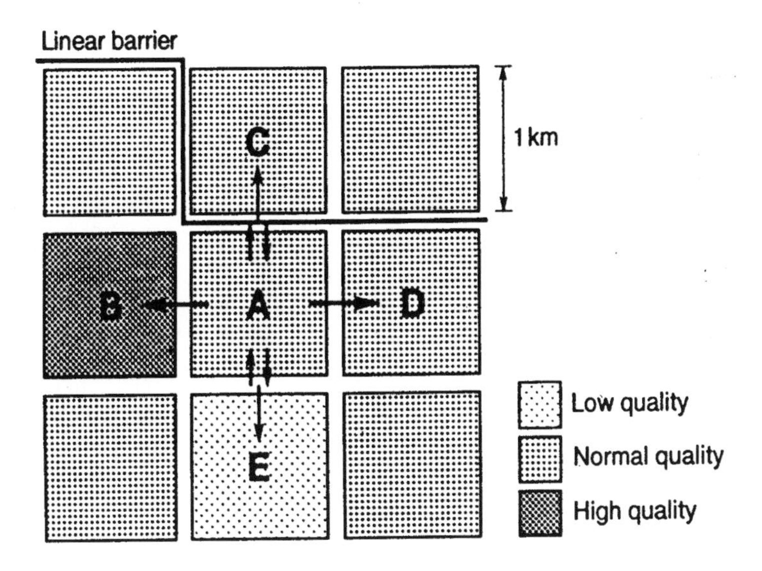

Fig. 1.46 - The quality of surrounding patches and barriers like motorways can greatly disturb the movements of animals and plants. Movements from patch A to patches B and D are very probable, but from A to E some reflection can be expected due to the low quality of the patch E. The presence of a barrier between patches A and C will prevent any movement (from Schippers et al. 1996).

Due to the functional properties of connectivity, landscape quality is quite difficult to measure. Many approaches such as the diffusion process, random walks and metapopulation models, are available for studying animals in a spatial arena (f.i. Johnson et al. 1992)

Schippers et al. (1996) have used a GIS-based random walk model (GRIDWALK) to predict dispersal capacity and habitat connectivity of badgers (*Meles meles*) in the central part of The Netherlands. The model was based on the probabilities of different movement choices. In order to determine dispersal behavior, the following parameters were considered: movement within one landscape type, movement between different landscape types, mortality, interactions with linear barriers like roads, and connectivity between populations. This model seems very promising for evaluating the dispersal capacity of animals that require different types of patchy habitats in a human disturbed system.

1.8.5.1 Patchiness and resource allocation

In a heterogeneous mosaic the probability that the composing patches change in quality and in the resources offered is quite high. When the quality of the composing patches changes, organisms may adopt at least one of the following strategies:

stay at home and wait until conditions are more favourable
move away by seasonal dispersal
move across the landscape and search in the best habitat patches
finally use more-than-one patch-habitat contemporarily (see f.i. Kozakiewicz & Szacki 1995).

From insects to mammals, individuals moving across a heterogeneous media often use field cues to create cognitive maps. Animal movements have been studied with the application of a dynamic state variable model by Baker (1996), who tested the model on northern bobwhite (*Colinus virginianus*). The author considered both the spatial qualities of the habitat and the state of the animal. The method, developed in the 1950s, was employed mainly in inventory scheduling and production control. The dynamic programming allows the animal choices to be optimized, reducing the computation time.

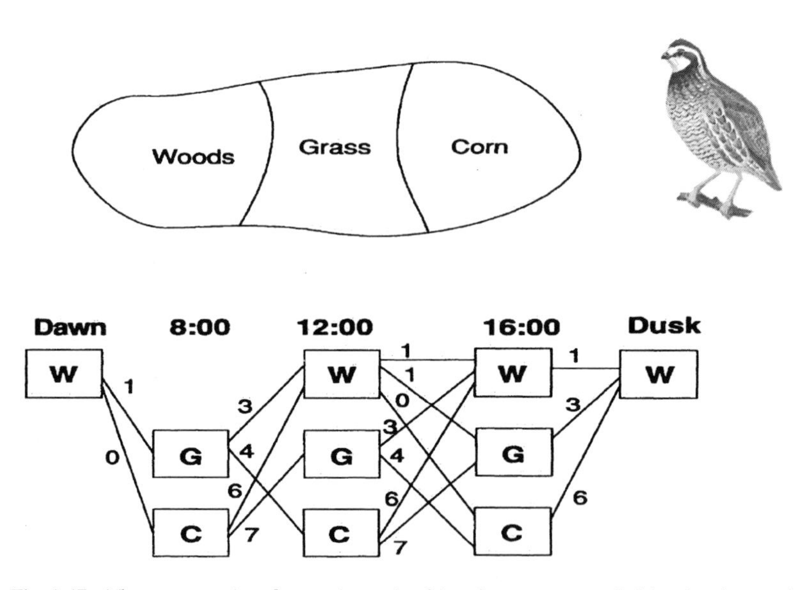

Fig. 1.47 - The energy gain of a moving animal in a heterogeneous habitat has been calculated assigning a different score for each habitat patch according to the different energy requirements of the species. The optimal solution is calculated selecting the best score according to the three possible uses of the habitat patches. This example has been based on bobwhite (*Colinus virginianus*) behavior (from Baker 1996).

In Fig. 1.47 an example is presented. In this case, a deer can move from woodland to grass and corn fields throughout a day. Starting at the end of time sequence it is possible to calculate the best patch (patch selection) for a bobwhite during three time sequences into which the day has been divided

(8.00, 12.00, 16.00). The best path in this example is represented by the sequence W+C+G+C+W with an energy gain of 17 (W=wood, G=grass, C=corn).

1.8.5.2 Meta-community concept

The term meta-community refers to the distribution of a community type across patches in a landscape, though it is not a common literary term (Lidicker 1995). It is related functionally with a land mosaic across the functional groups that compose the assemblage. The geographical scale is quite large. In each sub-community it is possible to find a patchy mosaic occupied in a different manner differently by the composing species. The resolution used to identify a meta-community is by necessity coarser than that used for a community. A population may be found in different communities across a large area and the density of each population may be reflected in the different interactions with the different species found across the meta-community.

1.8.5.3 Patch selection

Patches offer different services for a species and this variability may depend on the patch itself (seasonal availability of resources) or on temporal changes in the physiological demands of organisms. Patch selection depends on these factors but also on the intrinsic quality of the patches and their modification under different disturbance regimes. The combination of two habitat patch types may positively influence a species. For instance, a dense woodland surrounded by crop lands provides an optimal mosaic for many frugivorous birds. The selection of optimal patches is a common feature of vagile organisms and from this perspective, the distance between favourable patches and the degree of hostility of the matrix are very important for the dispersion success of organisms (Diffendorfer et al. 1995).

1.8.5.4 Organism density and heterogeneity

The abundance of organisms in heterogeneous environments is difficult to predict without considering the whole system in which organisms live. In fact the density of organisms in the landscape is affected by many variables and in some cases the mosaic geometry is important for determining the species assemblages (Haila 1988). Often, animals such as birds have an adaptive behaviour related to mosaic conditions which affects their distribution. Blue tits (*Parus caeruleus*) and great tits (*Parus major*) can be found at higher concentrations in breeding areas when abundant food resources are juxtaposed with few breeding sites.

1.8.5.5 Foraging behavior and patchiness

Fragmentation is increasing world wide and organisms are generally negatively affected by this new situation of the human disturbed regimes (Lord & Norton 1990), which often seem to be an important driver of prey-predator relationships.

The climatic changes that are occurring at all latitudes create a dramatic crisis for many organisms and related processes. For this reason an in-depth knowledge of fragmentation processes seems to be extraordinarily important as a predictive function of ecology. Natural patchiness is not only an attribute of plant populations but is present also in many groups of animals, however the difficulty of detecting such a pattern often reduces the possibilities of exploring this issue in detail. The spatial distribution of pelagic organisms is more complex than supposed and a good indicator of such patchiness are large predators such as Antartic fur seals (*Arctocephalus gazella*) (Boyd 1996).

In the pelagic landscape the patchiness of prey is likely to be copied by predators. The fur seal utilizes two spatial-temporal scales to track krill. At a fine scale (0.18-0.27 km) the fur seal moves from one krill patch to another in less than 5 minutes. At a mesoscale (1.3-1.6 km) travel duration is longer than 5 minutes. However, these values changed during the five years of study, demonstrating the capacity of this predator to activate an adaptive behavior according to the food patch distribution.

The two types of behavior are indicators that krill is clumped at a fine scale, but that this aggregation appears to be patchy at a large scale. This study provides a new view of the prey-predator relationship and offers a new approach for predicting the behaviour of large predators in a changing seascape (Fig. 1.48).

1.8.5.6 Predatory-prey interactions and habitat fragmentation

Fragmentation of habitats, "patchiness", has a strong influence on species interactions (Klein 1989, Herkert 1994, Kattan et al. 1994). Some authors have found stabilizing effects between prey and predator in fragmented landscapes, though this must not be considered a general rule, as outlined by Kareiva (1987).

This author experimentally fragmented an homogeneous field of goldenrod (*Solidago canadensis*) infested by the aphid *Uroleucon nigrotuberculatum* which feeds only on *Solidago* and overwinters at the base of the host plant. One of the main predators of this species is the lady bird (*Coccinella septempunctata*).

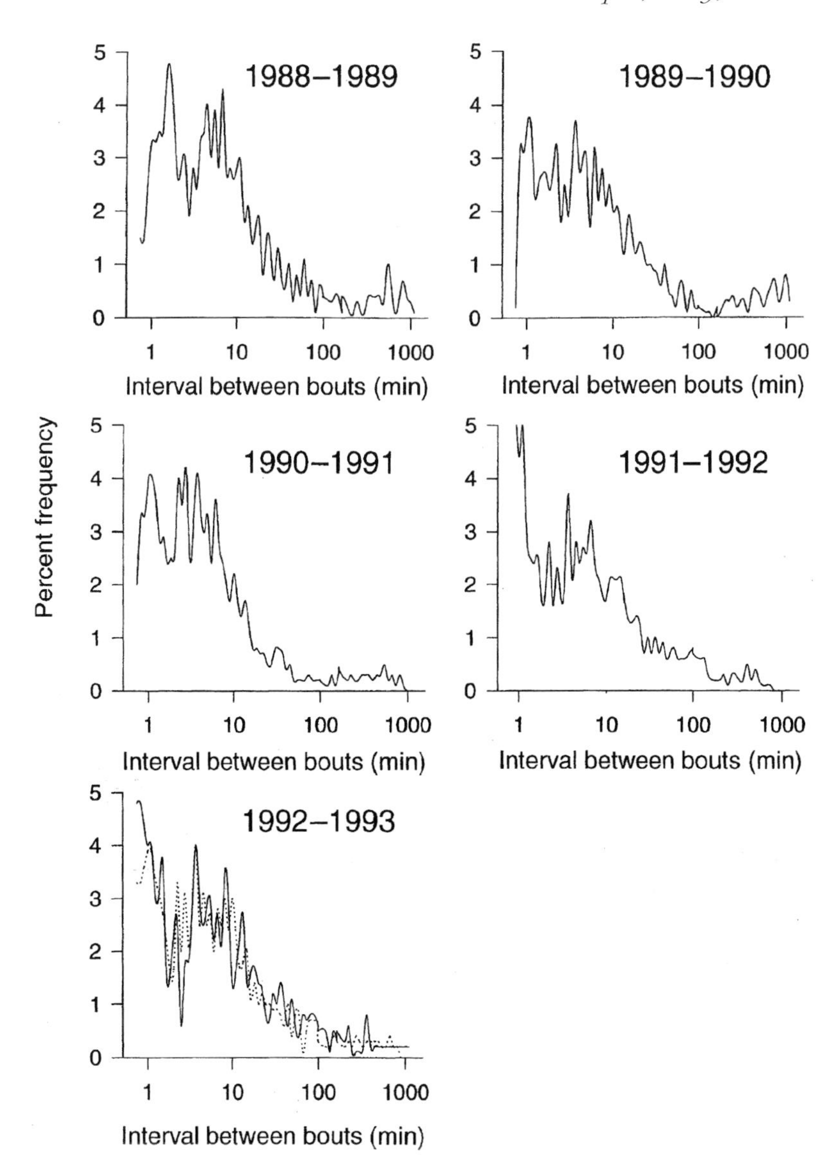

Fig. 1.48 - Lactating females of Antartic fur seals show a variability in foraging bouts in different years, and according to a large scale patchiness of krill (from Boyd 1996).

From the fragmentation experiment and a comparison with untouched *Solidago* cover, it emerged that fragmentation increased aphid density and consequently also the density of the ladybirds. However, fragmentation also increased the aphid outbreaks which decreased the predator pressure of ladybirds. Despite the general assumption that fragmentation should stabilize the relationship between prey and predator, this effect may not materialize as it is connected with the behavior of the species and their relationships. Again, exploring a process like the prey-predator relationship at the landscape scale we find new information and new surprises.

1.8.6 Fragmentation versus aggregation

Fragmentation is a world wide process that reduces dramatically habitats and biodiversity (Saunders et al. 1991) increasing the risk of extinction (Burkey 1995) and, in some cases, modifying the social and economic structure of human populations (Skole et al. 1994). Fragmentation encourages the invasion of alien plants (Brothers & Spingarn 1992) and affects the distribution, abundance and dispersal of many organisms (Galli et al. 1976, vanDorp & Opdam 1987, Gibbs & Faarborg 1990, Hinsley et al. 1995, Margules et al. 1994, Newmark 1990, Norton et al. 1995, Ostfeld 1992, Pearson 1991, Rudis 1995, Stouffer & Bierregaard 1995, Villard & Taylor 1994, Villard et al. 1995) and their cycles (Roland 1993). In addition fragmentation increases the risk of predation (Wilcove 1985).

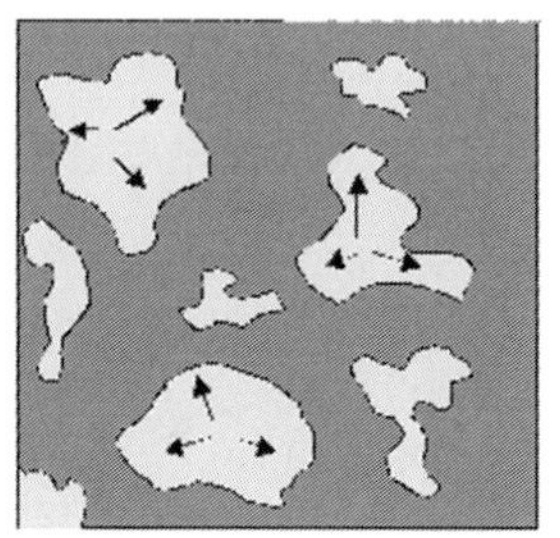
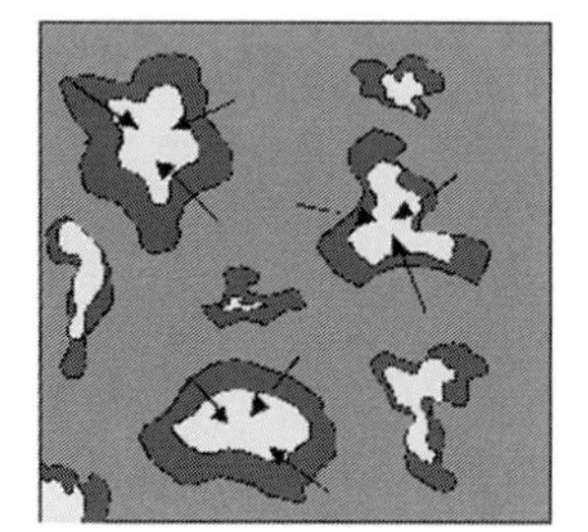

Fig. 1.49 - Aggregation and fragmentation are processes that develop from an "intact" matrix (A). Fragmentation opens holes (gaps) (B), and aggregation fills such gaps (C).

Aggregation is the opposite of the fragmentation process; it occurs when a disturbance that has created fragmentation is stopped and vegetation recovers, regaining the open space maintained by the disturbance process.

Aggregation is very common in nature and actually widespread due to land abandonment in rural landscapes (Fig. 1.49). This process is not well studied because most of our attention is absorbed by processes that destroy "valuable" landscapes like fragmentation. Indeed, aggregation is a relevant process that in many cases has the same negative effects of fragmentation; in fact, under aggregation landscape heterogeneity and (regional) diversity decrease. Aggregation occurs under many different conditions, the most common of which is represented by the recovery of woodland after clearing. Like heterogeneity, aggregation is also characterized by different shapes of aggregating patches according to different time sequences.

No detailed studies have been made of this relevant process though the consequences of aggregation may be severe for plant and animal communities. In many cases, aggregation creates new interior areas and a loss of edges. From this view point, the process is important for species that require large patches of the same type.

1.8.7 Homogeneity and heterogeneity

Heterogeneity is a characteristic of a pattern, homogeneity represents the characteristic of a functional unit (Fig. 1.50). Homogeneity is not a structural quality of a patch, but a functional attitude. When we find a patch represented by the same type of pixels we call this patch homogeneous, but homogeneity is a matter of scale and does not exist in nature as a structural pattern. In fact the resolution of a "homogeneous" patch is increased immediately a new heterogeneous characteristic appears.

For instance homogeneity is the characteristic of the functional landscape unit, the ecotope. In this case, we assume a particular configuration of land cover, vegetation, morphology, etc., to be homogeneous from a functional point of view where it has a precise role for a life trait of an organism at a precise spatio-temporal scale. We can find homogeneity at the level of a unit, for instance a pixel is by definition homogeneous, but when we aggregate several pixels they must be considered part of a heterogeneous patch.

Heterogeneity is composed of different categories of units and by the different positions of such units in the patch. So we have at least two levels of spatial complexity connected with heterogeneity: the type and the position. Both concur to create heterogeneity. Such a configuration, if observed from a distance, again disappears and we are induced to consider such an entity homogeneous.

1.8.8 Disturbance

Disturbance may be defined as any relatively discrete event that disrupts an ecosystem, community, or population structure and changes resources, substrate availability, or the physical environment (White & Pickett 1985).

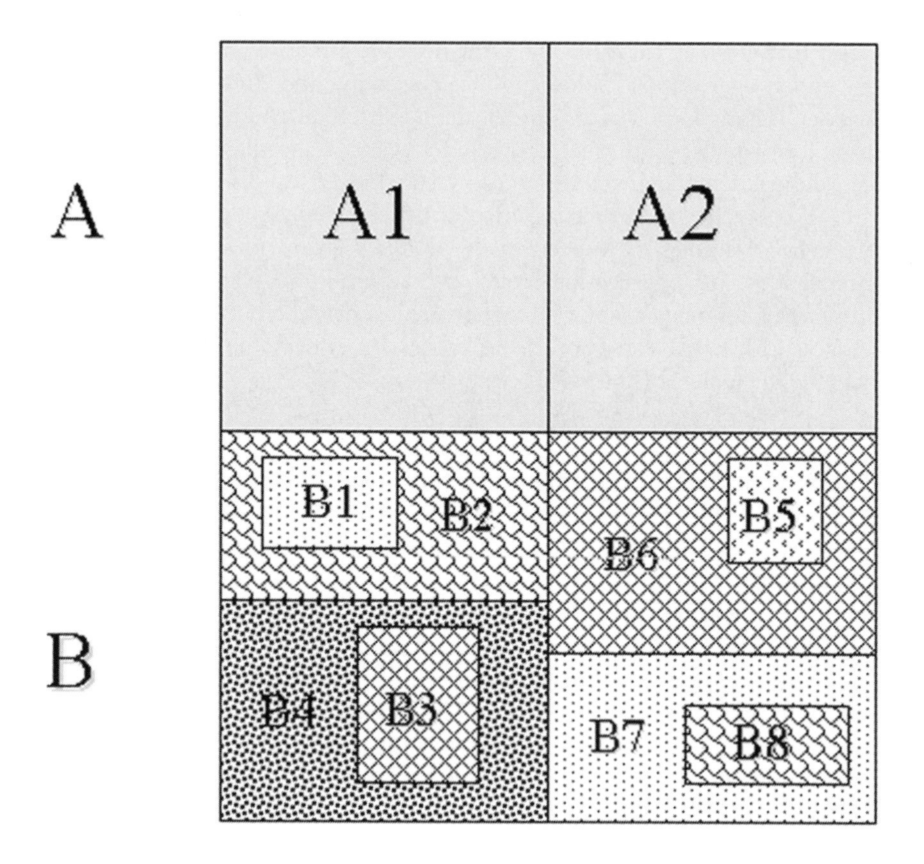

Fig. 1.50 - Representation of homogeneity (function) (A) and heterogeneity (pattern) (B) in a schematic habitat composed of two functional patches (A1: roosting, A2:foraging), and different heterogeneous patterned patches (from B1 to B8).

In a fragmented landscape, large disturbances are rare due to the low contagion capacity of the remnants to propagate disturbances such as fire or pest diseases.

Small or large scale disturbances are major changing forces shaping landscapes. Disturbance is a necessary shaping force. Fires, flooding, hurricanes, insect outbreaks, and tree fall dynamics, modify and influence landscapes differently according to their scale (Fig. 1.51). Disturbance may be the main factor influencing the presence and persistence of species or communities (Fig. 1.52).

Fig. 1.51 - Wind disturbance is one of the more frequent causes of damage in the forested landscape affecting the dynamics of a forest at a multiscale, from vegetation cover to soil structure (Pine wood, Monterey coastal range, California).

Also, events like earthquakes can have a relevant impact on the landscape at different spatial scales as documented by Allen et al. (1999) in the temperate montane forest of New Zealand. These authors found that the effect of such disturbance on trees was linked to the scale of resolution. Many trees were damaged by landslides, but at a broad scale many trees appeared damaged by other effects linked to this event, such as modification of underground water availability. The effects of such injuries are similar to those produced by hurricanes and pathogen infestation.

Fig. 1.52 - Tree fall by wind storm or by disease creates gaps in the forest cover, producing local diversity in plants and in animals attracted respectively by more light, nutrients from debris, and a greater food supply. Dogwood (*Cornus mas*) fruits, and Blackcap (*Sylvia atricapilla*), a common warbler in forest edges.

Disturbance can guarantee the survival of species sensitive to competition. This may be the case for *Salix erbacea*, a dioecious, prostrate dwarf shrub living on high mountains across Europe. Due to it's low stature it requires a regular disturbance regime (e.g. exposure, snow cover, soliflux and grazing) to survive more competitive species (Fig. 1.53). This is a good example of how disturbance may play a foundamental role for species that are not able to compete with more vigourous and taller species (Beerling 1998).

Fig. 1.53 - *Salix erbacea* can survive in relict areas in South Europe only on the highest mountains and in landscapes which severe disturbance (frost, snow cover, poor soil) prevents and reduces the competition with other more vigorous plants.

1.8.8.1 Disturbance and diversity

Disturbance and diversity seem to have contradicting relationships. In some cases disturbance increases diversity, as in cultivated meadows, but in other cases disturbance seems independent of diversity.

Small scale disturbance due to tree fall in a closed-canopy forest in the neotropical forest of Panama was studied for 13 years by Hubbell et al. (1999) (Fig. 1.54). The results suggested that the diversity was unaffected by this disturbance and so was due to other events. Tree diversity seems due to other events. Apparently canopy height changes are not related to richness per stem; a result which opened new uncertainty on the role of local disturbance for total plant diversity. These authors concluded that gaps are a diffuse and very frequent phenomenon in old growth tropical forests. There is no gap-to-gap predictability of species richness in gaps. This process is common in all forests but richness can be quite different from one forest to another.

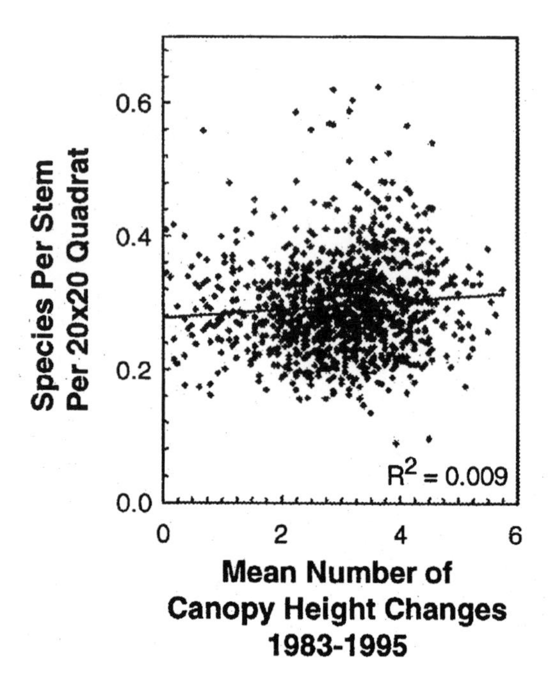

Fig. 1.54 - Species diversity was not found to have a direct relationship with canopy height changes in a neotropical forest, demonstrating that at least in some neotropical forests forest gap are not the main drivers of high plant diversity (from Hubbell et al. 1999).

1.8.8.2 Disturbance processes important for large scale landscapes

Physical processes such as glaciers and changes in sea level, hurricanes, wind storms, flooding, landslides and volcanoes, are common physical processes that occur at the larger scale and are independent of any type of biological interactions. Other forces, like lighting, are connected with biological heterogeneity. In some areas, such as the Florida peninsula, lighting is a very common process and the main cause of fire spread before Colombian's time. Vegetation in this peninsula has been exposed to natural fire regimes for thousands of years, though fire connectivity has now been broken by roads, artificial channels and urban areas. Thus, biodiversity derived from such regimes is at risk and the loss of species cannot be stopped by local mimicking of large scale processes.

Many biological processes have a great relevance for the large scale landscape. For instance, guano dispersion by colonial birds in the mangrove system, or guano deposits by colonial bats in caves which is then transported by water to the surroundings.

Fig. 1.55 - Large roads are considered important factors of habitat fragmentation and reduction of connectivity across landscapes. Highway crossing in Los Angeles area (South California).

Large scale effects have been recognized for beavers (*Castor canadensis*) (Johnston & Naiman 1987) which retain 1,000 time more nitrogen for impoundment activities than in streams not occupied by this species. Many organisms produce disturbance by grazing, trampling, depositing manure, and by changing soil properties, bio-chemical characteristics, and vegetation cover (Butler 1995). Human activity is another important factor responsible for deep changes at a broad landscape scale (Fig. 1.55). One of the major effect is the increase of heterogeneity by fragmentation, the reduction of ecotone complexity and often their substitution by linear borders characterized by high contrast and strong filter effects.

The main effect of human impacts is the reduction of energy and nutrient fluxes across large areas. Human activity reduces populations and communities into more and more restricted regions. All species suited to the forest interior are dramatically reduced, there is an increase of alien and common species, but also a disruption of processes like nest competition, nest predation, nest parasitism, etc.

One of the most explicit effects of human-dominated landscape is the spread of pests like feral cats and dogs in New Zealand and Australia or weeds such as *Artemisia* spp. in Europe.

"Area-sensitive" species are suffering due to human alteration of their habitats and most of these are the highest level predators such as Florida panther (*Felis concolor coryi*). The reduction of these predators has favored middle size predators with dramatic effects on small animals like game species of birds and small snakes, which suffer a higher predatory rate.

PART II

Landscape Evaluation

Contents

2. Preface 112

2.1 Exploring evaluation procedures 112
2.1.1 Define the geographical context 112
2.1.2 Relate evaluation with processes\problems 113
2.1.3 Sort evaluation before political decisions 113
2.1.4 Assure representativeness 113

2.2 Landscape evaluation and scale 114
2.2.1 Spatial scale 114
2.2.2 Temporal scale 115
2.2.3 Assessing habitat suitability at multiple scales 115

2.3 Concepts and procedures of landscape evaluation 117
2.3.1 Evaluating integrity 118
2.3.2 Evaluating fragmentation 121
2.3.3 Fragility *versus* stability: A vision for evaluating the complexity of a landscape 121
2.3.4 Evaluating resilience *versus* resistance 125
2.3.5 Disturbance as an indicator 128
2.3.6 Evaluating keystone species 129
2.3.7 Biodiversity indicators in the regional landscape 132
2.3.8 Animal behavior for evaluating the spatial availability of resources 133
2.3.9 Evaluating landscape changes 137
2.3.9.1 Succession and disturbance 139
2.3.9.2 Land use changes 140
2.3.9.3 Patterns in large scale vegetation changes 141
2.3.9.4 Changes in soil properties 143
2.3.10 Concepts of landscape health 145
2.3.10.1 Landscape health paradigm in action 147
2.3.10.2 Health criteria 148
2.3.10.3 Health indicators 148
2.3.10.4 Scaling landscape health 150
2.3.10.5 Patchiness and health evaluation procedures 152
2.3.11 Ecosystem distress syndrome 153
2.3.12 Examples of landscape pathologies 154
2.3.13 Evaluating forest landscapes 156
2.3.14 Evaluating cultural landscapes 157

2.3.15 Other criteria of landscape evaluation 159
2.3.16 Some additional considerations of landscape evaluation 161
2.3.17 Discontinuity, synergism and ecological debts 162

2.4 Tools for evaluating the structure and functions of land mosaics 164
2.4.1 Neutral models 164
2.4.2 Landscape metrics 168
2.4.2.1 Metric at individual patch resolution 171
2.4.2.2 Spatial arrangement 176
2.4.2.3 Proximity indexes 178
2.4.2.4 Diversity indexes 179
2.4.2.5 Core area indexes 181
2.4.3 Fractal geometry in landscape metrics 183
2.4.3.1 Application of fractals to ecology and specifically to landscape ecology 184
2.4.4 Fuzzy theory in landscape metrics 190

2. Preface

There are many parts of the planet that have recently experienced an impressive urban development, for instance most of the Mediterranean coasts (from Spain to Italy, in Europe) and along the California coast (from San Francisco to the border with Mexico). Urbanization in general organizes the human landscape affecting structures and processes. Human population density and road distance seem two good descriptors of human landscape organization and changes. In fact, changes in land use and the way in which such changes occur can be detected by using only land cover maps combined with infrastructure maps (like road maps, urban distribution, power facilities, etc.). In human dominated landscapes, changes occur at the fringes between urban and rural land use. This represents an important ecotone along which most of the changes like fragmentation processes or, the opposite, aggregation processes can be observed.

The environmental variables that can usually be utilized to evaluate changes in land use are both natural (as in sloped elevation) and human related (population density, distance to the closest roads, distance along road to the main urban center, etc.).

Evaluation methods are the basis for any information procedure in a target area. Landscape evaluation is generally carried out at a large scale (the human landscape scale), that ranges between a few kilometers to hundered kilometers.

Landscape evaluation can also be done at a smaller scale but often at this smaller scale we loose many important components of the human-interacting landscape. The starting point for a good evaluation is to fix a reference system based on a cultural model or on a real landscape. Diversity, integrity, fragility, resilience, and resistence are some of the commonest processes (working like indicators) that can be used to explore the functioning conditions of a landscape. These elements have the capacity to describe, in a very synthetic way, many structural and functional components of a landscape and these processes have been applied in large scale landscapes such as the human perceived landscape.

Processes like water drainage across a watershed can be used as guide line to develop an evaluation procedure, and river systems can be usefully employed as an indicator of large scale quality. Why do we evaluate a landscape, what is the area to be investigated and what are the tools for such an investigation ?

We need to reply to the first question. Evaluation is necessary as an investigation tool before any political decision, forming the objective basis on which final decisions are produced. The purpose of the evaluation is to formulate detailed information to utilize either for conservation or for developing policies (management, restoration, mitigation actions). There are several tools for pursuiting evaluation procedures in a scaled landscape from landscape metrics, that will be explained later in this chapter, to remote sensing procedures and GIS (Geographic Information Systems), coupled to spatial statistics.

2.1 Exploring evaluation procedures

Any evaluation procedure assigns a score to the landscape according to a criterion (such as animal suitability, vegetation diversity, landscape amenity, etc.), centered on different targets, but it is often only the human perspective that prevails thus masking other genuine approaches. Some important steps have to be followed before starting evaluation procedures.

2.1.1 Define the geographical context

To evaluate an "assigned" landscape we must formulate precise questions to which we need to reply. It is a nonsense to move into a landscape and decide to evaluate it on an empirical basis. The choice of the area under evaluation and it's extent are questionable. It is not possible to select a part of a land mosaic at random and then try to evaluate the status. The first step of the evaluation is the selection of a problematic area that may be delineated by geographical boundaries (watershed, mountain catena, lake district), by a functional unit (bio-region, eco-region, etc.), or determined by an administrative boundary like a county, a province or a parish. This selection is important for determining the scale of analysis. After the determination of the extent, the selection of the grain size of the area assumes a central importance. Grain size must be scaled in relation to the dimension of the area. Large areas like a European region (Tuscany, Italy is Km2), must be analyzed at least at a grain size of 400x400 m cell. A province (Massa Carrara Province, 1156 Km^2) can be studied at a resolution of 200x200 m. A small watershed (100 $k.m^2$) may be studied at the resolution of 10x10 m size. Extent and resolution are the basis for the matrix on which to transfer the information requested for developing the evaluation procedure.

2.1.2 Relate evaluation with processes/problems

Any evaluation procedure must be related with a process (function/problem). For instance, an evaluation action may be asked to measure the logging patterns that allow the maintenance of an economic income and that at the same time assure the presence of logging-sensitive species (Fig. 2.1).

2.1.3 Sort evaluation before political decisions

The evaluation is important if independent of the political decision on land use or on land transformation. Often evaluation and impact assessment procedures are confused and carried out after the definitive decision to make an intervention.

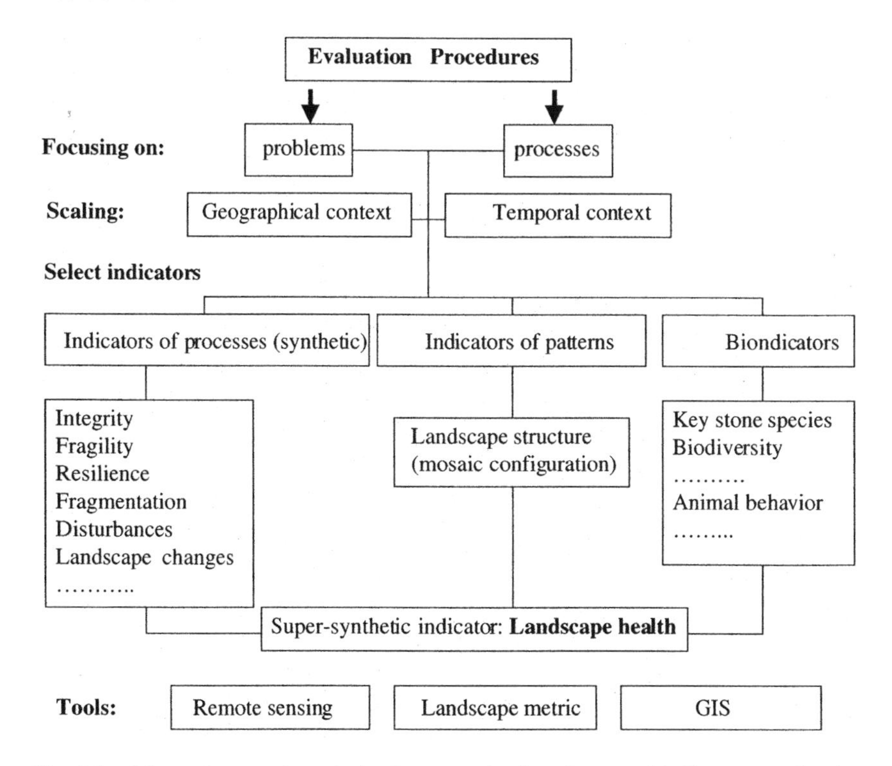

Fig. 2.1 - Schematic procedures in landscape evaluation. Steps and indicators are listed according to priority.

2.1.4 Assure representativeness

Historical biases due to our cultural model can create constraints in evaluation procedures. In general, the more fertile soils have received a higher impact and modification than the areas with a low commercial value. For this reason, some significant remnants are often not considered in a protection plain, but on the other hand, less productive and "marginal" areas are well represented. So, if the minimum target to protect at least 10% of a region has been achieved, this value may not be enough when we use representativeness as a major criterion.

The identification of dominant land form vegetation classes originally present in a district was coupled by Awimbo et al. (1996) with the main land use activities. The first criterion assured representativeness, while the second was useful to understand the environmental pressure affecting the remnant patches of original land cover. In the case presented by these authors, despite the protection of 78% of the South Island (New Zealand), the landform-vegetation classes were still under-represented. These authors suggested a stratified multi-scalar approach in defining protected areas in which spatial scale as well as representativeness have to be considered.

The case presented above is very common in many countries. Most efforts to conserve nature (biodiversity, ecodiversity, healthy conditions, scenic beauty) are concentrated in remote upland areas in which climatic constraints limit the richness of organisms and the complexity of ecological processes. But it is along lowlands such as delta regions, that we find the highest value of biodiversity in most of countries. Unfortunately, human competition is removing many biological entities from these areas.

2.2 Landscape evaluation and scale

Landscape evaluation is a fundamental procedure for assessing the conditions and quality of a system. In general, the evaluation of a landscape is the first step towards planning or managing actions. It seems relevant to work at the right scale, but often this is not enough to describe the status of a complex system and a multi-scalar approach is required for a more accurate landscape evaluation.

In the evaluating process it is important to have in mind the spatial and temporal scale. In fact diversity, integrity, fragility, stability, resilience and resistance are all landscape attributes that can be evaluated differently thus changing the spatial and temporal context in which they are considered.

2.2.1 Spatial scale

Evaluation is generally carried out at a large scale (called also landscape scale, but we have seen that this terminology creates confusion, a landscape may have a different extent). Nevertheless, it is true that the relative scale is enlarged around an organism or a process when we are dealing with a landscape.

The heterogeneity of the landscape creates variations in the density of populations across a broad range of environmental factors, conditioning many population related processes. For instance, heterogeneity influences nomadic predators that move from highly populated patches to others. Distance decreases dispersal processes and environmental synchrony. In fact, in general, spatial correlation decreases on the increase of distance. Some less mobile species are probably affected by other de-synchronized processes.

Environmental perturbations probably affect synchrony, an effect known as the Moran effect. Despite the evidence, there are few data corroborating the hypothesis that at the large scale climatic factors are the main causes of the de-synchronization of biological cycles.

Also the outbreak of population density weaves observed in voles confirm that spatial distribution of phenomena are more asynchronous when the geographical distance increases.

By enlarging the scale of investigation studies on population density we acquire new insights into the factors responsible across the geographical range of a species. So, today, we can move from the study of the patterns affecting an individual population to the fluctuations of many geographically distinct populations. One other important factor is represented by the temporal window by which the correlation is tested. For instance, in Canadian lynx (*Lynx canadensis*) the correlation between sites may change from highly positive to highly negative by moving the temporal window (Ranta et al. 1997). The same trend has been found by Koenig (1999) on the spatial structure of acorn production in central coastal California using a sliding time window of 10 years.

2.2.2 Temporal scale

Often the temporal scale is scarcely considered during the evaluating procedure, but time assumes a central position for understanding the potential direction of environmental evolution. It is often the temporal comparison that allows us to understand the ecological integrity of a system, as is the capacity to recover after a severe perturbation, and to discover pathologies that are masked by ephemeral attributes.

2.2.3 Assessing habitat suitability at multiple scales

Landscape pattern and ecological processes are linked but it is quite difficult to find the connection between such patterns and functions, and yet to apply landscape principles it is necessary to interpret such patterns ecologically (see f.i. Turner 1987a, 1989, 1990, Turner et al. 1989a,b, Turner et al. 1991). Habitat suitability is generally carried out by using an evaluation of within-site habitat quality but this analysis is only a partial component of the specific ecotope requirement (see Witthaker 1977). Spatial patterns contribute to an evaluation of the total species niche.

Fragmentation is one of the most severe threats to species ecotopes and an increase of fragmentation increases the risk of species extinction. Species react differently to habitat fragmentation according to their specific ecotope requirements (habitat, home range size, perception of the mosaic).

The ecotope is scaled to the species and as a consequence every species has a different ecotope-oriented response to landscape complexity. Riitters et al. (1997) presented a good example of the application of a multiscale and multifunctioning landscape assessment.

Since it is not easy to find a good umbrella or key species to describe the landscape requirement, it is often enough to start with an explicit model (Turner et al. 1995) to distinguish species of forests, species of open fields and species living at edges.

This broad classification enables easy classification of most remotely sensed images or the rescaling of land cover maps employed in planning activities. The selection of a sliding window scaled according to the home range of a selected species, in this case forest, open field and edge species, is moved across the map and the character of the central pixel of each window is reported on the map. In this way we are progressing from a pattern map to a process map. Such a process reduces the resolution of the area according to the size of the window. After this step it is necessary to establish a filter to select the windows that distinguish suitable from unsuitable habitat. We can select a threshold such as 90% for woody windows, or 11% for edge windows (f.i. Riitters et al. 1997).

Some measurements can be done on suitable habitats such as area, percent of total area, connectivity (calculated as percentage of neighboring pixels of suitable habitat), number of patches, largest patch size, average patch size, contagion (see Li & Reynolds 1994), and fractal dimension (overall image). Fig. 2.2 shows an example of three types of habitat in Chesapeake Bay Watershed.

The home range of a species is a more complicated issue than the simplified square of a sliding window, but the model is enough for the first modeling. Of course, this analysis is timeless and the changes in habitat scaling across seasons are not considered. In actuality, many organisms show a shift in home range according to the different seasons and this varies species by species. Some species that defend territories during the breeding period are more evenly distributed within this season than than outside of it (Farina 1985).

This change in behaviour is important in flocks for increasing the possibility of finding more concentrated resources, for transmitting "culturally" the location of the best sites and reduce the risk of predators.

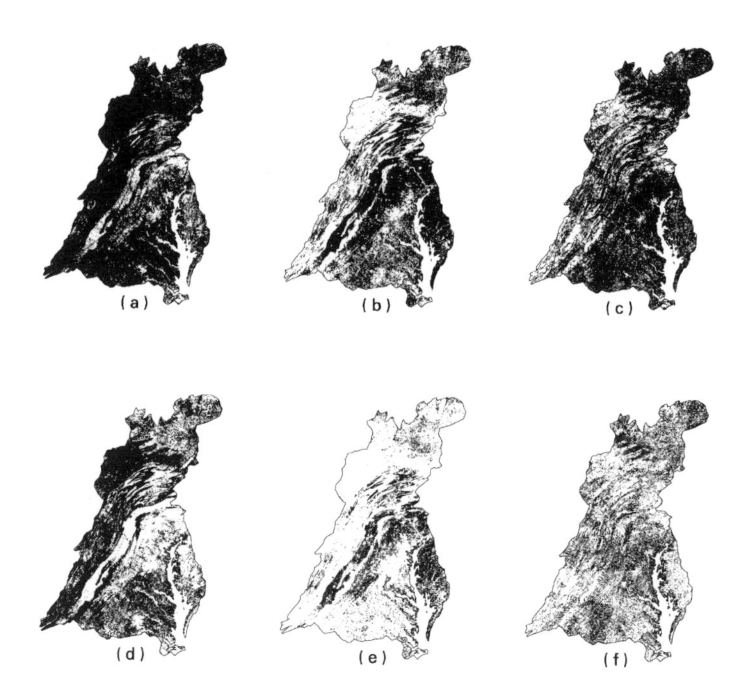

Fig. 2.2 - Regional distribution of woody habitat (a), herbaceous habitat (b) and woody-edge habitat (c) in Chesapeake Bay watershed. From d to f the same categories but land cover has been evaluated using a threshold value of 90% (d,e) and 11% for (f) (from Riitters et al. 1997).

2.3 Concepts and procedures of landscape evaluation

What has to be evaluated in a landscape is a vell discussed issue (see f.i. Hargis et al. 1997). Generally, discontinuities and some emerging characters have to be evaluated. In fact, our evaluation is largely based on the visual perception of patches like woodlots, lakes, river buffers, and cultivated fields (Lucas 1991) and appreciation is based on the contrast between the different "objects". Discontinuities can be seen as ecotones, corridors or simply the meting point between different cover types (vegetation, soil attributes, water type, temperature and air composition). Discontinuities exist at all scales and are perceived by organisms in different ways according to their species-specific caliper.

Often human intervention creates simplified landscapes, for instance a maize cultivation, or very complex mosaics in which the dimension of habitat patches is so small that these patches have lost the function of individual habitats.

In a fragmented landscape it is possible to assess high quality to a large fragment than to a small one, but this analysis seems inadequate for ensuring the maintenance of a species in a larger area.

Patches are not always habitat patches; often if they are too small or too isolated they are devoid by a particular species although the intrinsic patch characteristics appear suitable. This is exemplified by small open fields which are avoided by larks that live in "larger" open fields.

A criterion for evaluating a landscape is the measure of the basic matrix by using as attributes such as dominance of a cover type, diversity of the cover type, aggregation, diversity in patch size, shape of the patches, length of the edges, level of contagion, and spatial distribution of the patches. This last attribute measures the uniqueness of a landscape. In fact the spatial distribution of patches is the most variable depending on the object considered and the area covered. The comparison between a theoretical landscape (neutral model) and real landscapes is necessary in order to move from a simple description to a more functional characterization of the landscape considered (Dale et al. 1989). Especially in regard to human landscapes the need to preserve the actual mosaic dominates compared with studies of future scenarios. Generally, an historical approach is applied to interpret the actual landscape (Farina 1991,1995a).

A general paradigm has to be found exploring new areas of integration between ecosystem and landscape approaches.

In general terms, landscape evaluation is conducted at the scale of the human landscape perception and consequently the patterns and processes described have to be considered roughly as indicators of many other overlapping and interacting processes.

2.3.1 Evaluating integrity

The decline of biodiversity alters the performance of ecosystems as experimentally demonstrated by Naeem et al. (1994). For instance, CO_2 is consumed more in a species rich community than in lower-diversity community. Plant productivity has higher values in higher-diversity plant communities. While diversity is a collective property of an ecosystem, the concept of integrity represents a synthetic property of a system and has been defined as "the capability of supporting and maintaining a balanced, integrated, adaptive community of organisms having a species composition, diversity, and functional organization, comparable to that of natural habitat of the region" (Karr & Dudley 1981). Assessing biological integrity is not an easy task; the influence of processes occurring at multiple organizational levels and at multi-spatio-temporal scales must be assessed. In addition, it requires a set of indicators to measure it (Angermeier & Karr 1994, De Leo & Levin 1997).

Considering the hierarchical structure of complexity, integrity at one organizational level is maintained by processes occurring above and below the selected level (O'Neill et al. 1989). Due to the difficulty of assessing integrity by the use of processes, in practice species and their aggregations are a realistic evaluating approach. For instance, to assess integrity in a lotic system, some of the criteria utilized are species composition and diversity, trophic composition, population density, and tolerance to human impact (Angermeier & Karr 1994). Tab 2.1 represents an example of factors responsible for the organization of ecological systems that can be used as framework to assess integrity in aquatic (A) or in terrestrial (T) systems (Angermeier & Karr 1994).

Integrity in some instances, can be restored successfully by actions that mime natural recovery after perturbation. Such an approach is extremely important in aquatic as well as in terrestrial systems. Restoration actions focus on processes like fire regime, flow regime, grazing regime, etc (see Part III in this book). Restoration assumes a central role for integrity recovery but needs an in depth knowledge of the processes involved in each specific case and must be related to the real world by an appropriate spatio-temporal scale.

Despite an increase of funds to protect species, biodiversity is declining year after year. This largely relates to a strategy which is focused more on the elements of biodiversity than on the processes that generate and maintain all of the elements.

Integrity seems a more efficient and scientifically "realistic" goal when focusing on the conservation of the landscape matrix.

To achieve this policy, an appropriate geographical large scale which includes most of the relevant processes to be conserved, is required. An important example is presented by the riparian and floodplain landscape. In these strategic zones where the landscape paradigm of ecotones has been fully described, integrity can be ensured only if both aquatic and terrestrial components are considered (see Naiman & Decamps 1990).

In many regions integrity is strongly menaced by habitat destruction but also by alien species invasion, pollution, over exploitation and diseases. Wilcove et al. (1998) reviewing the threats that imperil species in the United States, have found that the main cause of biodiversity decline is due to habitat destruction, followed by the competition of alien species with native elements and an increase of "management debt" like fire suppression.

Tab. 2.1 - List of the five main classes of factors that structure ecological systems and that can be used to evaluate integrity (A in aquatic, T in terrestrial systems) (from Angermeier & Karr 1994).

Class	**Factors**	
Physiochemical conditions	Temperature PH Insolation Nutrients	Salinity Precipitation (T) Oxygen (A) Contaminants
Trophic base	Energy source Productivity Food particle size	Energy content of food Spatial distribution of food Energy transfer efficiency
Habitat structure	Spatial complexity Cover and refugia Topography (T) Soil composition (T) Vegetation height (T)	Vegetation form (T) Basin and channel form (A) Substrate composition (A) Water depth (A) Current velocity (A)
Temporal variation	Diurnal Seasonal Annual	Predictability Weather (T) Flow regime (A)
Biotic interactions	Competition Parasitism Predation	Disease Mutualism Coevolution

Management debt represents the main concern in many developed countries where agriculture reduction is dramatically threatening most species which have adapted to human disturbance.

Land abandonment affects the organization of land mosaics at the small scale so increasing the "management debt" whilst, at the large scale, the fragmentation of habitats and the increase of physical constraints like road infrastructure, reduce the functions of disturbance processes creating a "large scale disturbance debt". In land abandonment, local recovery is isolated by infrastructures like roads and railways thus the new local system cannot develop by the necessary large scale disturbance processes.

2.3.2 Evaluating fragmentation

One of the commonest processes affecting the integrity of landscapes is fragmentation. Much data are available on this subject. Fragmentation may be measured as the level of patch dispersal in the landscape or in terms of dynamic processes. Both approaches can be used. Patch size and shape, patch spatial arrangement, distance between patches (Fig. 2.3), and hostility of the interpatch environment are the main attributes usually used to measure fragmentation (see later: landscape metrics). Simulating landscape patterns under fragmentation processes seems a very efficient system. For this neutral models and other spatial aggregations have been used extensively (see f.i. Gardner et al. 1987, 1992, Gardner & O'Neill 1991, Hargis et al. 1997).

Neutral models create a landscape without local constraints based only on the probability of a pixel occupying a certain position in the landscape.

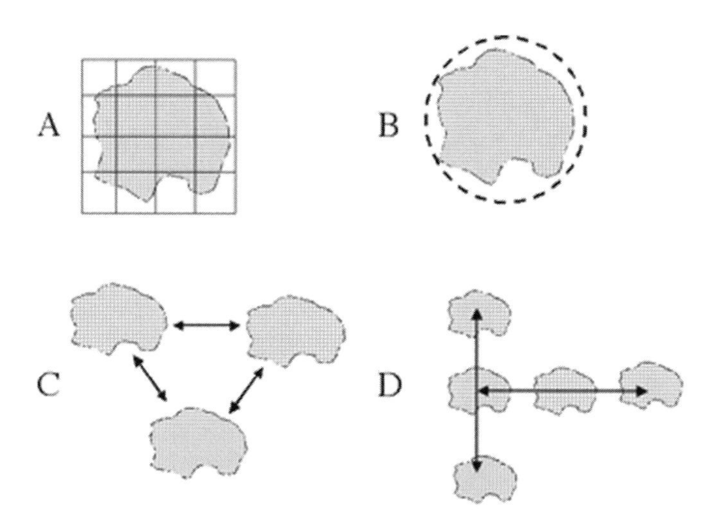

Fig. 2.3 - Schematic representation of the main attributes that can be measured in a fragmented landscape. A= Size, B=Shape, C=Distance, D=Spatial arrangement of fragment.

2.3.3 Fragility *versus* stability: A vision for evaluating the complexity of a landscape

Fragility is the attitude of a system which is modified when disturbed (Nilsson & Grelsson 1995). The rapid change in species composition and abundance is an indicator of fragility. Fragility is a timing attribute of a system, and can be evaluated only by adopting a long term time scale. There are landscapes and/or ecosystems that are intrinsically fragile, for example low productivity systems or systems located in areas in which the disturbance regime is particularly severe as on high mountains.

Most human dominated landscapes are fragile systems and they change plant/animal compositions and land mosaics, according to the disturbance regime adopted. The complexity of a landscape creates problems when evaluating the degree of fragility. This attribute can change according to the process or the species considered. Fragility/stability are scale dependent. Thus, a forest fire may produce a big change in a short scale period but, if the temporal scale is enlarged, the system appears more stable.

Rivers are very dynamic, unstable systems over a short time period, but conserve their major patterns (stability) over a long period.

Lakes are an opposing example; they seem extremely stable over a short period, but over a long period are destined to be deeply modified by a progressive inorganic and organic sediment filling that transforms water bodies into marshes. Fragility is enhanced by extinction and invasion, both of which processes are responsible for species turnover. Extinction is produced by modification of internal characteristics like forest fragmentation and a decrease in connectivity that transforms a suitable land mosaic into an unsuitable landscape. Invasion generally depends on external conditions like availability of seeds or demographics explosion of neighboring populations.

The attitude to invasion is strictly linked to the richness or complexity of the system (ecosystem/landscape). The more complex a community is, the less probability there is of adding new species as empty niches are limited. Extinction and invasion increase the fragility of a system. Extinction may be produced by mechanisms such as:

Extirpation of a species by over-hunting or over-harvesting
Habitat destruction
Introduction of pathogens
Introduction of predators
Introduction of new competitors
Release of pollutants

Invasion is very active in areas where species have no defensive capacity, or in areas where the system experiences severe disturbances, for instance along the river beds. Invading species require temporarily empty niches.

In general, the extinction of a species has more severe consequences for the system when compared with the addition of a new species, though this rule has many exceptions. For example the introduction of rabbits and wild boar in Australia has produced a deep change in the system. Land abandonment in many European upland agricultural landscapes has created favorable conditions for wild boar invasion. This species, which lives in many types of shrub-lands and forests, is one of the main causes of recent changes in prairie plant dynamics.

The removal of a predator component in a community can cause the extinction of many components of the community, so the role of key species is important in landscape fragility assessment.

A system in which plant recovery progress only slowly, may be considered fragile. Human disturbed systems are more fragile than undisturbed systems, a pattern which is evident in agricultural field successions which change diversity and heterogeneity according to different disturbance regimes (Fig. 2.4). The process of land abandonment that allows the recovery of new vegetation types apparently moves from a fragile (dynamic) system to a stable one.

Conservation strategies and fragility evaluation are important terms for effective action. Often our awareness grows when we are dealing with a rare species living in a restricted habitat. It is not always true that a rare species is more prone to extinction. Nevertheless if a species is restricted to a small area, the risk of extinction can be very high.

Fragility increases in systems in which productivity is low due to poor soil, low or hot temperatures, water shortage, etc. (deserts, tundra, high mountains). These systems rapidly change the diversity of living organisms when perturbed and can only recover the original species composition after a long time lag.

Fragility is an important attribute of every system and detection of such an attribute can be done indirectly using the attitude of a system towards perturbation. Perturbations are produced by many disturbance regimes but most such regimes may be modeled in a landscape at a specific scale.

This provides tremendous possibilities for analyzing areas at different scales and to arrange predictive multipurpose maps. For example, it is possible to produce a map of flooding risks, landslide falls, fire ignition, etc.

We have to consider that, though we can predict the probability of an unfavorable occurrence, this doesn't help us to know the real effects of such a disturbance. However, spatially explicit models can help in this analysis. The prediction of disturbance effects can be carried out using key or umbrella species though in reality these species are not enough to describe the complexity of a system. The use of guilds seems a good enough system but again monitoring is restricted to a group of selected species that only partially express the complexity and behavior of the system.

The use of processes instead of species seems to better cover the issue of understanding the behavior of a system. Processes act at a broader range of scales than organisms and can capture, when analyzed across the different scales, more information.

Fig. 2.4 - Cultivated fields have very high fragility, changing soil properties (microbial, fungal and animal composition) and vegetation according on the cultivation patches. Vegetables (above) and corn (below).

A process being used as an indicator of fragility must be scaled properly according to the geographical scale of interest. The wide capacity of a process to be tracked using physical measures makes their use quite realistic.

For instance, fire occurs at a broad range of scales, from individual trees to the entire forest. The severity of fires depends strongly on the extent of the observed area. The increase of an area enhances the relationship of fire with land heterogeneity, tree diversity, diversity of age class, and with many other relevant patterns and processes of a forested mosaic. The increased forest size reduces the severity of the fire according to the scale extension and, at a certain scale, the effects of burning can be easily predicted according to morphology, slope aspect, slope acclivity, and natural and man-made corridors and barriers. The behavior of fire assumes a more stochastic character when the scale is enlarged.

The effect of burning or flooding on plant and animal populations is important and is often the main driver of the evolution of vegetation mosaics and animal communities.

It is possible to attempt a broad classification of fragile sites:

> Areas affected by frequent natural intrinsic disturbances like fire and flooding, and by an active secondary succession.
> Areas dominated by extrinsic periodical disturbances like - landslides, avalanches, lavic cover, etc.
> Sites in which human intervention is increasing and in which the environment suffers from a pathological disturbance (intensive agricultural stands).

Lakes and lagoons which are affected by acid deposition and pollution concentration may also be considered highly fragile. Such impacts of human origin are not intentional but the effects disrupt the complexity of plant and animal communities.

2.3.4 Evaluating resilience *versus* resistance

Resilience is the property of environmental systems to recover after disturbance. This property is extremely important for the maintenance of patterns and processes over a long time period, allowing evolutionary processes to take part in maintaining stability. Willows in a river bed act in a resilient way during flooding. Their branches are flexible and folding, and regain their normal position after the stress event. Resistance is a less common property of natural systems, but is very common in human "engineered" landscapes. Most human landscapes have resistant characters and after perturbations that surpass the threshold of resistance, they progress in another directions. We can consider two types of resilience: an "engineering resilience" and an "ecological resilience" (Peterson et al. 1998) (Fig. 2.5).

Engineering resilience measures the rate at which a system returns to a single steady state after perturbation (Holling 1996). Ecological resilience is a measure of the amount of change or disruption necessary to transform a system into another (Holling 1973). According to this last definition we assume that a system has self-structuring and self-reinforcing mechanisms that create stability.

The scaling capacity of a system increases ecological resilience. For instance, the predatory process can affect a prey at one scale but becomes inefficient at another scale which is used by the prey for a different life trait. Also, the redundancy of processes increases the resilience capacity of a system. Again the scaling capacity of species and process characteristics are important to ensure the resilience of a system. Modern management of systems reduces the diversity of species and processes, reducing many ecological functions and consequently reducing the resilience capacity. This increases the risk of abrupt changes to the system with catastrophic consequences in term of species turnovers, species extinction, and environmental instability.

The presence of different functional groups within a scale and the repetition of functions across a range of scales reinforces the resilience capacity of a system. Ecological resilience may present different patterns according to the scale. We have described these patterns using examples of mountainous vegetation along the Northern Apennines (Italy).

A system may be unstable at small and at broad scales of perturbation (low engineering resilience).

Generally this system can be found in organisms that are geographically rare and ecologically exigent. This is the case for *Salix erbacea* in the Mediterranean basin, a species which survives as a relict at the top of the highest mountains in small, very isolated populations. A severe (local) disturbance of a population (like the building of a ski cabin) can suddenly cause the extinction of that population. Also, a change in the grazing regime of sheep may cause extinction, with no possibility of recovery in another place.

Stable at small scales of perturbation, but unstable at broad scales of perturbation (snow vegetation).

Another example is that of mountain vegetation which was adapted to living in depressions where snow accumulation covers the soil for long periods. Where some human disturbance such as soil removal for a ski facility produces local extinction, vegetation can find other micro-sites in which to develop because such microsites are not rare. However, large scale climatic change, such as a temperature increase, can affect the survival of such vegetation.

Unstable due to local perturbation, but stable at larger scales.

The group of *Sempervivum* is represented on Southern European mountains by three species : *Sempervivum arachnoideum*, *S. montanum* and *S. tectorum.* which are living as pioneers in recently formed rocky habitats, suffering for competition with other species during the recolonization process; but their presence is so widespread and soil disturbance events (f.i. landslides), so frequent, that at large scale these species maintain stable populations.

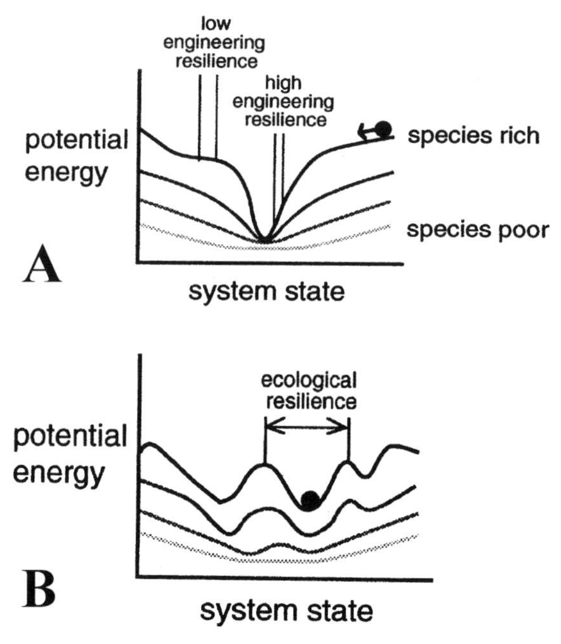

Fig. 2.5 - Relationship between species richness and the stability of ecological function (A), and between potential energy and the system (B). Different landscape topographies exist according to the different levels of species richness. Engineering resilience is high where there are steep slopes of potential energy. Ecological resilience changes according to the potential energy and the system state. The stability of the system depends on local conditions and different topographies of the system can exist (from Peterson et al. 1998).

Stable at local and large scale perturbations.
The blue berry (*Vaccinium myrtillus* and *V. uliginosum*) is an example of this system.These plants are very common on mountains above tree level (from 1500 to 2000 on the Northern Apennines). Local and broad scale of perturbations are unable to affect the distribution of these plants because they are dominant, very common and distributed along a broad elevation range.

Evaluating the resilience property of a system is not an easy task and it is necessary to have knowledge of the components, structure and functioning of the system. It is possible, from the predicted behavior, to create broad classes of resilience for each part of the system and then classify the entire system using this procedure. Resilience properties increase enlarging the spatial scale of observation and this behavior creates difficulties for assessing affordable resilient values to a select area. To overcome such a constraint, maps of resilience must be provided at different scales.

2.3.5 Disturbance as an indicator

Ecosystem functioning may be reduced in efficiency due to human disturbance, especially in highly dynamic systems like rivers, coastlines, and high mountain slopes. Disturbance may be a good indicator and a landscape can be classified as undisturbed, lightly disturbed, moderately disturbed, or heavily disturbed.

These categories must be considered as a "fuzzy set" and relative to the regional context in which the landscape is embedded. The level of disturbance from human activity can be measured by human population density, length of roads and other linear barriers, fossil energy involved in the productive system, quantity of chemical products in agriculture, quantity of fertilizers spread on fields, level of environmental noise from traffic or other activities, amount of logged biomass, amount of computerized services, etc.

Invasion seems a relevant process both in disturbed and undisturbed ecosystems (see Mack & D'Antonio 1998)(Tab. 2.2).

Invaders (plants, animals, bacteria and virus) have deep effects on the functioning of a system thus they are ideal as indicators.

Invaders can interact with the physical forces of disturbance like fire, erosion or biotic disturbance.

Grass invasion can facilitate fire regimes due to newly available undecomposed biomass acting as a fuel load. We have several examples from the abandoned fields on most of the European uplands. In other cases, the presence of invaders can reduce fire frequency due to the thickness of canopies and reduction of the fire prone grass layer.

When an invader occupies a highly dynamic system, for instance steep slopes, and has a weak root system, the slope may be exposed to rapid erosive processes which are more severe than under natural conditions. In other cases, for instance in Mediterranean Europe, the utilization of black locust (*Robinia pseudoacacia*) to consolidate eroded soil surfaces affected by natural and man-made disturbance, reduces the natural dynamics of the system with no clear stabilizing effects.

Large and medium size mammals can produce severe changes in disturbance regimes by a combination of trampling, biomass removal and manure deposition. Many studies have focused on the effects on rare or endangered plants but scant information is available on the changes in the disturbance.

For instance, the removal of the grass understory by grazing alters the nutrient retention causing an increase in decomposition of soil organic matter.

New evidence is required because the alteration of disturbance regimes by invaders has not been described enough and many effects are far for being fully understood.

Tab. 2.2 Some effects of invaders on disturbance regimes (from Mack & D'Antonio 1998).

ALTERATION	EXAMPLE
Fire enhancement	Grasses increase rates of fire frequency, spread, and/or areal extent
Fire suppression	Trees decrease fine fuel load and fire spread in grassland or open woodland
Increased erosion	Trees increase stream bank collapse in riparian zones
Decreased erosion	Plants stabilize mobile substrates Pigs increase soil disturbance
Increased biotic disturbance	Pests or pathogens cause stand-level or population-level dieback
Change in consequences of disturbance	Invader changes susceptibility of community to physical forces of disturbance

2.3.6 Evaluating keystone species

The word "keystone species" was coined by Paine (1966, 1969) and from that time to the present the idea has been discussed by several ecologists (for a recent review see Mills et al. 1993). The keystone species concept, as often happens in ecology, has been adapted for a broad range of environmental conditions and roles though originally Paine used this term to describe the effects of predators in the rocky intertidal zone. At least five keystone categories can be described: predator, prey, plant, link and modifier (Tab. 2.3).

Despite a general reconnaissance of the role of keystone species in a community, few data are available to demonstrate the effective importance of a species in an ecological system. One way is to manipulate such species in order to evaluate the effect of experimental removal. However, this concept was at times been abused within the conservation arena with some species being considered key stone only because they are large, occupy broad areas and, often, are predators like grizzly and wolf. Both have negligible effects on most of the wild fauna living inside their territories whereas other species such as large ungulates, play a real role as keystone "engineering" species for several organisms. I believe that the "modifier" category is the most important when we analyze the environmental mosaic at a large scale (Fig. 2.6). The same role could be applied to many plants that have strict linkages with dependent pollinator and seed disperse animals. In this way the keystone concept is fully applied and verified.

According to Power et al. (1996) keystone species occur in most ecosystems around the world, but I agree with Mills et al. (1993) that the term keystone species is "poorly defined and broadly applied".

Many organisms have a very complicated niche in which more than one patch type is required. In this case the presence of a species is an indicator of a landscape complex. Such species can function as landscape indicator.

A keystone species differs from a dominant species as their effects are much larger if compared with their abundance (see Power et al. 1996). They don't always occupy high trophic status, and they exert their effect not only by consumption mechanisms but also through processes like competition, mutualism, dispersal, pollination, diseases, etc.

The recognition that a species assumes the role of keystone according to the ecosystem (landscape) context in which it operates is also of great relevance. Often the local conditions present inside the community or inside the land mosaic determine the role. Apparently, there is quite a good (inverse) relationship between diversity and keystone species; the more diverse community is, and more species will have similar functions and traits. With the impoverishment of a community, remnant species, that originally shared functions with other species, will assume greater control and become keystones.

The status of key species is not easy to detect and often indirect measures are required. According to Power et al. (1996), an index of community importance can be used to establish if a species is a keystone or not. This index is fomulated as follows:

$$CI = \left[\frac{(Tn - Td)}{Tn} \right] \left[\frac{1}{Pi} \right]$$

where *Tn* is a quantitative measure of a trait (e.g. diversity), *Td* is the measure of a trait when the species is deleted and *Pi* is the abundance of species *i*. When *CI* is > 1, this indicates that the species is a keystone species. For the bison of the Konza prairies (USA), *CI* ranges from 6 to 25.

Fig. 2.6 - Beaver (*Castor canadensis*) and buffalo (*Bison bison*) are two key stone species in forest and grassland systems respectively. Beavers modify water flux along streams by dam building and buffalos affect grassland mosaics by grazing and wallowing (figures from Barker 1987).

Tab. 2.3 - List of some relevant keystone categories and the effects when removed from a system (from Mills et al. 1993).

KEYSTONE CATEGORIES	EFFECT OF REMOVAL
Predator	Increase in one or several predators/consumers/competitors, which subsequently extirpates several prey/competitor species
Prey	Other species more sensitive to predation may become extinct; predator populations may crash
Plant	Extirpation of dependent animals, potentially including pollinators and seed dispersers
Link	Failure of reproduction and recruitment in certain plants, with potential subsequent losses
Modifier	Loss of structures/materials that affect habitat type and energy flow; disappearance of species dependent on particular successional habitats and resources

2.3.7 Biodiversity indicators in the regional landscape

Biodiversity indicators at the regional landscape level represent one strategy for evaluating the "biodiversity" status of a large area.

Often the use of species indicators is not sufficient to describe the trends of complexity. Again we are forced to move across different hierarchical levels in which, at each level, the focus must be oriented to a specific trait of complexity (for more details see Noss 1990). According to Noss biodiversity can be detected at genetic level, at species-population level, at community-ecosystem level and at regional landscape level. I agree with this view but we must remember that a landscape level may range from a few millimeters to thousands of kilometers. If we use the regional (large scale) landscape level it is possible to describe some important spatial variables which are able to function as indicators of biodiversity at the landscape scale. Regional landscapes can be considered a mosaic of geographical patches of vegetation, animal distribution, and land use with an area ranging from 10^2 to 10^7 Km^2. The distribution of habitat patches in a landscape mosaic is vital for organisms that use a different variety of resources in different geographical areas. The complexity of a landscape can be monitored using remotely sensed satellite images, aerial photographs, land use maps and field surveys. The location of ecotones is an important indicator of vegetation responses to climatic changes. The analysis must be carried out at time lags that fit the functioning of the system. Variation of some processes like erosion, fire regimes, human fragmentation, patch persistence and patch turnover rate, are some of the possible indicators of landscape "biodiversity/ecodiversity".

Tab. 2.4 shows the different variables used to monitor biodiversity at four levels of organization, subdivided by compositional, structural and functional components. Landscape ecology provides a powerful tool for assessing environmental indicators through the use of a multiscalar approach, remote sensing techniques including a modest investment in field observation, GIS, GAP analysis and a set of indices of landscape structure like fractal complexity, ecotone length, interpatch contagion, etc.

2.3.8 Animal behavior for evaluating the spatial availability of resources

The distribution of species abundance is a good indicator of the capacity of a system to fit the niche requirement of a species. Across the geographical range of a species, variation in abundance is a common characteristic of many species. Most species show "hot spots" of abundance in which the majority of the total abundance, as indicated by Brown et al. (1995) is concentrated. This variation depends on the niche requirement of a species.

The environmental conditions that we find in areas of high abundance seem important for defining the components of a specific niche. This approach is very useful for detecting areas in which environmental conditions are unfavourable, and for applying procedures to improve locally unfavourable conditions.

Tab. 2.4 - Categories of indicators at four levels of organization, distinguished according compositional, structural, and functional components. A list of inventory and monitoring tools is provided (from Noss 1990).

INDICATORS				
	Composition	**Structure**	**Function**	**Inventory and Monitoring tools**
Regional Landscape	Identity, distribution, richness, and proportions of patch (habitat) types and multipatch landscape types; collective patterns of species distributions (richness, endemis)	Heterogeneity; connectivity; spatial linkage; patchiness; porosity; contrast; grain size; fragmentation; configuration; juxtaposition; patch size frequency distribution; perimeter-area ratio; pattern of habitat layer distribution	Disturbance processes (areal extent, frequency or return interval, rotation period, predictability, intensity, severity, seasonality); nutrient cycling rates; energy flow rates; patch persistence and turnover rates; rates of erosion and geomorphic and hydrologic processes; human land-use trends	Aerial photographs (satellite and conventional aircraft) and other remote sensing data; Geographic Information System (GIS) technology; time series analysis; spatial statistics; mathErmatical indices (of pattern, heterogeneity, connectivity, layering, diversity, edge, morphology, autocorrelation, fractal dimension)
Community-Ecosystem	Identity, relative abundance, frequency, richness, evenness, and diversity of species and guilds; proportions of endemic, exotic, threatened, and endangered species; dominance-diversity curves; life-form proportions; similarity coefficients; C4:C3 plant species ratios	Substrate and soil variables; slope and aspect; vegetation biomass and physiognomy; foliage density and layering; horizontal patchiness; canopy openness and gap proportions; abundance, density, and distribution of key physical features (e.g., cliffs, outcrops, sinks) and structural elements (snags, down logs); water and resource (e.g., mast) availability; snow cover	Biomass and resource productivity; herbivory, parasitism, and predation rates; colonization and local extinction rates; patch dynamics (fine-scale disturbance processes), nutrient cycling rates; human intrusion rates and intensities	Aerial photographs and other remote sensing data; ground-level photo stations; time series analysis; physical habitat measures and resource inventories; habitat suitability indices (HSI, multispecies); observations, censuses and inventories, captures, and other sampling methodologies; mathematical indices (e.g., of diversity, heterogeneity, layering dispersion, biotic integrity)
Population-Species	Absolute or relative abundance; frequency; importance or cover value; biomass; density	Dispersion (microdistribution); range (macrodistribution); population structure (sex ratio, age ratio); habitat variables (see community-ecosystem structure, above); within-individual morphological variability	Demographic processes (fertility, recruitment rate, survivorship, mortality); metapopulation dynamics; population genetics (see below); population fluctuations; physiology; life history; phenology; growth rate (of individuals); acclimation; adaptation	Censuses (observations, counts, captures, signs, radio-tracking); remote sensing; habitat suitability index (HSI); species-habitat modeling; population viability analysis
Genetic	Allelic diversity; presence of particular rare alleles, deleterious recessives, or karyotypic variants	Census and effective population size; heterozygosity; chromosomal or phenotypic polymorphism; generation overlap; heritability	Inbreeding depression; outbreeding rate; rate of genetic drift; gene flow; mutation rate; selection intensity	Electrophoresis; karyotypic analysis; DNA sequencing; offspring-parent regression; sib analysis; morphological analysis

The study of intraspecific abundance has many possibilities which could be adopted as a methodology for evaluating the quality of an area by using a comparative scoring program. This approach could also be used to evaluate areas more prone to exotic or pest species. Organisms from bacteria to mammals are distributed according to the displacement of resources (food, nesting and roostings places). This distribution is often strictly related to the temporal availability of such resources. In particular, food is a discriminant factor in the spatial arrangement of organisms.

Resources act as attractors in animal distribution and consequently the distribution of animals may be used as a bio-indicator of resource availability. Often, when we are dealing with resources, we consider only some types of prey, but we know quite well that this is not enough for animals like birds or mammals; other factors act contemporarily to create favorable conditions.

For instance, birds forage in suitable patches but this suitability is the result of a combination of food as well as physiographic conditions (e.g. the structure of vegetation, size of the patch, distance from other patches, etc.). In heterogeneous areas especially the size and distance of suitable patches are very important. For instance, tree larks (*Lullula arborea*) and sky larks (*Alauda arvensis*) both prefer open habitats, but the former can be found in small clearings close to trees, and the latter (sky lark) need open, wide areas far from trees (Fig. 2.7).

Foraging behavior of herbivores is a typical example of the integrated, multiscalar information that an organism has with an heterogeneous neighboring of resources availability and resource quality. A strategic point is represented by the scale at which organisms perceive the environment. When a landscape pattern is outside the scale of a species this pattern becomes indifferent. The selection of an appropriate scale is the main concern when we investigate the environmental relationships between a species and the habitat in which it lives. It becomes increasingly clear that the study of ecological relationships between an organism and its environment is not enough if we investigate at one selected scale only.

The heterogeneity of a landscape is intercepted by organisms in a species-specific way and is linked to body size. For instance, during foraging behavior tortuosity and time of movements are both important indicators of landscape quality. To test the null hypotheses that the turtuosity of foraging paths and velocities are not controlled by landscape heterogeneity and that animals with similar body sizes perceive the landscape in the same way, Etzenhouser et al. (1998) conducted an investigation of white-tailed deer (*Odocoileus virginianus*) and Spanish goats (*Capra hircus*). The application of fractal geometry seems a promising approach; in fact fractals are scale invariant over a discrete spatial range. Two measures of fractal complexity were used. The perimeter:area fractal (using the computer software FRAGSTATS (McGarigal & Marks 1995)) is an index ranging from from about 1 when the shape of patches is regular, to 2 when the shape is complex with a large perimeter.

The second index is the grid count fractal based on measurement scales ranging from 10 to 1000 cm. The animal foraging path was characterized by the 'dividers' method of fractal estimation, plotting the log of path length and the log of the measurement scale in a range of scales (2.5 to 780 m). A fractal dimension close to 1 indicates linear movements, so animals spend less time in a site.

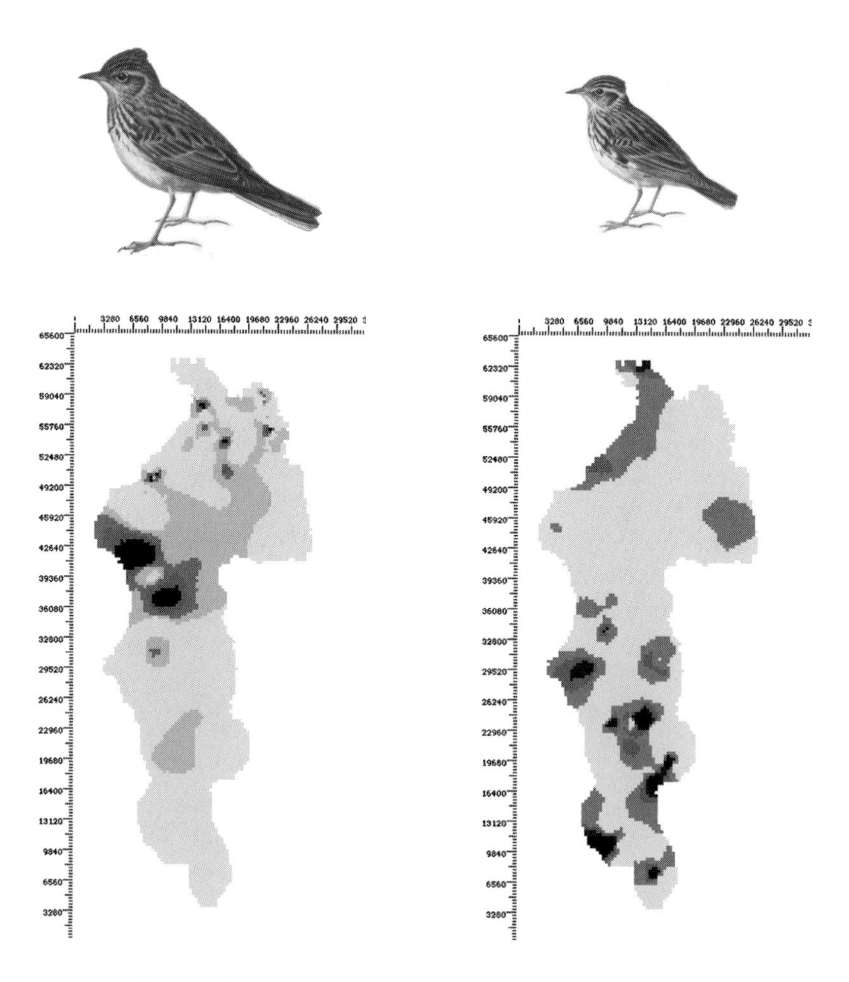

Fig. 2.7 - Distribution of skylarks (*Alauda arvensis*) and tree larks (*Lullula arborea*) on an upland ecotone (Logarghena, Northern Apennines, Italy) during the breeding season. Skylarks are located in open spaces far from edges and woodlots. Tree larks prefer small clearings into wooded mosaic (see Fig. 3.17, pag. 247).

On the other hand, a fractal dimension close to 2 means that the animal path is very tortuous, travelling more in small areas. Differences in foraging behavior were found; deer had more linear paths compared with goats, although both goat and deer 'viewshed' had an equivalent amount of shrubs; deer are more sensitive to the shape of shrubs than goat; and both have a different reaction to the presence of specific plants. In other words, the two species have a different perception of the same landscape.Spatial distribution of animals during their foraging activities is one of the best indicators of resource quality, location and seasonal availability, and this in turn may be used for landscape evaluation.

2.3.9 Evaluating landscape changes

Historical analysis is the basis of a landscape evaluation. It is not possible to evaluate the present conditions of a mosaic without knowing at least it's recent history. It is only by considering the evolution of a landscape that it is possible to understand the level of reaction to different perturbations.

Landscape change is an important process of environmental complexity. The main difficulties of studying landscape changes are represented by the different qualities and lack of information when we go back more than 50 years. In fact, the lack of data, especially at the spatial level, creates barriers for a comparative analysis. However, the new technologies of remote sensing have created the condition for a more regular and homogeneous survey of most of the globe for future comparisons. No landscape is stable but, they all have an homeorethic capacity to respond to climatic changes, and changes in land use and disturbances regimes. The composition of a mosaic, dimension and shape of patches, and quality of the matrix are some of the attributes that can change in a landscape. All these changes, which can occur separately and with different spatio-temporal scales and frequencies, influence the evolution of a species, their aggregations, patches, and mosaics.

The final product is represented by a mosaic of patches of different size and shape, with different ecosystem functioning. Changes are the integration of many abiotic and biotic processes and are basic components in every ecosystem. The behavior of a system is observed by analyzing the rate, intensity and frequency of changes. Changes occur at different levels of species aggregation and dynamism (extinction/colonization), species importance (dominance), and interspecific interactions.

Small scale systems have a higher rate of changes than large systems, though this is a result of the time scale. As a consequence, the study of changes is not confined to a unique scale but can be spanned, for instance, from the micro-landscape of an epyphit colony to the entire planet mosaic. In the study of changes it is important to have a reference system by which to carry out comparative studies.

The evaluation of changes occurring in large scale landscapes (at the human scale) seems, today, very popular and strategic for their economic consequences (f.i. Turner 1987a, Foran & Wardle 1995, Foster et al. 1998). This scale is useful for understanding the changes occurring within the land mosaic. Procedures and techniques for analyzing changes in the landscape by using aerial photographs, land register maps, and satellite imagines are available and well described (Fig. 2.8). Particular care is used to describe changes in the function of the different land configurations and social/economical scenarios are compared with landscape patterns. Landscape changes have increased in intensity and frequency during recent decades, characterized by the globalization of social, economic and information processes. In some regions such changes are incredibly wide involving entire countries like Brazil (Skole et al. 1994) (Fig. 2.9). In other parts, such as Europe, land abandonment has modified entire upland regions with profound consequences on the living style and economy of the people.

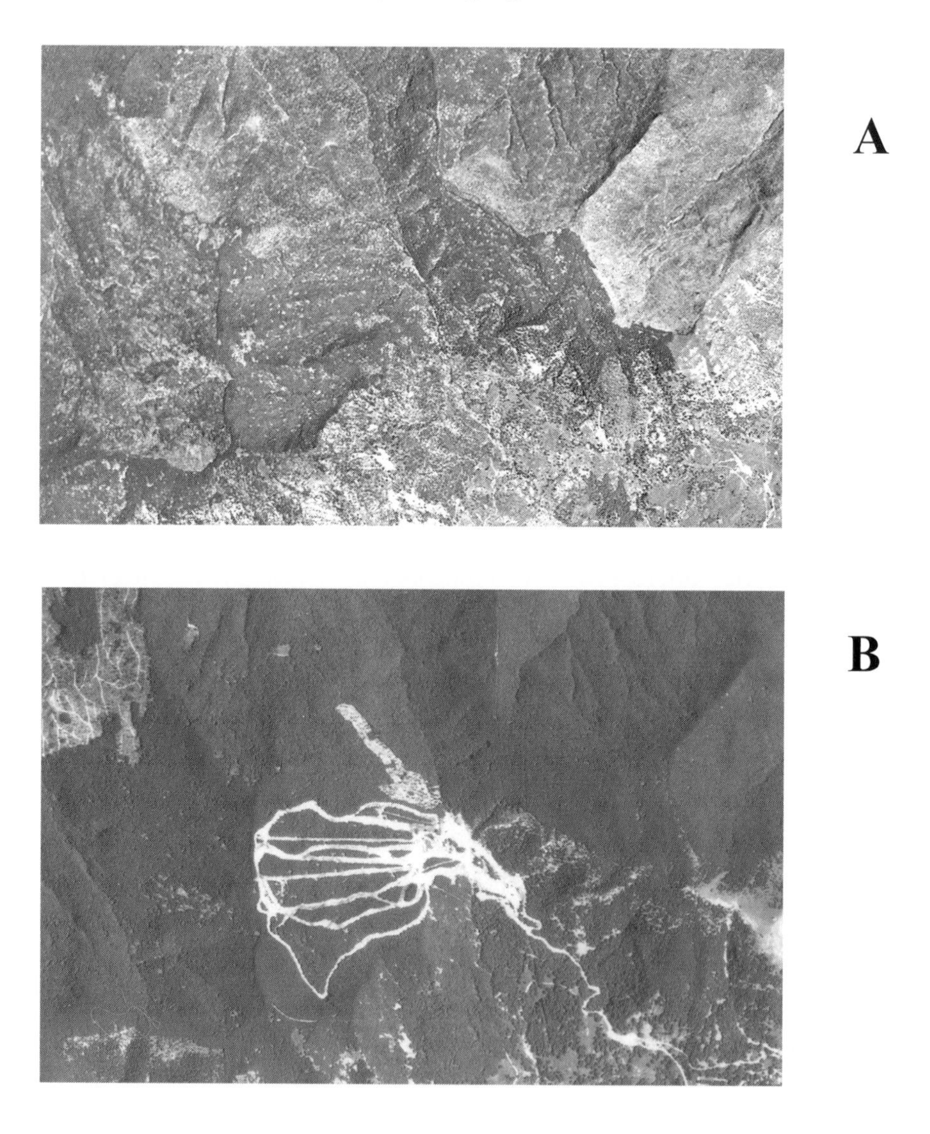

Fig. 2.8 - From 1950 until the present (1989), dramatic changes occurred along the mountain ranges of most of Europe. In the picture on the left (Zeri commune, Northern Apennines) the fragmented mosaic of 1950 (A) shows a through utilization of the mountain landscape for charcoal and permanent pastures. After land abandonment woodlands recovered and ski facilities were built (white areas) (B).

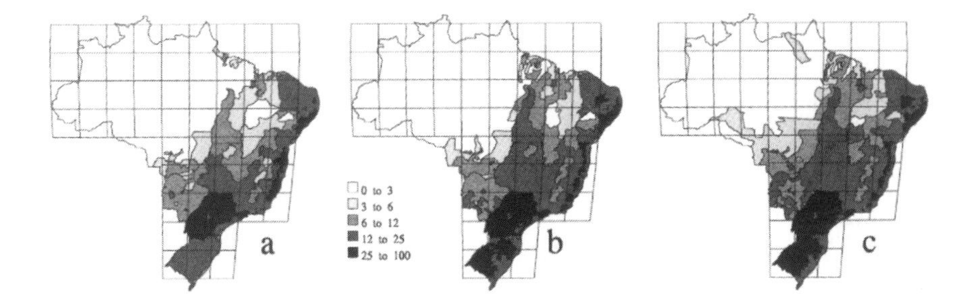

Fig. 2.9 - Time series of land changes across Brazil (A-1970, b-1975, c-1980) using a geographic information system, espressed as density of land cover transformed into agriculture (from Skole 1994).

2.3.9.1 Succession and disturbance

Dynamism is at the basis of all ecological processes. Changes across a landscape are due to natural factors like succession or disturbance. The first process behaves gradually, the second occurs suddenly and deeply modifies the structure and functioning of a system.

For plants, succession is considered to be the dynamic event that shapes diversity and structure of vegetation through time. Succession may be considered a process during which species aggregations change according to resource availability, competition game, dominance success and disturbance regime.

Animal populations and communities respond promptly to vegetation dynamism and their movements are coupled to vegetation changes. Animals produce a scaled disturbance throughout succession which retards or accelerates the successional process accordingly.

Secondary succession is a well known process common to most of the vegetational realm across the globe. Generally, such a process consists of changes along edges or ecotones of vegetation and animal communities or aggregation coherences. Moving towards a broader scale, changes can be observed across the mosaic matrix but evidence can only be detected if a suitable time-scaled window is considered.

2.3.9.2 Land use changes

Due to the diffuse presence of people, most changes that occur are produced by changes in the land use of a territory. Two opposite processes can be observed in today's landscape: a) the intensification of agriculture, and urbanization and b) land abandonment. Both have profound consequences on the structure and functioning of landscapes.

The intensification of agriculture produces a less heterogeneous area, thus decreasing the diversity of the land mosaic and it's spatial complexity. Edges and other buffer zones are drastically eradicated to simplify and optimize the mechanical work.

This simplification also has severe consequences on the cycle of water, nutrients and pollutants like phytossic sprays. This is the main cause of wildlife disappearance from "modern" agriculture farms (f.i. Smith at al. 1993, Andrews & Rebane 1994).

The reason for this change is simple; it is connected to the necessity to reduce the cost of production and to the impossibility of finding the necessary human and economical resources to handle the uncultivated land.

Foster et al. (1998), whilst studying forest dynamics in Central New England over the last three centuries, found a general "homogenization" of vegetation, disruption of vegetation-environment relationships and the formation of new assemblages (Fig. 2.10). This was apparently due to a massive new disturbance regime, a reduction of human interference during a reforestation period, permanent changes in abiotic and biotic components of the landscape, and a short period of forest recovery (100-150 years).This trend is also common to other regions such as the Mediterranean and Western Australia. Human pressure in the period considered, changed patterns across the landscape causing different pulses of clumping (urbanization) and spreading (farming) waves. However, this pattern is decreasing in importance due to a declining interest in natural resources by modern society.

This fact is well known in all developed countries where the productive compartment (agriculture) is separated from the non productive compartment (most of the hilly and upland regions).

2.3.9.3 Patterns in large scale vegetation changes

Changes of landscape occur due to external disturbances like fire, flooding, avalanches, and, especially today, due to human intervention.

Changes in vegetation cover were considered until recently, a matter of time (million years was the Reid's prevision)(Reid 1998).

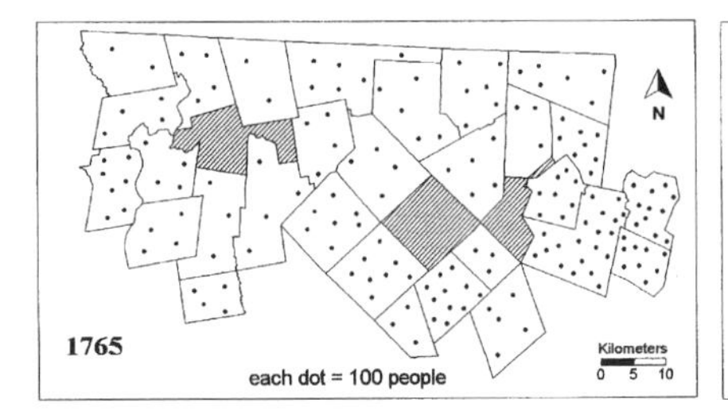

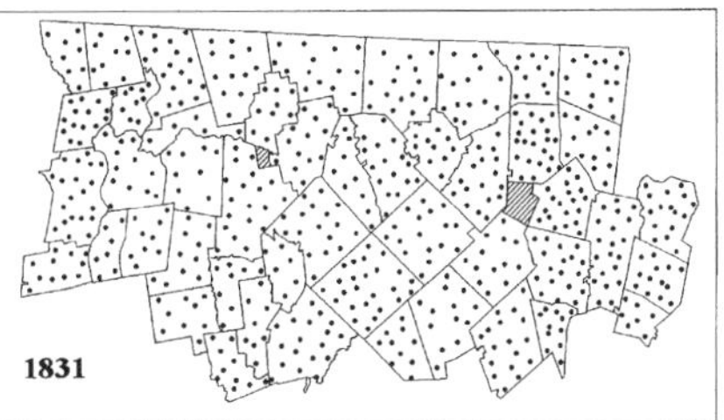

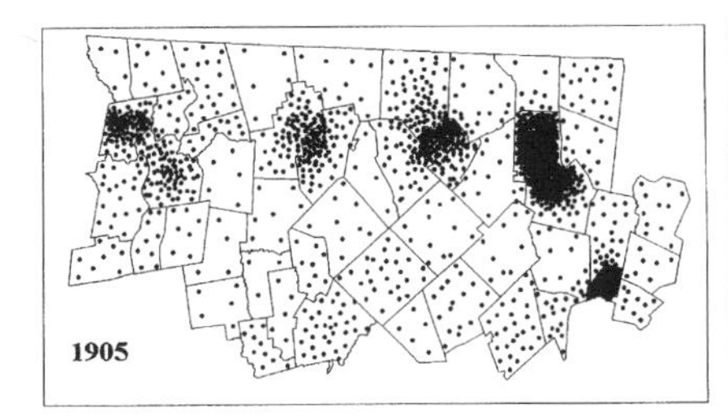

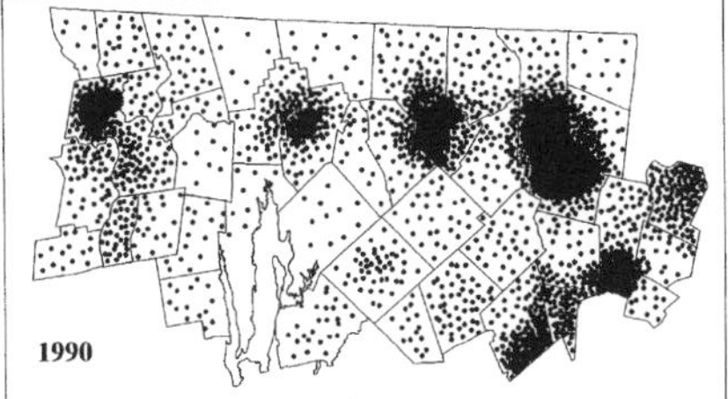

Fig. 2.10 - Changes in population distribution accurred in New England from the Colonial period to the present. After a peak of rural settlement in the mid 19th century, industrialization at the beginning of the 20th century produced a concentration in urban areas. A redistribution of population in rural areas is the today prevailing demographic process. Diagonal lines indicate towns without demographic data (from Foster et al. 1998).

Despite this long term view, evidence available today suggests a scale of 1,000 years is appropriate for the geographical migration of species (Clarks et al. 1998) (Fig. 2.11). Trees have been estimated to move 150 to 500 m/y in Europe and North America. This movement is conditioned by the direction of climatic changes; it is species specific and large barriers like lakes and large rivers can reduce the potential for expansion. Dispersal distance and growth rates are the factors determining the speed of dispersion.

The settlement of seeds in a new locality is in part controlled by fire and wind throw. Past knowledge about the intensity and scaling of disturbance regimes is very poor. The fast Holocene migration of trees linked to a non uniform climate, and probably interaction with animal dispersers, can't be compared with future changes of climate.

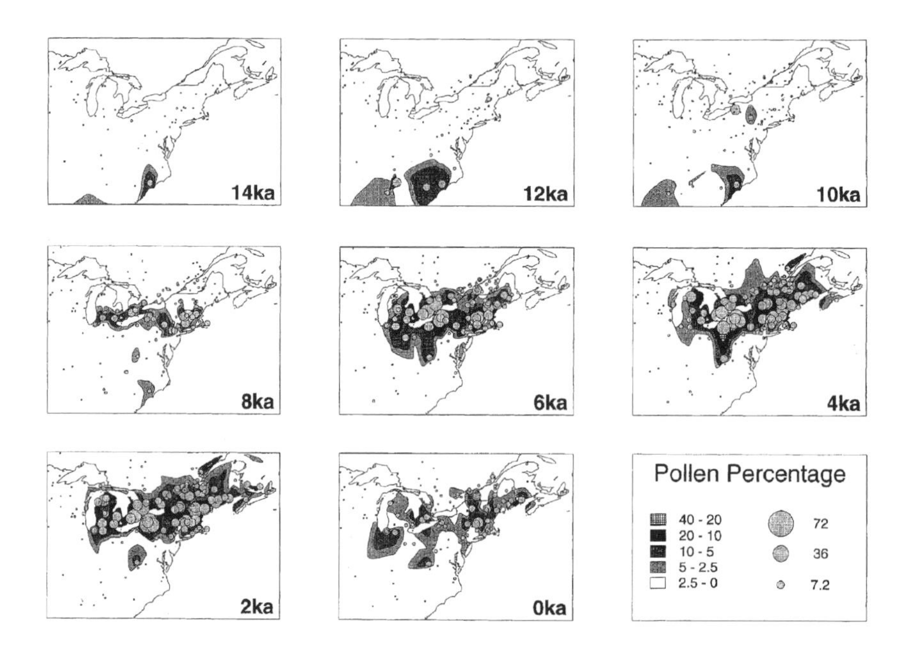

Fig. 2.11 - Spatial distribution of beech (*Fagus grandiflora*) pollen precentages from 14ka to 0ka (1ka=1000 radiocarbon years before present) in eastern North America. Circle size represent pollen percentage in the sediment samples and contours (5 categories from 0 to 40%) the interpolated abundances (from Clarks et al. 1998).

One significant difference consists of the different substrates in which seeds are reacting. The destruction of the natural environment, the presence of new structural and chemical barriers, and the reduction of large populations of birds and mammals, are probably factors that will reduce the migration rate of plants. The probability and the rate of dispersion of trees depends on many factors in which the interactions between a scaled landscape appears to be one of the more important elements. Spatially explicit models applied to micro-landscape studies appear as important as the dispersion models used to study plant movements at large scale. The combination of these two approaches could enhance comprehension of the migration process.

2.3.9.4 Changes in soil properties

The land mosaic is composed of discrete entities, i.e. patches, that contribute to the overall heterogeneity in different ways. It is especially along the border of each patch that changes occur and are more visible. Such a line is called an ecotone and plays a fundamental role for the entire land mosaic. Ecotones are particularly evident in a changing mosaic, as in most of the southwestern United States in which large portions of territories have changed from grassland to shrubland. These changes have been attributed to factors such as livestock overgrazing, drought, fire suppression, increase of rodents, or a combination of all these factors.

When environmental change occurs, the entire organization of biotic and abiotic components suffers profound modifications. In this way, new processes occur and new plant and animal communities replace the old ones. In this "eternal" play of changes, a dynamic equilibrium may be reached by a system for a discrete time period. The disruption of system can occur when new external perturbations modify the resilient capacities and the fragility of the system. In such respects, the investigation from Kief et al. (1998) in the Southwestern United States confirm the change in soil nutrient resource availability when the vegetation changes from grassland to desert shrubland (Fig. 2.12, 2.13).

Soil nutrients become more heterogeneous in space and time, and concentrated into islands of fertility. In effect these authors found a reduction in the soil profile of 3-5 centimeters from grassland to shrubland dominated by creosotebush, with an increase in calcium carbonate content at a lesser depth. In grasslands the sand content is higher, probably the change to shrubland has transported the soil from bare areas to creosotebush areas. In such a way, buried rocks and stones are exposed at the soil surface. The establishment of creosote modifies water circulation, sequestering most of the soil moisture under plants. Soil under plants, in grasslands rather than shrublands, has a higher nutrient content, but resources are more concentrated under creosotebushes. Soil resources show a greater temporal variation in shrubland than in grassland. The total resources available in grassland are higher than in shrubland due to higher plant cover (45% in grassland compared with 8% in creosotebush shrubland). With the change in vegetation typology, changes also occur in the soil physical structure resulting in a fine-scale relief with small mounds of soil under grasses. In creosotebush sites, the soil has more exposed gravel and deep islands under bushes. The transformation of grassland into shrubland has affected the redistribution of resources from a sparser pattern in grassland to a more concentrated pattern but with fewer locations in shrublands. This in effect produces a reduction in available resources when grasslands become substituted by shrublands.

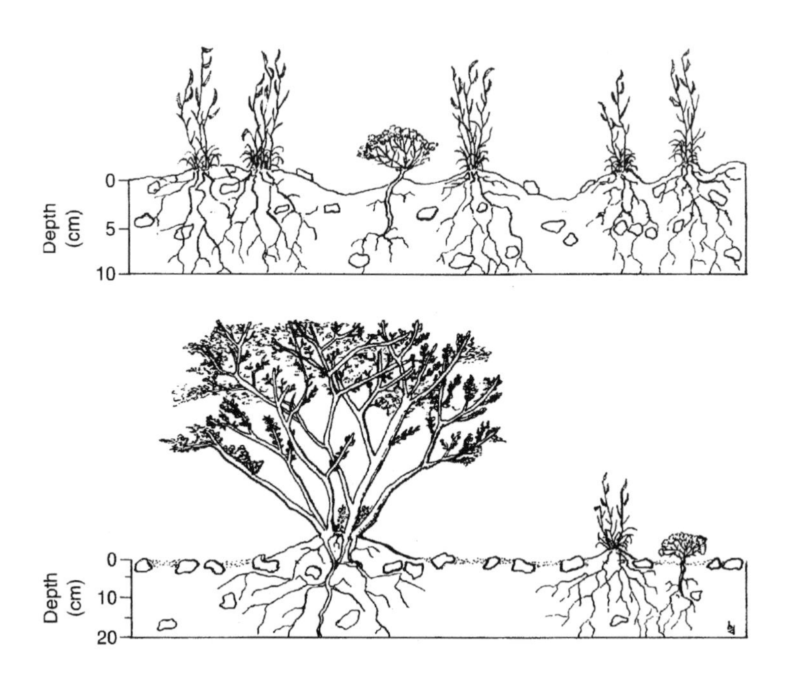

Fig. 2.12 Comparison between a soil profile and vegetation in grassland (upper) and creosotebush shrubland (lower). Although at different scales, it is possible to recognize a finer – scale relief in grassland transects with more vegetative cover. Creosotebush shrubland is characterized by a coarse-scaled relief, more exposed stones and gravels, and less vegetation cover (from Kief et al. 1998).

Actually, this evolution is considered a dysclimax but in other regions this phenomenon can't be considered as fragmentation or aggregation and therefore needs more attention as it is typical of many secondary successions.

Changing the type of vegetation also produces a change in soil fertility, as in humus forms (Vos & Stortelder 1992) and in the micro-physical properties of soils.

A morphological approach to soil texture could be extremely important for predicting the consequence of vegetation shifts after a change in disturbance regime.

A similar phenomenon has occurred on mountains prairies (above the tree limit) along the Northern Appennines in the 60yrs following complete abandonment of sheep farming.

Within a few years many slopes experienced landslides, apparently due to a change in vegetation cover from mixed grasslands to *Brachypodium pinnatum* dominated grasslands and to moorland (*Erica carnea*, *Calluna vulgaris* and *Vaccinium* spp.). Although there were no extensive studies, only empirical observations, it appears quite clear that the change in vegetation cover changed soil stability exposing the soil to erosion after snow melts.

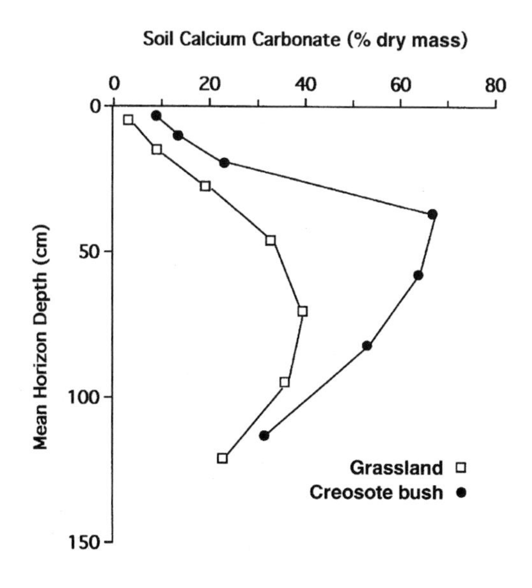

Fig. 2.13 - Difference in the presence of calcium carbonate in grassland and creosotebush along a soil profile (from Kief et al. 1998).

2.3.10 Concepts of landscape health

The health of the environment would apparently seem to be a reasonable concept to be encapsulated within a paradigm. Generally, health evaluation is carried out at the ecosystem scale. This means that the evaluation is conducted at a scale in which ecosystem functions are fully considered (Rapport et al. 1998). When we analyze the health concept it is, by necessity, in connection with the human occupation and use of land, and we can also express this in terms of environmental service. Such environmental services mainly consist of air cleaning, water quality maintenance, and renewable resources.

It is possible to use a multi-criteria approach to carry out an evaluation procedure. Such a multi-criteria procedure needs a multi-scalar approach starting from an individual and moving through different aggregations. In this procedure we must also consider processes and their aggregations. The difficulty of health assessment is the scalar design of investigation, the parameters to be considered and the capacity of every system, from organisms to geographical areas, to incorporate disturbance and to live for a long time in a pathological status.

When the health concept is applied to ecosystems and landscapes it assumes an immediate significance. A healthy status means that the different functions are working well and that energy flows within an organism are in good balance. Health is an integrated condition of a complex system like an animal or a plant. This concept can be also extended to a large area, such as a region.

Organisms are the first indicators in a landscape health evaluation (see f.i. Pollard & Yate 1993, Furness & Greenwood 1993, Jeffrey & Madden 1994). Their sensitivity to environmental changes is, in many cases, very high. For instance the quality of ornamental traits of some birds are used as indicators of environmental health. This field as stressed by Hill (1995) is just at the beginning, but several studies have confirmed that ornamental traits are important indicators of environmental condition.

In fact, there is a disproportional impact on ornamental traits produced by environmental perturbations compared with other morphological traits. The monitoring of ornamental traits also assumes particular importance because it is relatively inexpensive and because it captures the first signals of environmental pathology before irreversible damage (Fig. 2.14).

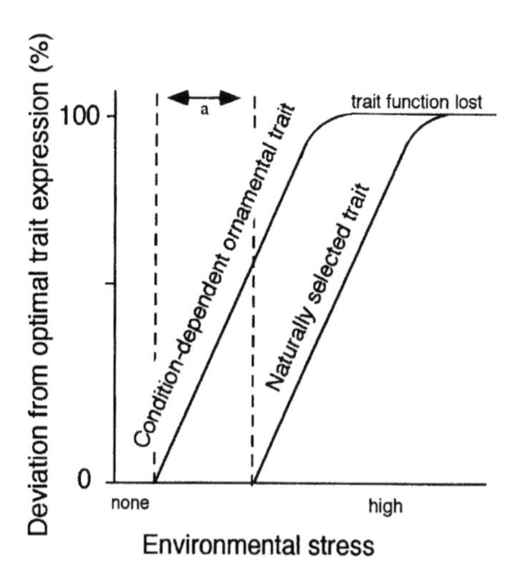

Fig. 2.14 - Relationship between environmental stresses and the deviation from the optimal trait (from Hill 1995).

2.3.10.1 Landscape health paradigm in action

As extensively discussed by Rapport et al. (1998), the health metaphor has received criticism because an ecosystem does not have the behavior of an organism. At the same time, a landscape is less "organismic " oriented than ecosystems (units of processes) and, according to a general theory, systems have weaker and weaker intra-relationship with an increase of the geographical scale.

Nevertheless, it is acceptable to collect general information on the functioning of an ecosystem or landscape and to find a synthesis that can express the capacity of a landscape to function properly (Fig. 2.15).

As in non systemic medicine, the view of the overall system in ecology is not popular and scientists try to move towards the proximate factors. As stressed by O'Neill et al. (1998), precise answers are often provided to trivial questions, rather than approximate answers to critical questions.

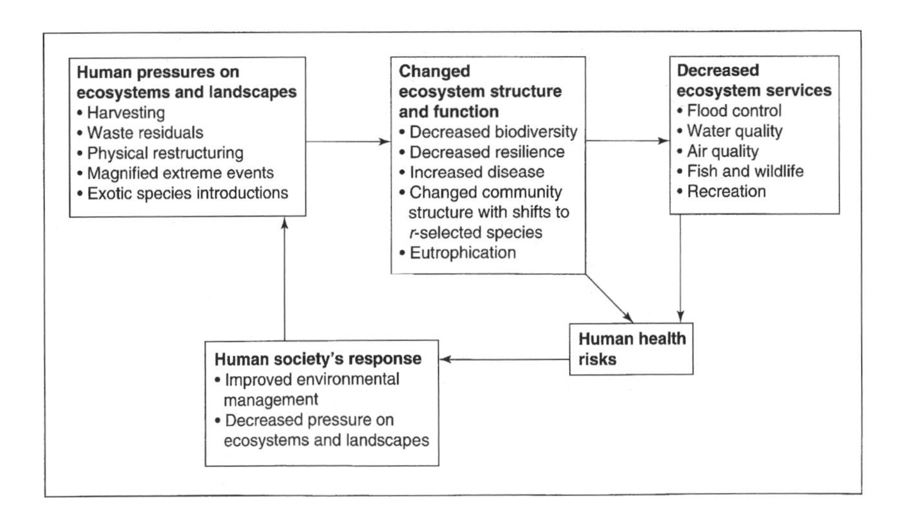

Fig. 2.15 - Connection between human activity, ecosystem changes and degradation, and human health (from Rapport et al. 1998).

Signs of dysfunction are the basis for evaluating the healthy status of a landscape. The capacity to implement preventive measures is really more convenient than to find solutions when damage has already occurred.

It is a general recognition with regards to the deterioration of most of the world's ecosystems that an improvement of the health of landscapes is strategic for the future of our ecosphere. This approach can be considered trans-disciplinary, for which the integration of information and approach from the different disciplines is considered in a unique framework. Of course, it is simpler to collect information on distressed behaviour than to find criteria for health indicators.

2.3.10.2 Health criteria

We can considered ecosystem health as a service furnished by a system providing energy, biomass (capital stock) or information. When the ecosystem service is considered as a criteria for evaluating ecosystem health, it may be defined as the capacity of a land mosaic to support organisms, human included, to incorporate inputs of chemical pollutants and as a source of water. The criteria for classifying the healthy condition of a landscape may be several; we will try to consider the most relevant but we can be sure that our analysis will be considered imperfect, and incomplete.

First of all it is important to define the geographical space in which the analysis is carried out. Generally the area is determined by a socio-economic context, for instance the administrative border of a county, of a province or of a parish. In other cases, it is the watershed area that is considered, but it is quite different if we consider the Mississippi (US) watershed (3,328,000 kmq) or the Po (Italy) watershed (70,000 kmq).

At every geographical scale different processes contribute to a health evaluation. The larger is the scale, the larger are the scale processes to be considered.

2.3.10.3 Health indicators

According to Rapport et al. (1988) the indicators of ecosystem health are: vigor, organization, resilience, level of fragmentation, degree of deforestation, amount of natural vegetation, quantity of development, number and quantities of human barriers, accurateness of management, rare and endangered species and the ecological debt. Ecosystem health may be also considered a super-indicator.

Vigor:

Vigor refers to activity, metabolism, primary productivity and nutrient cycling.

Organization:
It consider to be the complexity in life forms and biodiversity, and refers to the ratio of r-selected species to k-selected species, the ratio of short-lived species to long-lived species; the ratio of exotic to endemic species; the degree of mutualism; and the extinction of habitat specialists.

Resilience (Fig. 2.16*):*
Resilience is considered as the capacity to recover from natural perturbation or the resistance to natural perturbation. It is also the capacity to maintain structures and functions despite external stresses or disturbance regimes.

Level of fragmentation:
At a geographic scale, other useful criteria include the level of fragmentation of natural habitats, the degree of heterogeneity, the level of interference between natural structures and artifacts such as dikes, roads, electrical power lines, etc.

Degree of deforestation:
This indicator is good enough for small basins, but it looses importance when large basins, like the Mississippi, are considered. In this last case the river crosses different climatic regions that are in fact considered eco-regions in which forests and grasslands occur (Bailey 1995). Deforestation has a strong effect on many organisms, decreasing complexity in the system (Bancroft et al. 1995) and affecting forest dynamics (Bierregaard et al. 1992).

Amount of natural vegetation:
In general the quantity of natural vegetation is the prime criterion. Natural vegetation means vegetation which is untouched by human activity (no logging, no major roads and permanent settlements, no development). Such a situation is increasingly rare on our planet. From tropical to boreal forested areas, the degree of intactness of forest cover is the first criterion.

Quantity of development:
This criterion is especially relevant in human dominated landscapes and involves measuring the quantity of development and it's spatial arrangement.
In this case, a criterion of distance from natural hotspots seems very useful; a difference between evaluating a village along a river or far from a river.
Again the pattern of a developed area may be very important. It is better if a developed area occupies one part of a region uniformly, leaving another part untouched, than to have a sparse settlement.

Number and quantities of human barriers:
The number and distribution of human barriers such as roads and permanent fences (concrete walls, metallic nets, atc.) is another important crite-

rion. The dimension and spatial arrangement of barriers play a fundamental role in a landscape as their presence means the interruption of the main physical and nutritional fluxes. The worst situation is represented by roads surrounding the coastline of islands, marshland and rivers, covering and reducing natural ecotones/buffers. In this situation the exchange of material and organisms is more difficult or even prevented. For instance, roads parallel to a river have more negative effects than roads that cross the riverine system.

Accurateness of management in cultural landscapes:
The quality, intensity and accurateness of management assumes a fundamental importance in landscape evaluation. In fact, management action is the process that can create the ideal conditions for maintaining a landscape in good shape, or it can create degradation processes. This is typical of many cultural landscapes around the world that are maintained by an exacting stewardship according to practical rules imposed by the continuos feedback between natural processes and human disturbance regimes.
For instance, logging activity, if not well planned, is the main cause of species extinction. The management of forests can reduce, minimize or incorporate logging activity.

Rare and endangered species:
Though the species criterion is important at a local scale, it is less significant when we enlarge the geographical scale. Rare or endangered species can be entered in a regional survey and contribute to artificially improve the situation at a landscape scale.
It seems better to utilize the structure of different guilds as an indicator. In this case the diversity of guild composition and inter-guild diversity can be good estimators of landscape health.

2.3.10.4 Scaling landscape health

The different scales at which a landscape mosaic reacts to stressors and, especially, to external stimula, create a complication in the evaluation of health. The capacity of a system to incorporate a disturbance depends on the intensity of such disturbance but also on the dimension, frequency and scale of the disturbance. Small scale disturbance interacts at patch level and a stressed patch may be included in an healthy landscape mosaic. However, the number of stressed patches is an indication of the general impact on the entire landscape. If patches or their aggregations are considered the units of a mosaic, their health condition contributes to the overall health of the system.

Often multiple stresses act contemporarily and the relationship between each stress event and others becomes very difficult. A landscape is not an organism, the relationships that link together the different components are necessarily weak but, for this reason, no less important. In fact, the error that we may make when we consider a landscape mosaic is it has parts linked each to other with the same strength of a component system, ecosystem or organism.

Fig. 2.16 - Example of the incorporation of a disturbance (fire) into the trunk of a Turkey oak (*Quercus cerris*). The incorporation concept can be used to evaluate the capacity of a large scale landscape to reduce the effect of a stress like flooding or regional fires.

The relationships between parts become increasingly weak, with an increase in hierarchical level, and this is the main difficulty for finding strong management policies at large scales. There are many possible options and each option can change the system in a quite unpredictable way. At this level the system reacts not only to biological or physical stimulus but also to societal and economic ones (O'Neill et al. 1998). Enlarging the geographical dimension of a landscape, new processes interact and complexity is recognized by a weak system of relationships.

When the scale of human disturbance is coupled with the scale of primary processes like water flux, erosion and forest fragmentation, the effects on the landscape become evident. The saturation level of disturbances is progressive and often not easy to detect in advance. The mechanism of coupling between large scale human disturbances and natural processes cause catastrophic behavior in both systems. As pointed out by Haila (1998), the concept of a healthy versus unhealthy system may be applied only in the presence of a human disturbance regime and for a human context. In fact, in nature, every system can be viewed in a particular stage of its temporal trajectory, and the presence of feedback between natural disturbance processes and organisms is the limiting factor for degrading trajectories.

Another difficulty relating to the evaluation of a landscape is that often the processes operating in this geographical scenario have ambiguous scales and there are multiple scaled contexts.

When, at a large scale, we also consider the economic and social components, the boundaries between natural and human processes become weaker due to the multidimensionality of human activity. In this case, temporal scale (historical changes) plays a fundamental role.

2.3.10.5 Patchiness and health evaluation procedures

Patchiness in a landscape increases the complexity of processes. At fine temporal scales patches are more ephemeral than at long term scales, and so patchiness is perceived differently by various organisms. In each patch the relationship between species largely depends on the role of a single component, for instance, keystone species assume a higher importance for the maintenance of a patch.

Organisms living in a patch can modify it, or destroy it (for instance) by using it as an ephemeral food patch whilst organisms living in other patches can invade a patch producing a profound modification of the patch community. The so called "paradox of site" may be considered when evaluation procedures are in action. The conditions of a single site have a scant influence on the overall system, though the system is the result of inter-site relationships. The matrix in which patches are embedded acts as a cement of varying quality and a "biodiversity bank" according to the overall health conditions.

Patch dynamic in a mosaic probably is the basal mechanism that assures the survival of ephemeral successional stages that in turn support specialized organisms. Ephemeral patches could be considered not as marginal, unimportant events, but as the more important signals of a healthy matrix. Patches may be considered as a focal level of landscape architecture but different processes affect patch formation and the dynamics of the overall mosaic. In a forested mosaic the internal stand structure is affected by cutting practices but the stand configuration is created by management policy. Also, as reported by Haila (1998), cutting practice in the past occurred separately from management policy, whilst in a modern approach it must be considered part of a more integrated process in which local decisions are produced contemporarily with more bureaucratic management decisions.

In an originally forested landscape, the level of persisting forested patches is a good descriptor of the health condition of the area, but this is often not enough. In fact, with the same level of forest cover, it is the spatial arrangement of the remnants that can create a difference in the reaction of living organisms to the new conditions. There are very good examples in the Pacific US where the spatial management of forested patches has allowed the persistence of which are otherwise threatened (Harris 1984).

2.3.11 Ecosystem distress syndrome

Natural and human dominated ecosystems are subjected to increases in dysfunction and stress. Many ecosystems and landscapes are "unhealthy", which means unstable and unsustainable. This occurs at different scales from organisms (to which the term health is referred), to ecosystems, catchment areas and landscapes.

The Ecosystem Distress Syndrome as described by Rapport et al. (1985) and Rapport & Whitford (1999) represents the signal of an unhealthy system (land sickness). The distress syndrome reduces primary productivity and biodiversity, and alters the structure of communities so favoring short-lived species and decreases the capacity of population regulation thus hampering population fluctuations.

Many environmental systems suffer such a syndrome. For instance, most of the Ponderosa pine (*Ponderosa pine*) of the Western US is suffering from heavy grazing and fire suppression. This increases tree density and tree mortality due to intraspecific competition, thus increasing fuel load available and consequently increasing of fire hazard. This system is more prone to diseases and pest outbreaks.

A stressed ecosystem has a lower rate of decomposition and nutrient cycling. Biodiversity and landscape diversity are depressed, so affecting the aesthetic value. In a system stressed by heavy human pressure, Rapport & Whitford (1999) have recognized signals such as:

Progressive dominance by opportunistic species
Progressive invasion of non-local or non-native species
Shift in community structure
Loss of substrate stability
Disruption of nutrient cycling
Progressive loss of environmental service

Landscape degradation, which can also be expressed as a decline in the life-supporting capacity of the biosphere is strictly linked to human health and this last element represents a good indicator of the functioning of an ecological system.

2.3.12 Examples of landscape pathologies

I present three examples of unhealthy landscape conditions and disease spread which seem very active in different parts of the world and which are produced in a multi-cause context.

Lyme disease
The interaction between landscape changes as in the increase of forested area in urban settlements of north America follow abandonment of agricultural land, and human health are becoming explicit. Lyme disease, produced by ticks spread by rodents (*Peromyscus leucopus*) with the complicity of an increased density of deer *Odocoileus* sp. due to predator depression, is a good example.

Hantavirus pulmonary disease
Other diseases are potentially at risk of a "virulent" expansion, such as hantavirus pulmonary disease which is spread by the deer mouse *(Peromyscus maniculatus).*

The cholera case
The cholera bacterium may stay dormant in phytoplankton and zooplankton until environmental conditions become favourable for a new outbreak. The spread of cholera in these recent decades has been reviewed by Colwell (1996) and interesting relationships have been found between sea surface temperatures linked to El Nino events and cholera spread (Fig 2.17).

The pandemic behavior of this disease is strictly linked to other biological events like algal bloom concentration as detected by remote sensing approaches.

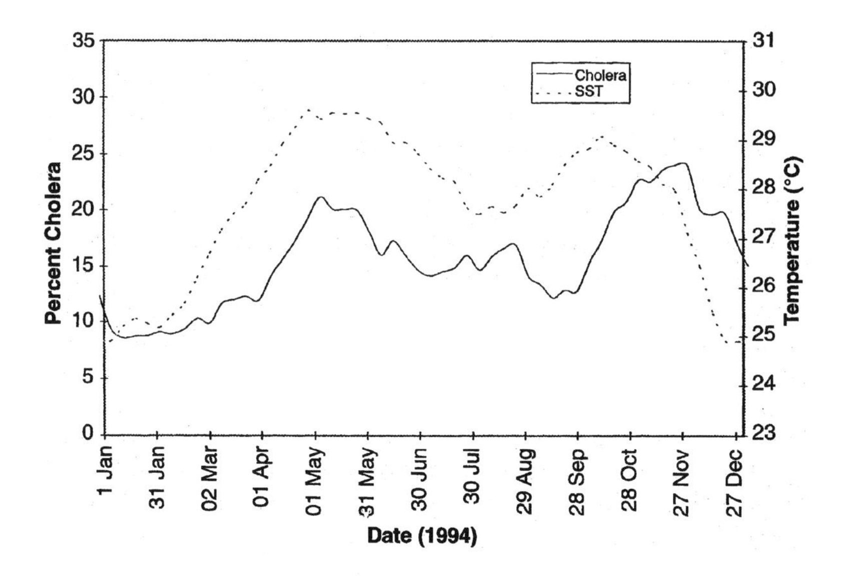

Fig. 2.17 - Cholera cases in Bangladesh have been found to relate to sea surface temperature (from Colwell 1996).

In fact, cholera is associated with plankton in many regions - *Vibrio cholerae* live in low salinity coastal water (5 to 10 p mil) and can survive in salt water in an inactive state such as spore-forming bacteria. Many species of *Vibrio* spp. are associated with chitinaceous zooplankton and shellfish.

Sea currents can easily transport this disease agent and the combined effect of human poverty, low hygienic conditions, brackish water eutrophication and natural episodic events like El Nino, can create predictable conditions for a new, dramatic spread. The study of coastal and off-shore systems in an integrated way, according to landscape ecology paradigms, seems a very promising way to investigate the oscillation of natural events and correlated pandemic diseases.

As reported by Rapport et al. (1998) open questions remain, such as to distinguish between human and natural perturbation regimes and related effects. Thus we might ask: what is the level of ecosystem survival in a perturbed system? What role can play the landscape configuration in terms of diseases spread such as Lyme disease, hantavirus and cholerae? What is the capacity of a landscape configuration to mitigate disease spread? Is it possible to keep systems from a pathological status? And, finally, what is the role of adaptation in altered ecosystems?

I believe that if a comparison between natural and natural-disturbed landscapes is important, it is also of fundamental importance to understand the way in which a healthy human-dominated system such as cultural landscapes, can be maintained by an economically-realistic stewardship.

In conclusion, there is evidence that small organisms seriously and dramatically affect large organisms (virus, bacteria versus trees, animals, man). These micro-organisms live in a micro-world in which the spatial and temporal patterns have a strong resemblance with large scale patterns. Micro heterogeneity is, for chemotaxic bacteria, important for growing procesess, but it is sea currents, at mega scale heterogeneity, that influence their geographical distribution. The large scale landscape is particularly adapted for investigating these pathologies because the scale in which the landscape can be recognized is the scale at which political decisions and management of natural resources occurs, i.e. at the level of river basins, forest districts and states (see also O'Neill et al. 1997).

2.3.13 Evaluating forest landscapes

In a forest landscape, evaluation is based mainly on the amount of logged surfaces and on the mosaic determined by repeated cuts. The assessment of clearing in a region is strategic to prevent soil erosion and loss of organism diversity. Some species, such as butterflies and birds are particularly sensitive to changes in forest cover and appropriate action is required to prevent their losses. Often clearings allow the presence of more visible edge species and the occupation of an area by new species can mask the lost of elusive interior rare species. Connection between remnant patches is important to support available populations of animals. An extended literature has been produced on this point (see f.i. Hansson 1991). Species specific connectivity is a relevant parameter in evaluation.

It is difficult to evaluate the threshold of a clearing that can be incorporated by a forested area without visible consequences on the structure, dynamics and diversity of the area. According to the principles exposed by MacArthur (1972) in the book "Geographical Ecology", many effects are linked to the geographical area in which a forest is located and by the disturbance regime in action.

It is not possible to present details on this point due to the enormous variety of conditions in different biomes and eco-regions. Any eco-region reacts differently to disturbance regimes and resilience capacities can be stressed differently, but a general rule can be presented: the presence of long term human disturbance regimes results in more severe deforestation than areas with low or no disturbance.

Temperate forests have a greater capacity for maintaining diversity after clearing than tropical ones. Yet at the same time the diversity of organisms in tropical areas is higher than in temperate areas and the effect of deforestation has stronger consequences for tropical forest because there is a strict relationship between plants and animals. Many examples and models have been produced to analyze clearings in the tropics as well as in boreal forests.

In temperate regions animals are less connected with a specific plant and have a broader niche. This is probably related to a longer history of interactions between human disturbance regimes and living organisms. Sensitive species probably became extinct thousands of years ago, following human civilization and the progressive occupancy of wild areas.

Prehistoric overkill is a general belief although in some cases there is evidence of long term enduring relationships between Pleistocene diversity and Palaeolithic culture. Also, it is reasonable to imagine that hunter gathering peoples had a deep respect for nature and natural phenomena coupled with religion.

2.3.14 Evaluating cultural landscapes

Healthy and unhealthy management can be a good approach to evaluating the landscape conditions. At the same time, when a landscape has lost the capacity to support any type of management this represents a clear indication of distressed conditions. A decrease in the ecological service of a system is often gradual and not easy to measure without a comparative analysis.

Human stewardship at a growing rate is often necessary to maintain some productions though this process may appear different when we are dealing with the cultural landscape. This landscape, which can be recognized by specific spatial patterns (Spanish dehesa, Tuscany "coltura mista") (Fig. 2.18), supports a high biodiversity but needs a large amount of human energy to plough, cut, harvest, prune and repair the land, whilst at the same time controlling water flows. All these activities have a role in social organization and culture, adding a non trivial contribution to the quality of people's lives and to creativity. In this case, the ecosystem service is larger than in natural systems and this new paradigm (ecosystem service) opens a door to a new research field. We consider such a landscape as a high profit investment because the stability of the system is very high, as is the overall fragility when the energy input is suspended.

Landscape management in any time period has been addressed towards better and larger production. In some cases, geo-climatic constraints were so strong that productivity reached a threshold with no possibility to surpass such a limit. This is very common in semi-arid or cold regions, where water or heat are the limiting factors. But in temperate and tropical regions the limit of the systems productivity has often been overcome by a surplus of human energy, and the diversity of life forms increases when compared with other types of landscapes.

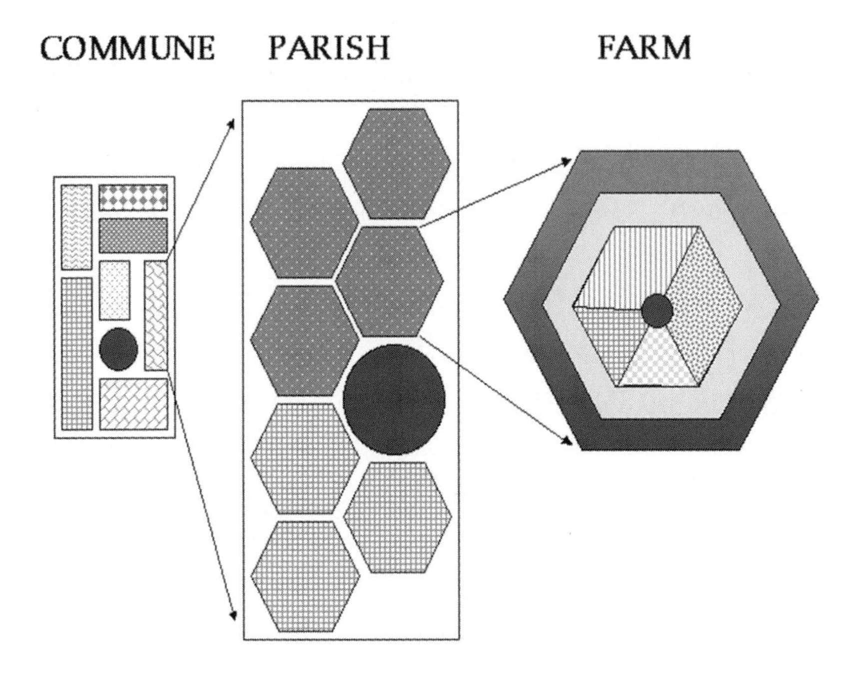

Fig 2.18 - Hierarchical organization of a Southern European cultural landscape. Black circles indicate urbanized areas (from farm buildings to the main city commune).

When human investment is reduced by socio-economic processes, the recovery of the natural path, namely secondary succession, occurs and the abandoned landscapes develop towards a more natural shape. This process can occur in more than half a century and may be interrupted by temporary investments such as the agricultural subsidies during the last sixty years in southern European rural areas.

The healthy condition of a cultural landscape can be easily detected when human stewardship is reduced. The system immediately shifts to other topological and chorological configurations. The matrix, which is not a hostile medium, allows the sprouting of new organisms and the organization of new biotic assemblages. Naturalness (sensu Anderson 1991, Gotzmark 1992) spreads to every part of the system. Naturalness is a scientific concept that allows us to evaluate and quantify the intactness or integrity of ecosystems.

When we are dealing with a cultural landscape, the consideration that the health condition in one part of a mosaic may worse in other parts becomes important. The "coltura mista" of the Tuscany landscape seems a good example. In fact, if we move from fields with scattered trees (oaks) and the rich hedgerows that border fields toward the streams and rivers of lowlands, we generally find that water is of low quality. This is because the water becomes polluted by manure and soil particles which migrate towards stream channels following disturbance during autumn ploughing.

2.3.15 Other criteria of landscape evaluation

Evaluating the mosaic

When we progress from an ecosystem approach (in the ecosystem is simplified as an homogeneous functional entity) to a mosaic of habitat patches, we have to consider, as a priority, the selection of the broad indicators that can embrace the entire mosaic considered. Hence, the first criterion may be the distance of such a mosaic from a mosaic with the same shape. This indicates the degree of isolation of the system in the manner as we calculate for a species. Attributes such as rare, uncommon, and common are important elements to consider. These attributes can be recognized using the spatial arrangement criterion and not just patch typology.

Spatial arrangement of patches

It is important to evaluate carefully the spatial arrangement of patches in a landscape, especially in human disturbed region. The position of a patch can increase or decrease the functionality of such a patch.

In fact we can recognize a cultural landscape by the spatial arrangement of "typical" patches. Patches are not located at random in a mosaic but have been "arranged" by nature-human interactions in precise positions because they have to accomplish specific functions. Such functional patches may be considered ecotopes. Size, shape, and position in a territory are fixed by multi-criteria such as soil type, exposure, slope steepness, economic constraints, social constraints, etc. The results of these mechanisms are landscapes such as the Tuscany "coltura mista" (Farina 1996a, Vos & Stortelder 1992) which has a centripetal organization around a farm house, with a repetition of similar patterns for each farm (type of cultivation, spatial arrangement of linear tree cover, extent of the farm, energy invested in production, infrastructures, and overall dynamics).

Focusing on patch function

Within each landscape (defined as a mosaic of patches), we can verify the patch conditions. A landscape can maintain a structure for a long time; for instance, an open field can remain without trees although it is not yet cultivated. However, it is possible to have composing elements with different degrees of functionality. This is the case of abandoned farmland landscapes where the changes of the last 50 years are visible only as different functions of the cultivated patches, whereas the structure of the overall mosaic has been maintained as that of a fine grained mosaic of small fields interspersed in a matrix of woodlands.

Changes of mosaic grain

Another criterion for landscape evaluation may be the rate of the change of mosaic grain, for instance from a patchy mosaic to a coarser grained mosaic. Generally, a coarser mosaic occurs during the abandonment process or during the "industrialization" of agriculture.

Water quality and river fluxes

The quality of water drained in the main stream may be an indicator of the capacity of the landscape to incorporate different land uses. When an aquatic system has lost the capacity to incorporate pollutants, signals such as a dramatic depression of biodiversity, algal bloom, and invasion of tolerant weeds appear. The presence of ecologically man-made or natural buffers can produce different results.

At the watershed scale, the main processes are water retention and flow across different patches until the main stream or river is reached. If we use processes linked to the water cycle as indicators of landscape functioning it is quite simple to assess structures such as dams, bridges, roads and urban settlements that intercept water fluxes. Their position is strategic for the movement of water and for the recharge of springs.

In many cases, the massive use of water captured along streams and rivers, for irrigation or for industrial use, reduces the amount of free flowing water in the natural bed. This is a good indicator of landscape stress. In fact the artificial reduction of water into streams and rivers has severe consequences on riparian buffer zones, on aquatic life and in particular on the above and under-ground circulation of water with long-term effects on the entire landscape.

The fuzzy aproach

The application of the fuzzy theory to the manipulation of ecological processes seems a good approach. In fact, it is the distance from a certain degree of health condition that we can to some extent measure, including a lot of intermediate stages between a completely healthy system and a completely degraded one.

Fuzziness (see later at 2.4.4). is a type of imprecision that characterizes classes which for different reasons, don't have sharp boundaries (Burrough & McDonnell 1998).

2.3.16 Some additional considerations of landscape evaluation

The indicators of landscape status consist of simple spatial metrics based on information theory and fractal dimensions. One such metric is the measure of dominance that to what extent a landscape is dominated by a specific land cover. Also, contagion is an important metric because it measures the degree of clumping between different land covers. Details about these metrics are indicated at the end of this chapter.

One indicator of landscape integrity is represented by land cover changes, using potential natural vegetation as a baseline, i.e. the vegetation that would exist if human intrusion was removed and vegetation was able to develop successionally. This approach has been used for evaluating areas with a higher value of natural vegetation retention.

Another indicator is the frequency distribution of distances between patches. Long distance between patches means fragmentation, which consequently reduces the integrity of the considered landscape.

The length of forest patches may be a good indicator for fragmented areas and for predicting the risk of predation for many species of animals.

It is possible to use the information about the habitat required by a species and by using an organismal view transformed into a spatial window that equals the specific home range. In this way it is possible to verify suitability across an entire landscape. These methods can be extended to guilds, populations and communities. It is important to scale the resolution of land information and the size of the species home range. Another method based on probability considers random movements of imaginary organisms across a landscape. At each step, which is scaled according to the species requirements, the quality of the land cover is valued.

Moving from an entire ecoregion to a watershed, it is possible to evaluate the integrity of water quality as an indicator of landscape dynamics. The more a landscape is occupied by agriculture and urban settlements, the more nutrient surplus (N, P), turbidity and high temperature are present in free flowing waters. Landscape properties and lotic conditions are strictly related.

Other indicators of watershed integrity suggested by O'Neill et al. (1997), are the quantity of riparian vegetation and the number of intersections between paved roads and rivers. Roads intercept river dynamics and represent barriers for many organisms. Banks, which protect roads from river erosion, and bridges alter the resilience characteristics of river beds. Also, hazard events (from a human perspective), such as floods, can be predicted by using landscape models based on digital elevation models of soil, land cover type, extent of riparian zone, etc.

When operating on fragmented landscapes, as in most of human dominated landscapes, it seems very useful to be able to apply the percolation theory to predict the rate of movement of many organisms. If combined with epidemiology theory it is also possible to calculate the probability that an epidemic event will spread across the landscape or becomes endemic (O'Neill et al. 1992). The assessment of the number of possible ecological scales can be an important indicator of landscape structure and function.

The loss of some landscape configurations can determine the loss of some animal guilds. In fact, Holling (1992) has demonstrated that animal size (body size) is related with opportunistically scaled landscape configurations.

The disappearance of some configurations produces the extinction of specific scale-related guilds. Large animals live in larger home ranges, encounter different habitat patches, and have a scale of reference which is directly related to the scale of resource distribution. Coupling remotely sensed data with the landscape paradigm seems the most efficient way to critically evaluate the integrity of large areas. This method is an integration of finer-scaled monitoring with coarse scale surveys.

2.3.17 Discontinuity, synergism and ecological debts

Many signals are necessary before assessing the health condition of a landscape, and these signals must be integrated in such a way as to produce a state of health without focusing on only a few indicators.

The state of functioning of one component of the landscape is apparently not enough to judge general conditions. Considering the non-linearity of a system and the complexity of the components, we can utilize the concept of ecological debt to describe the situation in general terms. There are different ecological debts depending on the local or regional context. In a cultural landscape we have a "stewardship" debt, where a decrease of human intervention gradually releases other processes like secondary succession. In fragmented and human dominated landscapes a "disturbance" debt can create highly unpredictable conditions in grasslands and forest remnants. The progressive simplification of complexity in nutrient cycles by species extinction creates a "biodiversity debt". In this way, many environmental crises and human related problems have been discovered later during their emergence, when most of the negative influences on the environmental structure and behavior are clearly visible. Related to this fact, which dominates the scene of ecological research (along with an orientation towards publishing papers and career advancement), a new research agenda has recently been proposed by Meyers (1995) which aims for a more innovative direction.

Two different families of processes are underestimated or barely recognized in ecological research: discontinuity and synergism.

Discontinuity

Environmental discontinuity occurs when a system absorbs stress for a long period without any apparent reaction and it is suddenly modified at an unknown threshold. This is the case of a forest exposed to acid rain. Apparently, a very low decline in tree vigor occurs year after year but above an unknown threshold, i.e. after an additional stress, a forest may manifest a rapid and severe injury.

Discontinuities can be observed in many stressed landscapes. For instance the demand for fuelwood can be satisfied by a forest even when the recovery of trees is no longer possible but the subsequent deforestation will be visible some years after the decline start and will probably be discovered too late for countermeasures. The same has occurred in most European uplands over the last fifty years. During this time, the emigration of a considerable part of the population from hilly and mountainous regions occurred, the long term consequence of which was agricultural abandonment. This effect was under-estimated because it was buffered by the older generations of farmers who remained in the countryside, and by huge agricultural subsidies.

Contemporarily, an increase of biodiversity occurred from a combination of lower human pressure on the environment and a release of vigorous secondary succession. Land abandonment signals appeared just 20 years after the beginning of the process; they are now well known and coupled with related problems such as: increase of fire risk, decline of biodiversity, and increase in social poverty. Most conservation policies in such areas were concentrated on protecting "wild areas" and individual biotopes of "scientific" interest, neglecting the importance of cultural landscapes as "strongboxes" of the majority of biodiversity. Conservationists changed their philosophy and actions later in the 1980's when most such valuable landscapes had already vanished.

The debt (sensu Tilman et al. 1994) that a landscape stores during a disturbance process may create unsustainability in local populations. For instance, fragmentation can reduce the dimension and extent of a population. Such a population can utilize a metapopulation strategy to maintain a genetic flux between sub-populations and to exploit the remnant patches. However, this condition is clearly fragile and a small increase in fragmentation and isolation can negatively affect this population and pave the way for a massive local extinction.

Synergism
Synergism is a phenomenon that appears when two or more environmental processes interact in such a way that their effects on ecological systems are multiplied and not added. Thus a problem is not a double problem but a super problem. For instance, a plant living in conditions of poor soil nutrients may be more prone to diseases when changes in climatic conditions occur. This is highly probable for the plants of most biota which, when subjected to the climatic stresses of global warming, may be more sensitive to pandemic diseases.

Discontinuity and synergism are two of the "surprises" that may result from our demographic expansion, and energy and service demands.

We should direct our research towards the unknowns, such as the effect of more than 70,000 synthetic chemicals that we have released into the environment. To pursue such an agenda, research on landscape structure and functions, seem very promising as a tool of defining the presence of many correlated processes that can be found when the spatial and temporal scales are expanded. For instance, the forest services of storing and releasing hydrological resources for lowland agriculture, regulating river fluxes and stabilizating soil erosion are often not considered enough by environmental policies which are oriented towards short term benefits for human populations, rather than a long term assurance of ecosystem functioning.

2.4 Tools for evaluating the structure and functions of land mosaics

2.4.1 Neutral models

A neutral model is used to generate an expected pattern in the absence of specific processes which affect the landscape such as topography, disturbance history, and ecological impact. Due to the impossibility of conducting experiments on large landscapes, neutral models in the form of simulations seem to be a very promising approach. For a review of the application of neutral models in ecology see With & King (1997).

Neutral models, also synonymous with null models, are used to determine which of the structural properties of a landscape (e.g. patch size, patch shape, connectivity, amount of edges) lie far from a theoretical (random). distribution Neutral models can also be used to evaluate the effect of known patterns on ecological processes such as seed dispersion, animal movement, nutrient movement, pollutant dispersion, etc.

Neutral models are extensively utilized in landscape ecology modelling due to their spatially explicit characteristic and the simplicity of the generation rules. These models are used to test if an observed pattern is similar or not to a neutral distribution. If there are no differences the observed patterns are probably randomly generated. These models are useful for understanding the spatial properties of a landscape and allow the association of observed patterns with ecological processes.

Neutral models are generated by using a simple metric calculation based on a matrix of $n*n$ cells. In this matrix a specific land-cover is randomly distributed using different abundance classes. This model is perfectly neutral because the cells or pixels occupied by a specific land-use are selected randomly.

If we consider P as the probability of a landcover being in the matrix, $1\text{-}P$ represents the amount of area without the selected landcover. In this model the pixels of the selected land-cover are considered to be packed together in distinct clusters. A cluster is considered when each pixel (in this case square in shape) has at least one side in contact with another pixel. Diagonal contacts are not considered for clustering. This rule is called the nearest-neighbor rule and when we use the four cardinal directions to establish a patch between pixels this is called the four-neighbor rule. An eight-neighbor rule also considers the diagonal directions (sub-cardinal).

Patch size and shape are dependent on the probability P of the presence or absence of the landcover considered. If P increases from the lowest number then the number of patches also increeases. However, when P is >0.3 then smaller patches will begin to coalesce into larger patches, thus decreasing in number. The number of large patches increases and the amount of edges reaches a maximum in proximity to $P=0.5$, after which the amount of edges decreases (see for instance Fig. 2.19).

By progressing through the P values it is possible to find a critical threshold close to $P=0.59$. At that point a small change in P produces large effects in the mosaic. For instance, a small change between $0.5< P < 0.6$ produces changes in the area of the largest patch (Pearson & Gardner 1997).

This effect is explained by the percolation theory (Stauffer 1985, Ziff 1986) which states that, if applying the four-neighbor rule 50% of very large maps randomly generated by a $P=0.5928$ value will have a single large patch spanning throughout the entire map. But if we use the eight-neighbor rule the critical threshold is equal to 0.4072. Generally these models are applied to animal responses to the distribution and abundance of suitable habitats. It is not correct to use the same neighbor rule for animals with a different level of vagility. A hedgehog (*Erinaceous europaeus*) has a different vagility to a fox (*Vulpes vulpes*) and in this case the hedgehog model should be based on a four-neighbor rule and the fox model on an eight-neighbor rule.

Fig. 2.19 - Sample random map with different proportions of *P* suitable (black) habitats. The map with *P=0.6* is considered a percolating map of suitable habitat.

However, we must remember that this is a rough example. No organism has an absolute scale of perception and interaction with it's habitat, but will change scale according to an internal and functional timer.

This is true also for processes that change scale across a patchy landscape. The distance between patterns exhibited by a land cover and a neutral model distribution is particularly useful for investigating the processes responsible for such difference.

It is also an extremely useful approach because when a land cover is close to the threshold value (*P=0.5928*) small changes due to logging, agriculture or urban development can drastically change the quality of an organism's habitat or the behavior of a process. For instance, a forest cover can quickly change from a connected system to a fragmented one and without a direct human perception of the changing threshold.

More evoluted neutral models have been generated recently using a multiscaled approach (Gardner & O'Neill 1991)(Fig. 2.20). The neutral models assume that, in a landscape, there are no significant processes affecting the spatial arrangement but often in a real landscape processes which act spatially vary according to the spatial extent of the considered landscape. For this reason hierarchical neutral models are required to test the null hypothesis that a landscape has different costraints at different scales.

A simple neutral model and a hierarchical neutral model describe the distribution of a land cover by using a bit map (0/1). In real landscapes, the suite of landcovers within a matrix rarely have abrupt changes so are expressed as a gradient.

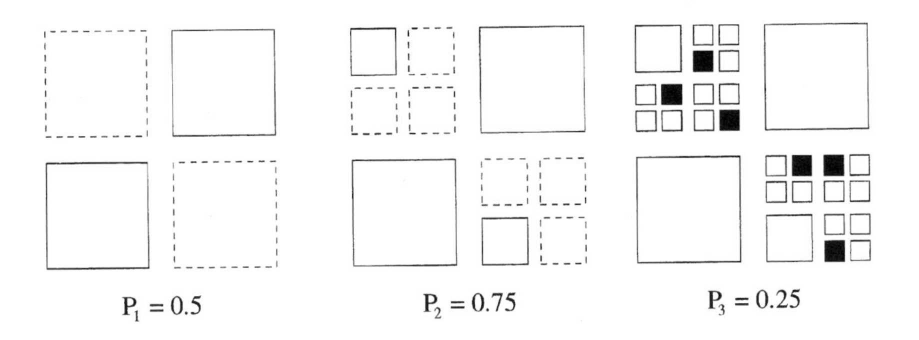

Fig. 2.20 - Example of the method of making a hierarchical neutral map. The dashed lines indicate randomly selected divisions that have suitable habitats in level 1 and 2. Shaded cells in level 3 contain suitable habitat (from Pearson & Gardner 1997).

One technique for producing these maps is the application of the Brownian Motion of a two dimensional random walk to create a fractal landscape. A random walk is produced by a sequence of steps *Xt*. If the next step is *Xt+1-Xt* and is drawn independently of the Gaussian distribution this sequence of steps is considered a Brownian motion. If we introduce the parameter H that control the correlation between the successive steps this is called Fractional Brownian motion. When $H=0.5$ the steps are not correlated, when $H<0.5$ they are negatively correlated, and when $H>0.5$ they are positively correlated.

Fractal maps can be used to simulate the distribution of an object, like a contaminant in a landscape. These maps have a different level of spatial autocorrelation (Gardner & O'Neill 1991).

2.4.2 Landscape metrics

In quantitative ecology, counting the number of species is the most popular approach for evaluating the complexity of a community (plant or animal). Other measures are based on plant density, on the diversity of vertical distributions of vegetation, and on the distribution of biomass across a study site (Mueller-Dombois & Ellenberg 1974). In landscape ecology the main subject being measured is the structure and complexity of the mosaic. We consider the mosaic at any level of spatial scale (from a few centimeters if we consider the mosaic of a moss settlement along a wall, to the distribution of forest patches along an elevational gradient on mountains).

Many metrics are available and from the literature many show redundancy, or if you prefer are not independent (Bogaert & Impens 1998). Assuming that such metrics are necessary to evaluate the complexity and dynamism of a mosaic (at any scale), I'll try to describe the more utilized and tested metrics, though many others are used for different conditions (see also Farina 1998). It is important to recognize that there is no fixed rule in the choice of indexes to apply, often the selection is linked to the specific requirement of the investigation.

The metrics used are based on the distribution, shape and spatial arrangement of the visible patches. A dedicated software, Fragstats (McGarigal & Marks 1995), may be employed in such calculations. Most of the indexes presented in this routine are from well known literature (Shannon & Weaver 1949, Krummel et al. 1987, O'Neill et al. 1988, Baker & Cai 1992, Gustafson & Parker 1992, Li & Reynolds 1994). Many of these indexes require data arranged in spatial formats (matrix) handled by Geographical Information Systems. Some preparatory routines for handling data are required especially when the criterion of distance is utilized (see f.i. Hulse & Larsen 1989, Tomlin 1990).

At least two different approaches are available for applying metrics to the landscape: landscape composition and landscape configuration.

Landscape composition

Landscape composition describes the quality and quantity of elements (patches) composing a mosaic. Landscape composition is not an explicit descriptor of a mosaic but nevertheless it is a good indicator of habitat suitability for some species that require, to some extension, a particular type of patch, or a particular percentage of mixed patches. Landscape composition is described only in a numerical way and not in a spatial way.
Many metrics can be used to measure landscape composition, such as relative importance of each patch type, patch richness, patch evenness, and patch diversity.

Landscape configuration

Landscape configuration describes the physical distribution of patches within the mosaic. The variables considered are patch isolation, patch contagion, landscape boundaries, and shape and size of core area. Patch juxtaposition and distance from patches of the same type are two other explicit indicators of landscape configuration. For example, boundary complexity can be measured either at patch type level or for the whole mosaic.

Spatially explicit models can be applied to incorporate information related to patch abundance, patch spatial distribution and organismal perception of the surrounding habitat. Some metrics, such as patch boundary complexity, are not explicit at landscape level and only have explicit information at patch level. For example, the extent of boundaries depends on the different amounts of patch type but also on their spatial relationship. In fact, an extensive boundary is typical of a heterogeneous mosaic and this attribute contains a combination of structure and configuration.

The mosaic can be studied at three levels:

at individual patch resolution

at patch class resolution

at landscape scale (calculating the interactions between different patch classes) (Fig. 2.21).

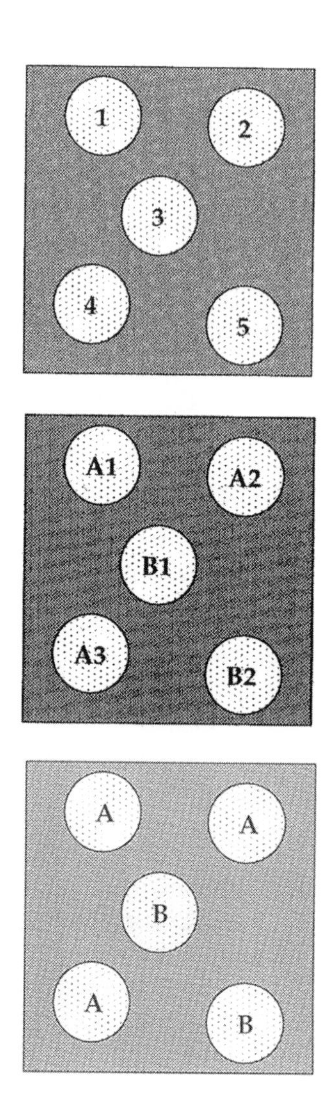

Fig. 2.21 - Representation of the three different approaches of studying the configuration of a landscape:
A = Collecting data at patch level
B = Collecting data at patch level class
C= Collecting data at landscape level (all classes considered).

The application of different metrics depends on the goal that we have. For example, patch resolution may be important in a forested landscape for evaluating the largest patches present. In other cases it is important to measure the diversity of the patch mosaic which has been assumed to be important for some groups of organisms (Fig. 2.22).

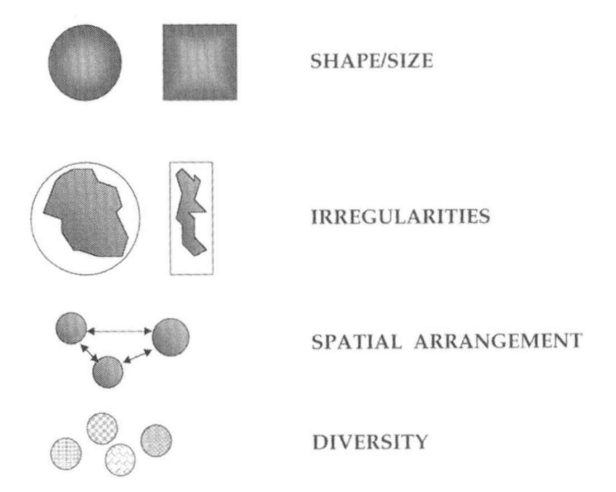

Fig. 2.22 - Schematic representation of metrics used to measure patches.

Most of the indexes proposed are presented in the form of a patch class resolution; for use with the other two resolutions (patch class, landscape class) it is sufficient to change the range of calculation.

2.4.2.1 Metric at individual patch resolution (Fig. 2.23)

Area of each patch *aij* **(***i*= patch type class; *j*= *n* of patches)

Perimeter of each patch (total length of edge) *pij*

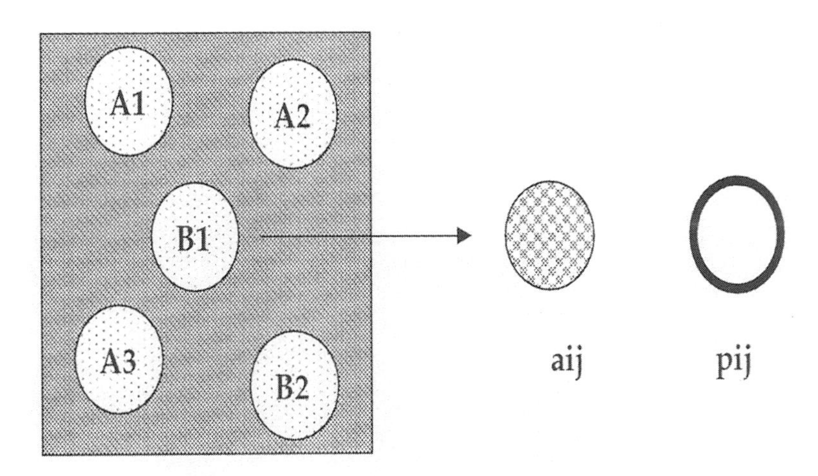

Fig. 2.23 - The simplest metrics: area of a patch *aij* and perimeter *pij*.

Patch Density

$$PD = \frac{ni}{A}(100)$$

ni= number of patches of type *i*, *A* is the total area

Mean Patch Size

$$MPS = \frac{\sum_{j=1}^{n} aij}{ni}$$

where *ni*= number of patches of type *i*, and ai is the surface of patch *ij*.

Largest Patch Index

$$LPI = \frac{\max(aij)}{A}(100)$$

where *max(aij)* is the largest patch in the landscape. *LPI* moves close to *0* when the dimension of patches is very small, and approaches *1* when only one patch occupies the entire landscape.

Patch size/patch shape (Compactness) (Bogaert & Impens 1998)

$$\phi_3^* = (1-\phi_3)\times(1-\frac{P\min(a)}{P\max(a)})^{-1}$$

where

$$\phi 3 = \frac{P}{P\max(a)}$$

P= Perimeter, *Pmax* is the maximum perimeter for an equal area a[L2], with L equal to the pixel sidelength This index measures some aspects of patch irregularity.

Perimeter Irregularity (Twist number statistics) (Bogaert et al. in press) (Fig. 2.24)

$$\Omega = \frac{\theta\max - \theta(n)}{\theta\max(n) - 4}$$

where

$$\theta\max = 2n+2$$

if there exists an integer i for which $n=4i+1$, then n is the number of pixels composing the patch or otherwise

$$\theta \max = 2n$$

if $n \# 4i+1$

This index counts the number of twists as a measure of edge irregularity. This index is independent of the fractal dimension, has a strict correlation with compactness indices (see Bogaert et al. in press) and ranges from *0* for very irregular shapes, to *1* for regular figures.

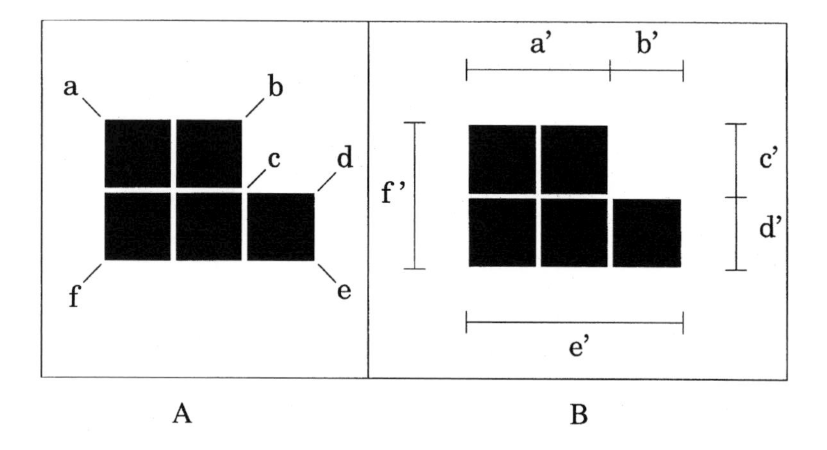

Fig. 2.24 - The perimeter irregularity (Twist number statistics)(Bogaert et al. in press)

Landscape Similarity Index

$$LSIM = \frac{\sum_{j=1}^{n} aij}{A}$$

This index is close to *0* when the patch type (class) is rare in the landscape, and *LSIM* is *1* when the entire landscape is composed of one patch type class.

Total Edge

$$TE = \sum_{k=1}^{m'} eik$$

Where *eik* is the length of the edge between patch type *i* and *k*.

Edge Density

$$ED = \frac{\sum_{k=1}^{m'} eik}{A}$$

where *A* is the surface of the entire area (landscape)

This is the total length of patch edge per unit area. This measurement can be done for each size category or for all. In a neutral landscape, edge density shows a well shaped curve for random pixels, enlarging patches and abutting patches. Buffered patches show a more distinct pattern.

Edge Contrast Index

$$EDGECON = \frac{\sum_{k=1}^{m'} (pijk * dik)}{pij} (100)$$

where *pijk* is the length *m* of edge of patch ij adjacent to patch type (class) *k*, and *dik* is the dissimilarity between patch types i and k. This dissimilarity must be fixed according to the different conditions.

This index measures the quantity of patch edges weighted by the value of edge contrast. Edge contrast is given different values according to an empirical evaluation. It ranges from *0*, when the landscape is composed only of a patch with no edge present, to *100* when the patch perimeter is adjacent patches which show the maximum contrast (dissimilarity).

2.4.2.2 Spatial arrangement

Mean Nearest-Neighbor Distance

This metric measures the average edge-to-edge distance between a patch and the nearest neighbor of the same class. The measure shows a rapid saturation curve and discrimination is possible only for cover values <0.20. The size of patches can produce different effects on this metric: large patches may produce larger values than many small patches of the same % covering.

$$MNN = \frac{\sum_{j=1}^{n'} hij}{n'i}$$

where *n'i*= number of patches in the landscape of patch type i that have close neighbours. *hij* is the distance from patch *ij* to the nearest neighboring patch of the same type (class) (see Fig. 2.25).

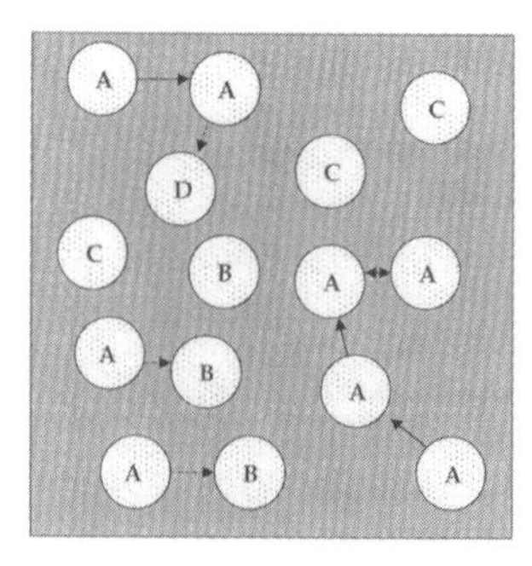

Fig. 2.25 - The Mean Nearest Neighbor Distance (MNN) index considers the distance of the nearest neighboring patch of the same type. Dashed lines indicate the closest distance from an adjacent patch of a different class.

Interdispersion and Juxtaposition Index

$$IJI = \frac{\sum_{i=1}^{m'} \sum_{k=i+1}^{m'} \left[\left(\frac{eik}{E} \right) * \ln \left(\frac{eik}{E} \right) \right]}{\ln(1/2)[m'(m'-1)]} * (100)$$

Where *eik* is the total edge length between patch class *i* and *k*. *E* is the total length of edges in the matrix; and *m'* is the number of patch types present in the landscape, including the matrix border. This index is close to 0 when adjacent patch types are unevenly distributed, and approaches 100 when all patch types are equally spaced around.

Contagion Index

Contagion measures the probability that two randomly chosen adjacent pixels belong to the same category (O'Neill et al. 1988, Li & Reynolds 1994). Contagion ranges between 0% to 100% of the maximum aggregation possible and has a value of 100% when the matrix is filled by only one type of land cover. Contagion is inversely correlated with the edge density.

$$CONTAG = [1 + \frac{\sum_{i=1}^{m} \sum_{k=1}^{m} \left[Pi \left(\frac{gik}{\sum_{k=1}^{m} gik} \right) \right] * \left[\ln(Pi) \left(\frac{gik}{\sum_{k=1}^{m} gik} \right) \right]}{2 \ln m}] * (100)$$

Where *Pi* is the relative abundance of patch class *i*, and *gik* is the number of adjacent pixels of patch types (classes) *i* and *k*.

If contagion is low, it means there are many small patches, and the percolation critical threshold is higher (0.65) compared with the predicted probability (Fig. 2.26).

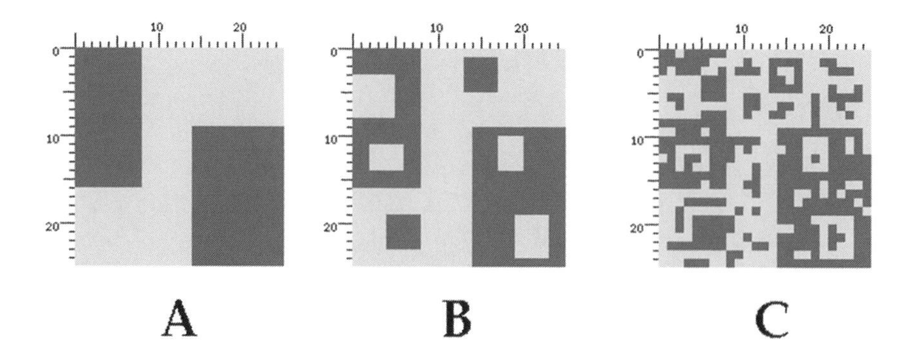

Fig. 2.26 - Different levels of contagion in random matrices. A) 37.34, B) 24.40, C) 3.50.

2.4.2.3 Proximity indexes

These indexes measure the distance between patches. It is possible to measure the distance of each patch from others of the same type or of different types. These indexes are important for evaluating species-specific connectivity.

Proximity Index

$$PROXIM = \sum_{s=1}^{n} \frac{aijs}{hijs^2}$$

aijs = Area of patch *ijs* within the specified neighborhood (*m*) of patch *ij*

hijs = Distance (m) between patch ijs (located within the specified neighborhood distance (m) of patch ij) and patch ij.

s= 1,... n patches within specified neighborhood

This index = 0 if a patch has no neighbors of the same patch type within the specified search radius, or >0 depending on the search radius (Fig. 2.27).

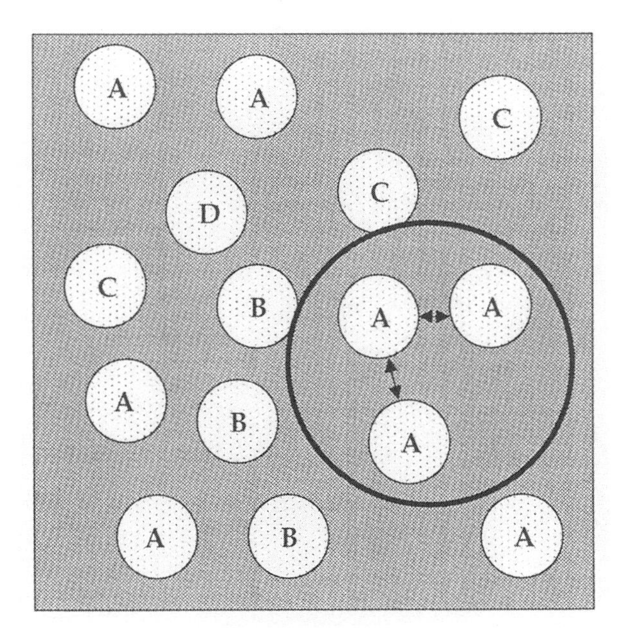

Fig. 2.27 - The proximity index measures the distance of a patch from all other patches of the same type within a specified neighborhood distance.

2.4.2.4 Diversity indexes

Measuring diversity means measuring the differences of a set of objects, organisms or processes. Diversity can be expressed as a number of different objects, in which case diversity is synonymous with richness. Of the numerous indexes for measuring diversity, two are particularly useful: the Shannon index and the Simpson index.

The first uses the product of the relative importance of the measured object and the logarithmic transformation of the same quantity. The Shannon index is a probabilistic index and is correctly employed to calculate the probability of finding a new species whilst tracing randomly across a sample. It is particularly sensitive to rare events.

The Simpson index, whilst based on the same probabilistic approach, is more sensitive to dominant processes. Both indexes can be used in a relative way to the relationship with the maximum diversity possible when all the objects composing the collections are present in the same quantity. This ratio is called evenness and is an important measure of the equi-distribution of an object or event. Diversity can be applied not only to objects but also to processes and to some relevant patterns.

Shannon's Diversity Index

$$SHDI = -\sum_{i=1}^{m} (Pi * \ln Pi)$$

where *Pi* is the relative importance of class patch *i*. This index is particularly sensitive to the rare events. In this case to the rare classes.

Simpson Diversity Index

$$SIDI = 1 - \sum_{i=1}^{m} Pi^2$$

Where *Pi* is the relative importance of patch class *i*. This index is particular sensitive to the dominant events. In this case, to the commonest classes.

Shannon's Evenness Index

$$SHEI = \frac{-\sum_{i=1}^{m} (Pi * \ln Pi)}{\ln m}$$

Where *Pi* is the relative iimportance of the patch class *i* and *m* is the total number of classes.

Simpson's Evenness Index

$$SIEI = \frac{1 - \sum_{i=1}^{m} Pi^2}{1 - (\frac{1}{m})}$$

Where *Pi* is the relative abundance of patch class *i*, and m is the total number of classes.

2.4.2.5 Core area indexes

A core area represents a selected area which is distant from the edge (Fig. 2.28). This distance is species specific, and there are no rules on this point. A core area is established according to the needs of the research or the application. Many indexes are based on the core area concept (see f.i. Fragstat software); considering the easy construction of such indexes I'll limit this presentation to only some of these, such as:

Total Core Area

$$TCA = \sum_{j=1}^{n} aij^c$$

where aij^c is the core area of the patch *ij* with a specified buffer width

Core Area Density

$$CAD = \frac{\sum_{j=1}^{n} nij^c}{A}$$

where *nijc* is the number of disjunct core areas of patch type *ij*, and *A* is the surface of the entire area (landscape).

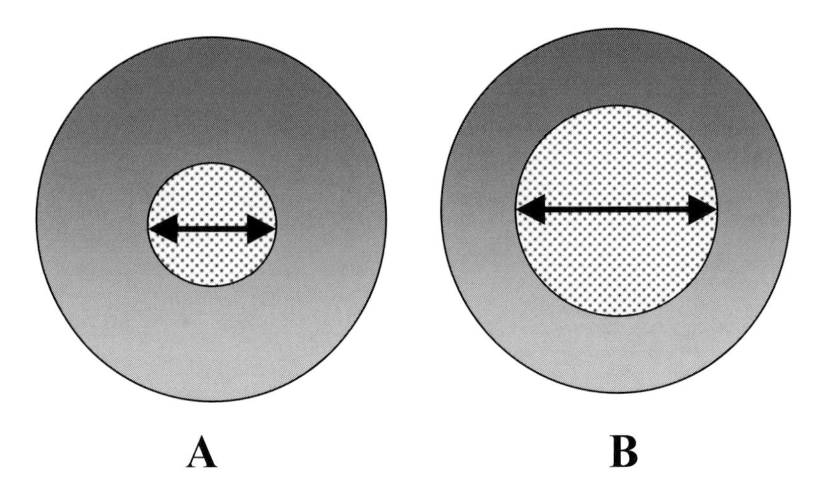

Fig. 2.28 - A core area is considered as the part of a habitat patch which is distant from the edges. Every species has a specific sensitivity to edges. For instance species in case A has amore restricted core area than species in case B. Metrics of core areas must be calibrated according to the species-specific habitat requirement, according to the functions/behavior, and according to the season.

Total Core Area Index

$$TCAI = \frac{\sum_{j=1}^{n} aij^{c}}{\sum_{j=1}^{n} aij}$$

where *aij* is the surface of patch *ij*.

2.4.3 Fractal geometry in landscape metrics

I have designated a chapter to fractal geometry because this new form of mathematics represents a formidable tool for investigating the irregularities of natural and human events and their patterns.

Mandelbrot coined the term "fractal" (Mandelbrot 1975) from the Latin "fractus" which describes a brocken irregular stone. Mandelbrot (1986) gave the following definition: "A fractal is a set for which the Hausdorff Besicovitch dimension strictly exceeds the topological dimension".

Fractal is a geometry of deterministic chaos which can describe the geometry of mountains, clouds and galaxies, but also the complexity of natural objects like ferns and cauliflowers. Fractal analysis can be considered as the study of regularities of the irregularities! Fractals are very common in nature and also in human culture and handcrafts; forms, size frequency distributions, spatial distributions, and paths or trajectories have fractal dimensions.

Fractal geometry is very intuitive but not easy for non-mathematicians to approach. The complexity of any "sky-", "sea-" and "land-"scape can be efficiently measured by using fractals. This new non-Euclidean geometry is very useful for many reasons. Fractals are completely irregular and their irregularity is shown with the same degree of extension across a range of scales.

The conceptual differences between fractal and Euclidean geometry consist of the capacity, in the fractal approach, to build complex objects by using simple and repetitive processes. On the other hand, Euclidean geometry uses simple blocks but requires complex processes.

The fractal approach can be used to describe objects and also time-series phenomena. One important key idea in fractal geometry is represented by self-similarity. In other words, a fractal object is composed of a whole which contains parts with the same shape of the whole. This property appears when changing the scale of resolution so that fractals can be considered to be scale-invariant or self-similar. The scale-invariance is expressed algebraically by power laws.

There are two categories of self-similarity: strict and statistic. Strict self-similarity means that a detail of a system has the same geometric structure as the whole. This property is not typical of natural fractals. Statistical self-similarity means that the magnification of a part is not a perfect representation of the whole though there is some invariance in the probability of describing a fractal object.

Fractals in nature are very common in organic forms from intracellular organelles to organisms and aggregations of organisms. Fractal attributes can be found in the turbulence of movements in water and air masses. Fractal dimension describes the complicated convolution of adaptive forms of life.

The soil has fractal properties and soil particles are organized across different sizes in the same matter. This geometry has strong effects on bacteria, and on arthropods and vertebrates too. The morpho-edaphic index assumes a particular importance when we evaluate the ratio between coastal shorelines and the water body mass of a lake. The importance of the convolution of the shoreline for the exchange of nutrients and organisms is well known.

Actually, a formidable literature is available and there are applications of fractal geometry to a wide range of abiotic and biotic processes (Mandelbrot 1975, 1982, 1986, Feder 1988, Milne 1991, Hastings & Sugihara 1993, Milne, 1997).

Fractal metrics are particularly useful when applied to the complexity of animal behavior (Loehle 1990, Fourcassié et al. 1992, Johnson et al. 1992, Russell et al. 1992, Wiens et al. 1993, Alados et al. 1996, Etzenhouser et al. 1998), and patch shape and landscape (Krummel et al. 1987, Rex & Malanson 1990, Milne 1991, Lathrop & Peterson 1992, Leduc et al. 1994, van Hees 1994, With 1994, Milne 1997). They also find application in a broad range of biological (Iannaccone & Khokha 1995, Smith et al. 1995, Campbell 1996) ecological (Frontier 1987, Sugihara & May 1990, Johnson et al. 1995, Underwood & Chapman 1996) and non-ecological disciplines such as hydrology (Pachepsky et al. 1995, Ichoku et al. 1996, Perfetc et al. 1996, Iturbe & Rinaldo 1997), geology (Acuna & Yortso 1995, Loehle & Li 1996), soil composition (Anderson & McBratney 1995, Perfect et al. 1993, Perfect & Kay 1995, Rasiah 1995, Perrier et al. 1995, Barak et al. 1996, Kozak et al. 1996, Perfect & Blevins 1997), microbial transport (Li et al. 1996), vegetation structure (Chen et al. 1992) and agronomy (Eghball & Power 1995).

2.4.3.1 Application of fractals to ecology and specifically to landscape ecology

I have selected only a few applications of fractal geometry to landscape ecology; detailed descriptions have been provided by Milne (1991, 1997), and by Farina (1998). Edge length, perimeter:area relationship, box and mass dimensions are outlined below.

The divider (caliper) method

It is well known that the length of a coast-line depends on the dimension of the caliper by which the measurement is carried out. Edge length is particularly useful for calculating how organisms perceive the complexity of edges. This metric is very important for estimating the potentiality of a habitat once we know the caliper.

The length of a convoluted line responds to a power law :

$$C(L) = bL^{-Dc}$$

where *C(L)* is the number of steps measured with a caliper of length *L*, and *b* is a constant.

The total length of such a line is calculated by multiplying *C(L)*L*

so:

$$Tot(L) = bL^{1-Dc}$$

This equation can be transformed into a logarithmic format and written as the equation of a line

$$\log Tot(L) = \log b + (1 - Dc)\log L$$

that assumes the form of $y=b+mx$, the equation of a line.

The logarithm of *Tot(L)* can be related to *log* L by using the linear regression $m=1-Dc$ and finally $Dc= 1-m$.
This method can be applied to studying the turtuosity of animal paths (see Wiens et al. 1993, With 1994) and vegetation complexity (van Hees 1994)(Fig. 2.29).

Perimeter:Area relationship

This is one of the most utilized metrics for evaluating the complexity of perimeters with the area of patches. There is much evidence that the complexity of edges plays an important role in species –specific patch suitability. The roughness of an edge can create the conditions for a new habitat, or can encourage/discourage the penetration of a species.

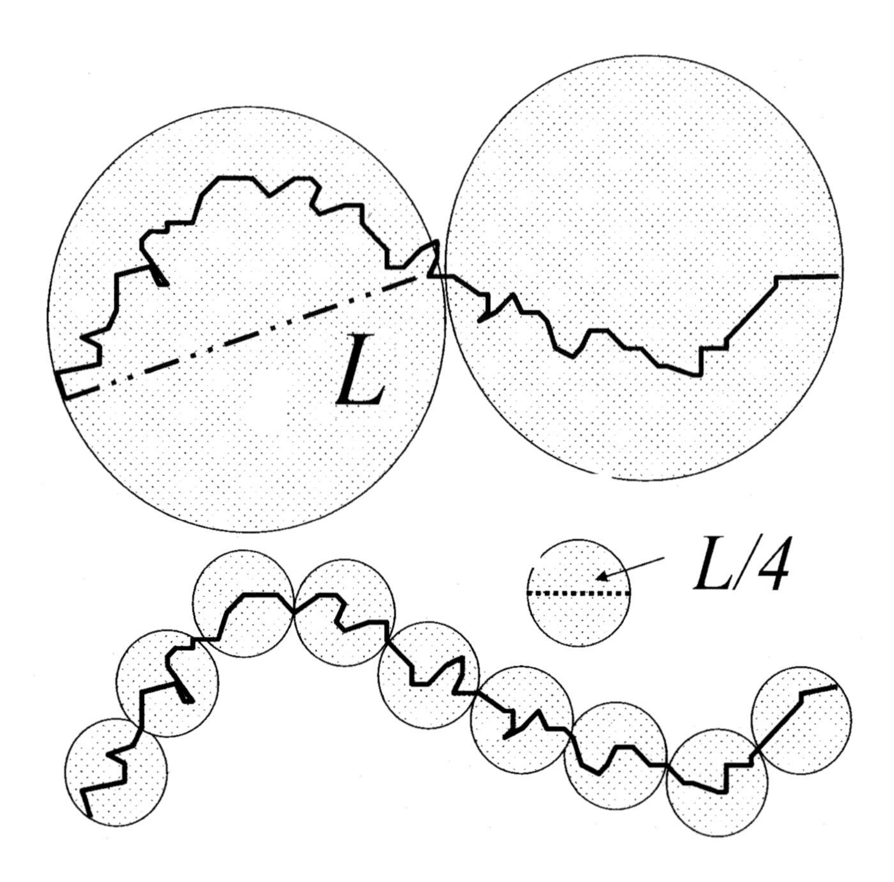

Fig. 2.29 - Representation of the divider caliper method. *L* is the length of caliper used to measure the turtuosity of this path.

This measurement must be applied carefully to organisms because it does not consider the way in which an organism perceives the complexity of edges but the method is important for comparing the complexity or the dynamics of two different mosaics.

The perimeter is related to the area of a patch according to the following equation:

$$P \approx \sqrt{A^{Df}}$$

For a simple Euclidean figure (circles and quadrats) *Df=1* and when the complexity of a polygon increases P is closer to 2 (f.i. Krummel et al. 1987).

Different equations have been used to calculate the dimension. For instance, Sugihara & May (1990), Kenkel & Walker (1996) used

$$Df = 2\frac{\log P}{\log A}$$

Milne (1991), Schumacher (1996) used

$$Df = \frac{\log A}{\log P/4}$$

and finally McGarigal & Marks (1995), Olsen et al. (1993) calculated Df as follows:

$$Df = 2\frac{\ln P/4}{\ln A}$$

Of these three equations, utilized by Bogaert (in press) for the quadratic Koch islands for which an exact fractal dimension is known, only the last has been found able to calculate the correct value. Other examples include Lathrop & Peterson (1992) who used fractal analysis for identifying structural self-similarity in mountainous landscapes.

Box (grid) dimension method (Mandelbrot 1982)

This dimension is particularly useful when we have to measure the availability of a specific habitat for an organism (Milne 1991).
For instance, there is a fractal power low for a line which is related to the number of boxes $N(L)$ that contain a target object (such as a forest patch, a breeding site, or a foraging animal) and the length of theses boxes. To estimate the box dimension the target object is covered by a regular grid of size L. The number of boxes occupied by the target object is calculated by changing the size L of the grid (Fig. 2.30).

$$N(L) = kL^{-Db}$$

This equation can be transformed by using logarithms

$$\log N(L) = \log k - D_b \log L$$

and Db is calculated as the slope of the linear regression. This dimension provides a representation of the spatial variation of patches. See Etzenhouser et al. (1998) for an application.

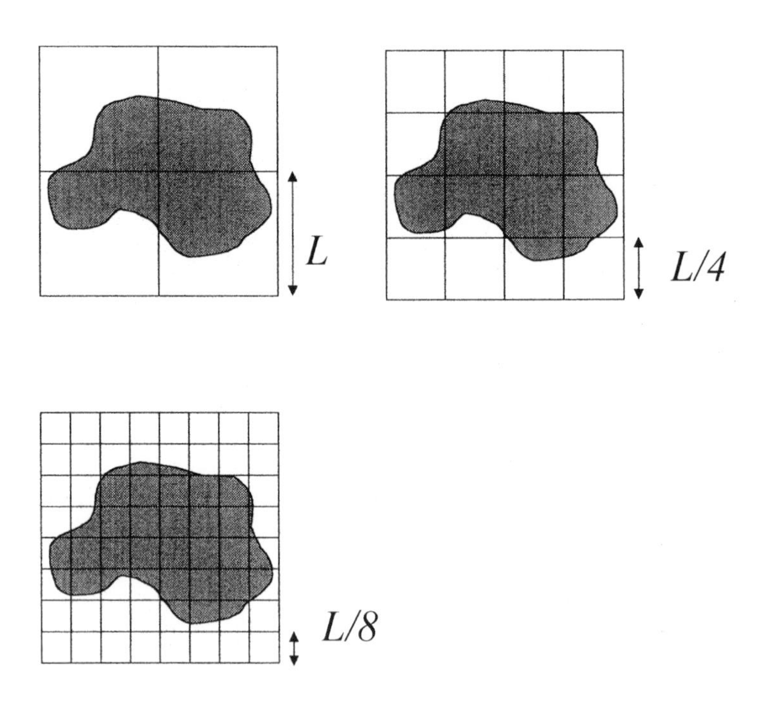

Fig. 2.30 - The box dimension method. L is the size of the box superimposed to the object.

Cluster (Mass) dimension (Feder 1988)

This dimension is particularly efficient for describing the availability of resource habitats available for a species. In fact, the measurement is conducted by using a sliding window around a target grain according to the home range of a species, or the functional requirements at that time of season.

The number of occupied cells *O(L)* within a box of length *L* increases as a power of *L* according to the equation

$$O(L) = KL^{D}$$

D is the amount of habitat grains expected to be found in a window of length *L* (Fig. 2.31). It describes the speed at which the number of occupied cells *O(L)* increases enlarging the window *L* and can be calculated as slope of a linear regression.

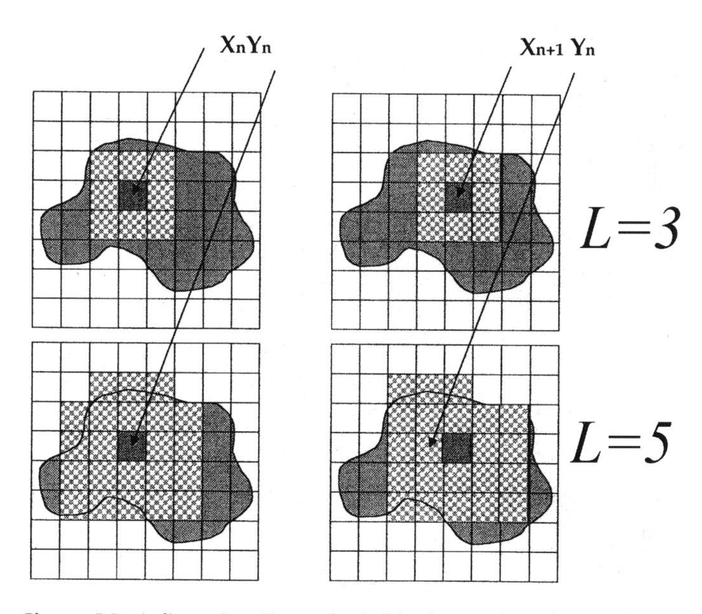

Fig. 2.31 - Cluster (Mass) dimension. For each pixel in the matrix are searched pixels of suitable habitat in a surrounding using different size windows. In the example the exploring windows have dimension l=3 and l=5 respectively.

This method is particularly useful for calculating the number of grains of suitable habitat according to the specific home range of an animal. There is a fundamental difference with the box counting method because mass dimension is calculated by moving from one grain to another and not by superimposing a fixed grid. This method tries to describe all possible states of the system.

When a map is composed of one point in a plane the box dimension is equal to 0. When patches are spread everywhere in the matrix the box dimension is equal to 2.

2.4.4 Fuzzy theory in landscape metrics

Fuzzy theory deals with a ambiguous evaluation of a phenomenon or attribute (Kosko 1993). Generally we consider an attribute using a binary code of 0 1 (true – false). This analysis reduces the possibility of comprising the variability that can be found in the real world within a few categories. The application of fuzzy logic allows us to create families of attributes or geographical configurations and to determine evaluation ranges which are useful for the specific goal.

The fuzzy approach can be usefully applied to soil surveys, land classification, vegetation mapping, pollution mapping and species overlapping. This approach allows us to transform crisp patterned objects into probable processes (Fig. 2.32).

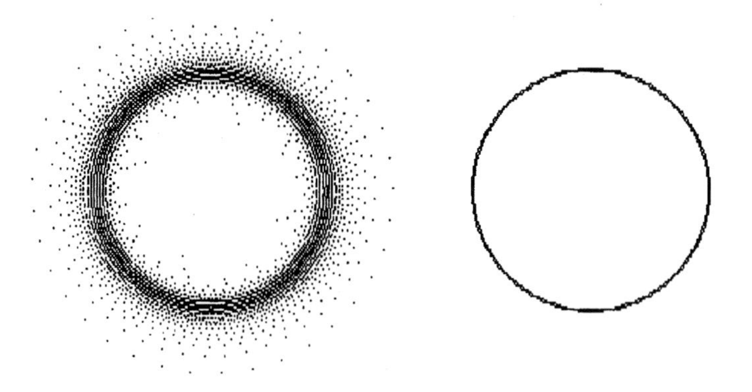

Fig. 2.32 - Representation of fuzzy sets (left) and Boolean (crisp) sets (right) (from Burrough & McDonnell 1998).

If we consider the dimension of an object or an organism we can easily describe the total size, such as human height. A person of 2 m may be considered tall, a person of 1.50 short but the intermediate conditions between the two extremes need another description. We can add another category such as the average and, if we consider people below or above this average, we will have created four categories. This procedure can be expanded as you like, thus creating many categories.

In landscape evaluation the suitability of a patch can be measured via classes that express a fuzzy system. Often the categories are not supported by precise measurements but by a logical combination of many descriptive factors. Fuzzy logic can be applied to describe geographical factors as biological patterns, and the human related view of landscape complexity. Fuzziness represents a degree of imprecision for classes that don't have sharply defined boundaries. These classes are called fuzzy sets. Fuzzy logic is not based on probabilism but on the assumption that a pattern has the possibility of being a member of a set. A crisp set is based on an assumption of true or false following a binary relationship. An element pertains to a crisp set and class boundaries are sharp.

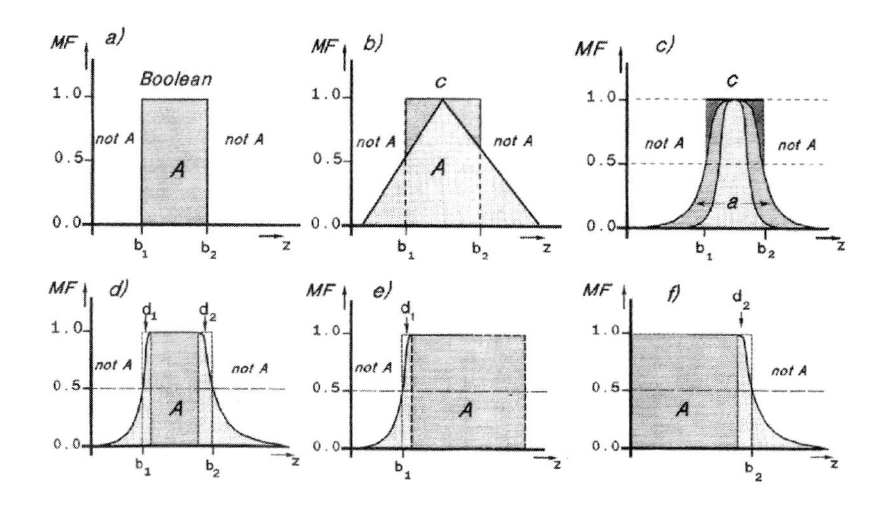

Fig. 2.33 - Comparison between Boolean and fuzzy membership functions using the SI method (from Burrough & McDonnell 1998).

Formally a fuzzy set is defined mathematically, i.e.:

Z is the space in which objects reside, the fuzzy set A in Z is a set of ordered pairs

$A = (z, MFfa(z))$ for all z and Z ranging from 0 (no membership) to 1 (full membership).

$MFfa(z)$ is the grade of membership of z in A. It is possible for z to pertain to other memberships.

The determination of fuzzy sets improves the quality of the geographical representation of a map and can also be usefully employed in the determination of ecotope classifications. This classification is based mainly on the combination of functional attributes which are important for an organism or for a process.

A fuzzy classification creates smooth borders between different ecotopes and seems very close to the perception capacity of organisms or to the scaled characteristics of the considered processes.

Two methods can be used to calculate a fuzzy set. The first method is called Semantic Import Approach or Model SI (Fig. 2.33). The second method follows procedure similar to numerical taxonomy and one commonly used method is the fuzzy k-mean.

PART III:

Management and conservation of landscapes

Contents

3. Preface 197

3.1 Some concepts in management, conservation 197
3.1.1 A new perspective on the landscape matrix 197
3.1.2 The keystone paradigm 198
3.1.3 The concept of refugia in land mosaics as biodiversity hotspots 200
3.1.4 Adaptive management 202

3.2 Landscape approach to management and conservation: Exploring new dimensions 203

3.3 Large scale landscape management 204

3.4 Multiple scale investigation and management 208

3.5 Constraints in managing 208

3.6 Types of management 210
3.6.1 Managing species: Land management and spatially explicit population models 211
3.6.2 Managing species: Multi-species management 211
3.6.3 Managing fragmented populations 212

3.7 Managing patterns 213
3.7.1 Matrix biodiversity versus patch biodiversity 213
3.7.2 Managing ecotones 214
3.7.3 Managing linear habitats 215
3.7.4 Managing remnant natural habitats 218
3.7.5 Forest management 218

3.8 Managing processes 220
3.8.1 Managing disturbance 220
3.8.2 Managing disturbance along river systems 223
3.8.3 Managing grazing regimes 225

3.9 Managing functional areas 231
3.9.1 Managing human dominated landscapes 231
3.9.2 The ecosystem service concept 233

3.9.3 Constraints in human dominated landscape management 234
3.9.4 Ecosystem-based management 236
3.9.5 Managing parks 239
3.9.6 Watershed management and water quality 239

3.10 Conserving the landscape 241
3.10.1 General principles 241
3.10.2 Nature conservation: Criteria which fit the landscape paradigm 244
3.10.3 Conserving hotspot biodiversity 245
3.10.4 Conserving riparian landscapes 248
3.10.5 The importance of preserving oases 252
3.10.6 From protected areas to diffuse conservation 253
3.10.7 Countryside (cultural landscape) heritage conservation 254
3.10.8 Soil conservation 256
3.10.9 Conserving biodiversity in managed forests 258
3.10.10 Conserving habitat fragments 260

3.11 Perspectives in management and conservation 260

3. Preface

The application of landscape paradigms in practice means the use of methodological approaches to manage, conserve and restore land mosaics (Hansson & Angelstam 1991, Naveh 1992).

From this perspective it would seen important to integrate the methods of landscape ecology with the commonest practices of conservation and management. Many principles of conservation were elaborated before a broad vision of environmental complexity was achieved and so need some revision to bring them up to date. Nevertheless, we must recognize the tremendous efforts made in these last decades by conservation and management sciences.

Landscape ecology may play an anbivalent role on this issue by serving as a scientific tool for conservation or by acting as the main science for managing, conserving, and restoring land mosaics across different spatio-temporal scales. To achieve both these points it is important to clarify some concepts and terms that, although popular and commonly used in a conservation context, assume a special significance when incorporated into landscape paradigms.

For simplicity I have sub-divided this chapter into three sub-chapters: Concepts, Management, and Conservation. It is clear that management can be considered an approach incorporating conservation, but to ensure an easier presentation I have separated the three different areas.

3.1 Some concepts in management and conservation

3.1.1 A new perspective on the landscape matrix.

Usually we refer to a matrix as a low quality area, often hostile to organism movements. This is true in some contexts but it is not a general rule. The role of the matrix should be reconsidered carefully especially when we find ephemeral patches in a land mosaic. Such patches cannot reorganize and conserve a high diversity and complex structure without a "source" matrix that provides new organisms and ensures the availability of nutrients. This vision opens new perspectives in management, as well as in conservation and restoration.

One example of the role of the matrix can be observed in lotic ecosystems. For example, flowing water along a river is populated by fish populations and such an area can be considered a matrix. At the confluence of streams into the main river, autocatalitic patches exist. In such patches the turbulence of water is less and water is deep. Riparian and hydrophil vegetation can develop providing patches suitable for breeding for most of the fish fauna. A very rich community of invertebrates is present and, additionally, amphibians, reptiles and birds are well represented.

Flowing water is a matrix that transport organisms, often leaving no possibility for self-organizing into permanent habitat patches, though there is a high possibility of temporary autocatalitic patches forming. Such patches need a strong intra-patch dynamism to survive because the young stages of succession are the dominant components.

In a landscape, the matrix is the medium in which an aggregation of objects is displaced. The short term (ephemeral) life of many of these objects is incorporated into autocatylitic processes initiated by the matrix. The matrix assumes a major role especially for fine grained landscapes such as the Mediterranean where the small sizes of patches means they are more exposed to external influences.

3.1.2 The keystone paradigm

If the main goal of conservation is preserve species, landscape conservation should aim towards the conservation of the dynamism of a system. This process is hierarchically more important than the conservation of a species so dynamics should be incorporated into models of landscape conservation.

Such a strategy allows the presence of hot spot biodiversity patches in areas that apparently have few chances of representing a permanent set of organisms. With this in mind the discovery of keystone species driving relevant processes that assure the dynamism of a system would be of great importance.

The keystone concept is extremely important for management, conservation and restoration, because the discovery of a keystone species in a community, ecosystem/ landscape allows the preservation of such organisms as the main actors of environmental processes, thus increasing the efficiency of the action.

In refining management policy, special attention should be devoted to the preservation of landscapes in which the role of some species is not obvious, so protecting a broad spectrum of organisms that may play a relevant role in ecosystem processes. Great attention also should be devoted to the introduction of alien species that can act as keystone species by conditioning the entire community. The lack of a precise protocol for recognizing keystone species appears one of the main gaps in present protection procedures, as stressed by Power et al. (1996). On the concept of keystone species, Knapp et al (1999) have presented new evidence of the effect of bison grazing on tallgrass prairies in Konza Prairie Natural Area in the Flint Hills of northeastern Kansas (Fig. 3.1). These prairies are remnants of a very widespread landscape (68 million of hectares) which has been almost completely destroyed from 1800's onwards, leaving just 5% of their original extent. In this area, in which prescribed fire regime attempts to save from extinction the tallgrass communities, bison has recently been re-introduced. Results after 10 years of grazing disturbance by this large ungulate show that grazing has increased the local heterogeneity, diversity and vegetation type.

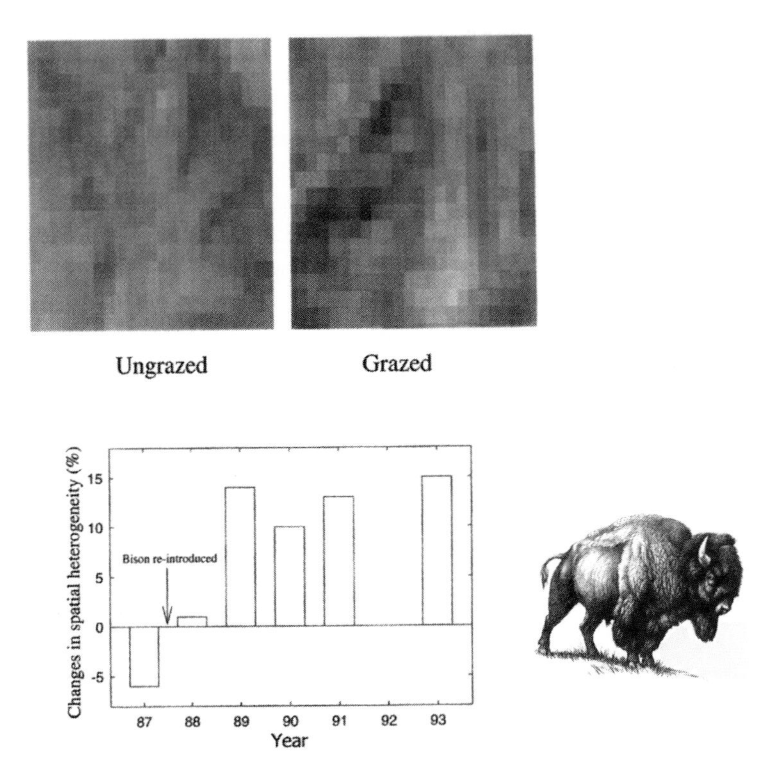

Fig. 3.1 - Changes in spatial heterogeneity after bison re-introduction to tall grass prairies in Konza Prairie Natural Area (Flint Hills, northeastern Kansas) (from Knapp et al. 1999).

Bison reduction of emerging biomass creates new cycles of nutrients and affects the C/N ratio. Urine and dung deposition, vegetation defoliation, wallowing, and bison carcasses are all fine scale disturbances which affect diversity and heterogeneity in Konza prairies. This affect is associated with the prescribed burns.

The management implications of adopting bison and fire are very important for the survival of these remnant prairies. Fire creates patchiness at a larger scale than grazing and their combination ensures a greater complexity and plant diversity within the land mosaic. In fact, before European settlement fire disturbance and bison herd disturbance probably worked at a greater scale than current situation. To day, management attempts to maintain complexity and diversity in tallgrass communities. Spatial patterning of burns associated with bison disturbance regimes creates a singular mosaic that is probably not too far from the pre-settlement situation.

Special attention should be devoted to the physical effect of wallowing. This activity creates an area of 3-5 m in diameter and 10-30 cm deep which is probably suitable for secondary succession. In addition, it may create a different water regime which, when the soil is compacted by trampling results in ephemeral humid patches. Plants that colonize these patches are different from the surrounding mosaic. Another relevant effect on the spatiality of vegetation is created by the degradation of bison carcasses. Actually this a very restricted phenomenon, but before 1800, when millions of bisons were living and dying, it was probable a relevant process affecting vegetation structure and composition. In fact, a short time effect of carcasses is the extirpation of vegetation below the carcass and immediately around. However, after three years the content of inorganic nitrogen is two to three times higher than that of the surroundings, and a luxuriant vegetation spreads.

The keystone effect of bison can be invoked for the transformation of C4 dominant grasses into C3 grasses and forbs with an increase of 23% of diversity, 38% of richness, and 13% of community heterogeneity.

3.1.3 The concept of refugia in land mosaics as biodiversity hot-spots

In long-term disturbed landscapes especially, there often exist small areas in which it is possible to find a great diversity of organisms, when compared with the embedding matrix. Such a matrix is generally composed of urban, developed areas or intensively cultivated land.

Such small remnants of naturalness are important refugia for biodiversity and are generally located along streams or rivers, in degraded marshlands, or in highly fragmented forested areas. Shape, size and composition may be very different and the potential biodiversity of these area may vary greatly.

A very clear case is represented by 40 ha of riverine landscape at the confluence of the Taverone and Civiglia streams into the Magra river (Northern Italy)(see Fig. 3.2). The area is enclosed by the highway, a bridge connecting the highway to Aulla center, by the railway (Parma-La Spezia) and by Statal Route N. 62. The urban area of Aulla and Terrarossa border the river.

This small area has experienced profound changes during the last 50 years. Open, devegetated areas can be seen on aerial images dating 1954 but for many reasons this situation changed resulting in development of the dense, rich riparian vegetation of today. An intense disturbance regime linked to seasonal flooding and to nearby gravel mining created new mosaics in just a few years. Various investigations on bird distributions have been carried out over the years in different areas (Farina 1981, 1993b).

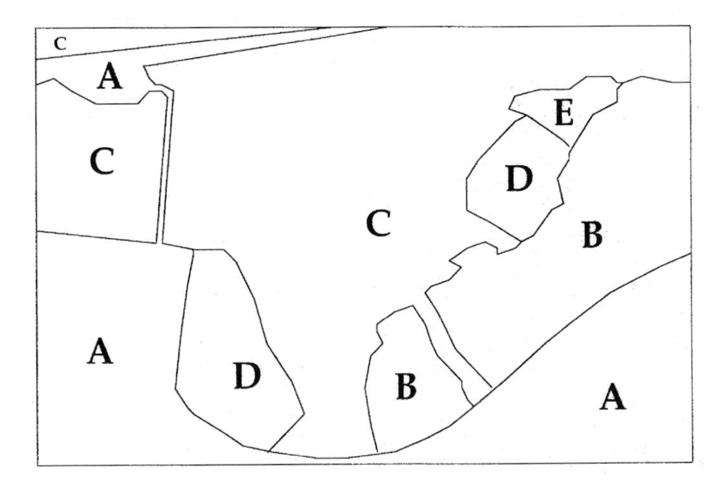

Fig. 3.2 - The confluence between Taverone stream and Magra river (Aulla, Massa Carrara province, Northern Italy) is an example of a biodiversity hotpsot embedded into a complex, human modified mosaic. A) Urban, B) agriculture, C) riparian area, D)recreation, E) land filling. The riparian area conserves a rich flora and fauna despite the highly disturbed neighboring patches (A,E,D).

Riparian vegetation has developed a dense stand dominated by *Alnus glutinosa*, *Populus* spp., *Salix* spp. *Platanus orientalis*, etc., whilst a dense undergrowth of *Rubus* spp. and invasive weeds such as *Artemisia* spp. and *Elianthus tuberosum*. dominate the lowest levels of this structured vegetation cover.

Wild boar (*Sus scrofa*) and fox (*Vulpes vulpes*) are the largest mammals and there are more than 30 species of breeding birds (Farina 1981, 1993b). The area is important for fishes, amphibians and reptiles (it is possible to find most vertebrates of the entire coastal region). *Lacerta sicula campestris*, *Natrix natrix*, *Natrix tessellata*, *Vipera aspis*, *Elaphe longissima*, *Coluber viridiflavus* are the commonest reptiles and *Bufo viridis*, *Rana dalmatina*, *Bufo bufo*, the commonest amphibia. Relevant fishes are: *Leuciscus souffia*, *Lampetra fluviatis*, *Leuciscus cabeda*, *Anguilla anguilla*. This list is incomplete but we can assume that biodiversity in this ephemeral area is very high. If analyzed at a seasonal scale, this relict landscape shows a surprising turnover of organisms and functions.

3.1.4 Adaptive management

Adaptive management is an approach used to guide ecological intervention in the face of uncertainty about the system. The main idea is that management actions are taken not only to manage, but also to learn explicitly about the processes governing the system. Monitoring is a key component and a plan for learning is fundamental (Shea et al. 1998) (Tab. 3.1).

Adaptive management mimics the feedback strategies that for thousand of years have been carried out by human societies in "cultural landscapes". To day monitoring action has substituted the practical lesson learned by individuals acting by trials and errors.

Tab. 3.1 - Details about the decision theory (from Shea et al. 1998).

What is decision theory?

Decision theory is a framework within which people responsible for management attempt to achieve explicitly stated objectives. This sort of theory is used extensively by other professionals, such as engineers and financial advisers, to make decisions. In some disciplines, decision theory uses complex mathematical tools that generally go under the banner of 'mathematical programming'. However, decision-theory is much broader and it includes qualitative methods, such as 'multi-criteria assessment' or 'ecological risk assessment'.
We should think of decision theory as a framework for objective management, rather than as a set of mathematical tools. In a disciplined approach to decision theory we need to define the following:

. A clear statement of management objectives, or at least a list of indicators of policy performance.
. A list of the management options, which can be expressed in terms of control variables.
. Variables that describe the state of the system (e.g. population size).
. Equations that describe the dynamics of the state variables.
. Constraints that bound the decision variables and state variables.
. All the parameters needed to describe the above, preferably with some notion of the variability around each.

3.2 Landscape approach to management and conservation: Exploring new dimensions

To protect means to preserve from destruction and assumes that process dynamics must be assured over time. A watch can measure the time only if all the wheels are moving. All actions aiming towards management and active protection, moving from species to land mosaic scale, need an environmental context in which to place such actions. To do this it is necessary to have an in depth knowledge of the natural history of a region and also a good knowledge of the socio-economic and historical constraints.

Conserving the mosaic means to preserve hotspots for biodiversity, source habitat patches, and keystone process patches. On this point we have to move from species level (population, metapopulation) to guild, to community and finally to land mosaic level.

The effects of managing actions can be measured at different levels from species to land mosaic. The use of keystone processes, species, guilds, communities, habitat patches and land mosaics, are all important approaches.

We don't know in advance the best strategy to preserve the complexity of our world though we are confident that the maintenance of diversity in processes and species is useful and important for the biological survival of the entire planet. We have no precise parameters to measure such a level, especially as this level varies geographically. The application of policy strategies to the environment has a lot of risks because human models are often quite distant from the reality of ecological processes.

The only way to act and react to maintain positive feedback of the environment, is to mimic natural processes and to ensure the broadest dynamics to ecological processes. For this reason the landscape approach assumes a special importance for carrying out this action. The landscape approach recognizes the complexity of the species, habitat, and land mosaic interactions.

It recognizes the context within which processes behave differently and the necessity of flexible actions when we operate in such a system. It avoids dogmatic positions and anedoctal "knowledge", and bases it's actions on functional units such as species-specific habitat patches, or process mosaics.

It describe the locally instable component of large scale stability, but recognizes large scale instability over a long time scale. Finally, it uses the concept of hierarchy extensively, as a mechanism which carries processes across the different spatio-temporal scales.

Adopting a landscape approach to preserve diversity means recognizing that this is possible on the condition that the environmental matrix has enough diversity to ensure autopoietic processes in "autocatalitic" patches. Autocatalitic patches may be defined as relatively distinct patches embedded in an active mosaic and in which communities find a good place to interact.

They function within a " context" of disturbance regimes in which intensity and frequency have a different impact according to the status of the patch.

In addition, they have many favourable attributes for a long list of species, and so act as a multiuse patch. These autocatalitic patches are easily recognized because the number of species is higher compared with the surrounding matrix. I believe that this model can exist only if the matrix has a high variety of species and has processes that ensure active and passive diffusion of organisms and resources.

3.3 Large scale landscape management

The management of large scale (human) landscapes is not an easy task because there are often social and economic implications beyond the interrupting actions of the environmental impact. Management means a direct action within a mosaic and such an approach can be processed in at least two different ways: by generating a rule of behavior, or by moving directly into the field and using a specific disturbance regime (logging, grazing, burning, etc.). Small areas can be managed directly but large areas need rules because the heterogeneity of the mosaic, different ownerships, different administrative tenures and a differentiated land use produce a prohibitive (economically) challenge for direct management.

In many countries the management of parks and protected areas is quite distinct from the management of all other parts of the territory. This difference in approach creates visible barriers between protected and non protected areas. Landscape ecology could also serve to reconcile this important form of management constraint that, in the last analysis, is not useful for a general conservation strategy.

It is not possible to maintain many organisms in parks and nature reserves because they need large spaces in which to move (i.e. large home range) or because there is not enough space for natural dispersal and spatial arrangement according to the more suitable habitats.

Management may be the main focus in a specific large scale landscape (e.g. Tuscany landscape, Georgia Piedmont landscape, Ohio river basin), or in a small scale mosaic (e.g. rivers, lakes, ponds, marshes). It seems, today, very promising to copy natural disturbance regimes as a formidable tool to reduce the "management debt" created by human intrusion and by natural disturbance depletion. According to the typology of the considered landscape, actions and prescriptions may be very different; in a forested landscape, the typology of logging (size, retention rate, frequency and spatial arrangement) should have priority only secondarily do afforestation practices assume a relevant importance in coniferous stands.

In cultivated areas the conservation of woodlots, marshes and riparian vegetation are a priority. Linear elements are the result of a combination between the necessity for open space to cultivate and the advantages of wind breaks and surplus nutrient interception buffers. In urban fringes the presence of green spaces is in growing demand.

In "today's" landscape, fragmentation and isolation appear to be dominant processes and reducing this trend is an ongoing challenge. The new paradigms of landscape ecology could be very useful for applying the integrated, multi-purpose policies of landscape conservation and soft (sustainable?) development across an informative evaluation and an ecologically oriented (based) management.

In the application of management rules the different cultural footprints must be taken into account. In every country, different values attributed to nature and natural components deeply influence actions towards the environment. The view of the environment in a developed country is completely different to the vision of a country undergoing development, with each facing two completely different action contexts. For a developed country the environment is considered a resource to preserve, for a country undergoing development the environment is still considered a resource from which to capture energy and products.

3.4 Multiple scale investigation and management

Hierarchical theory is increasingly used as a general paradigm for improving efficiency in ecological investigations. Studies conducted at one (space/time) scale are often incomplete and so collect information on a limited trait of species ecology. A multiscale investigation seems, today, a very promising approach to better understand the relationships that species and their aggregations develop. This is particularly important when the main goal of the study is the transfer of outcomes to management.

The importance of spatial scales has been demonstrated recently by Saab (1999) working on breeding birds in riparian forests along 100 km of the South Fork of the Snake river in southeastern Idaho. He identified a microhabitat (local vegetation characteristics), a macrohabitat (cottonwood forest patch characteristics) and "landscape" (composition and patterning of surrounding matrix, vegetation and land use). Details about habitat variables were reported in Tab. 3.2. At microhabitat, tree, ground cover and stem density were counted. The macrohabitat scale was considered by Saab(1999) as the dimension of the cottonwood stands, considering chorological characters like patch size, perimeter:area ratio, patch-edge contrast index, core area, etc.

Tab. 3.2 - Application of three spatial scales (microhabitat, macrohabitat and "landscape") to the measurement of habitat variables for riparian birds in a cottonwood forest (from Saab et al. 1999).

	Variables	Description
Microhabitat	Stem density of trees and shrubs (no./ha)	Recorded by structure (dbh size class: ≤2, >2-5, >5-8, >8-23, >23-38, >38 cm) and species composition.
	Tree canopy cover (%)	Recorded for all plant species (structure) and individually by plant species (composition).
	Ground cover (%)	Recorded as herbaceous, shrub, down, log, or bare ground (structure).
Macrohabitat	Patch size of cottonwood stand (ha)	
	Perimeter to area ratio (m/ha)	Index of amount of edge.
	Length of cottonwood patch (m)	
	Width of cottonwood patch (m)	Averaged over three measurements.
	Patch edge contrast index (%)	Percent of edge involving the sampled cottonwood patch weighted by degree of structural and floristic contrast between adjacent patches; equals 100% when all edge is maximum contrast (e.g., cottonwood vs. agriculture) and approaches 0 when all edge is minimum contrast (e.g.,cottonwood vs. aspen).
	Core area (ha)	Amount of core area of each sampled cottonwood patch defined by eliminating a 100 m wide buffer along the perimeter of each patch.
Landscape composition	Percentage of landscape	Percentage of landscape (within 1.0 km of edge of sampled cottonwood patches) composed of corresponding patch types.
	Simpson's diversity index (0-1)	Represents probability that any patch types selected at random (within 1 km of each sampled cottonwood patch) would be different patch types; the higher the value the greater the likelihood that any two randomly drawn patches would be different patch types (i.e., greater diversity).
	Relative patch richness (%)	Patch richness as a percentage of the maximum potential richness, which includes 10 patch types within 1 km of each sampled cottonwood patch.
	Simpson's evenness index (0-1)	Represents distribution of area among patch types within 1 km of each sampled cottonwood patch; larger values imply greater landscape diversity; maximum diversity for any level of richness is based on an equal distribution among patch types.
	Interspersion index (%)	Measures extent to which patch types are interspersed; higher values result from landscape in which patch types are well interspersed (i.e., equally adjacent to each other and greater landscape heterogeneity), whereas lower values characterize landscapes in which patch types are poorly interspersed with a disproportionate distribution of patch type adjacencies.
Landscape structure	Distance to contiguous cottonwood forest (m)	Mean distance from edge of sampled cottonwood patches to edge of contiguous cottonwood forests adjacent to river.
	Distance to nearest cottonwood neighbor (m)	Mean distance from edge of sampled cottonwood patches (mature stands only) to edges of nearest cottonwood patches (including young, mature, and old stands of cottonwoods).
	Landscape edge contrast index (m/ha)	Density of edge involving all corresponding patch types (within 1 km of sampled cottonwood patches) weighted by degree of contrast between adjacent patches; approaches 0 when all edge is minimum contrast.

At the "landscape scale" the characteristics of the surrounding mosaic (in terms of composition and structure) within 1 km of a cottonwood riparian forest were considered. The best predictors of bird species richness were found in natural and heterogeneous landscapes, in large cottonwood patches, close to other cottonwood patches and in microhabitats with open canopies. There is evidence that the surrounding mosaic is important for establishing the distribution and abundance of birds in a riparian forest. According to a general paradigm, all linear structures are deeply influenced by the surrounding matrix, due to the presence of a large contact surface between the linear structure and it's surroundings. The characteristics of the land mosaics seem to be the main factor in bird occurrence, followed by macrohabitat and microhabitat features. In conclusion the distribution of breeding birds in cottonwood forest patches is correlated with a hierarchical set of environmental factors. This confirms the validity of the proposed model of chapter 1.3.1.7 (pag. 22-24) in which the transient characteristics, from a topological perspective to a chorogical perspective, are included in the ecological traits of species.

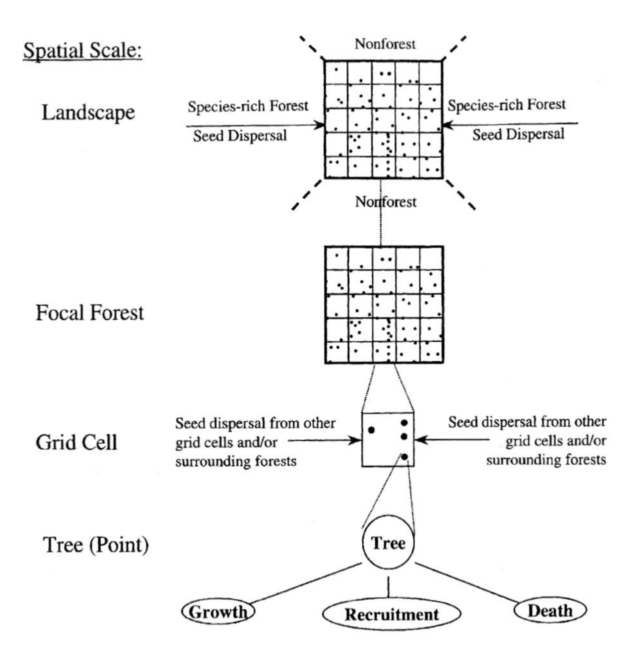

Fig. 3.3 Example of the utilization of a spatial multiscale approach to model tropical forest demography (from Liu & Ashton 1999).

The multiscale approach has been used by Liu & Ashton (1999) to develop an individual-based, spatially explicit landscape simulation model (FORMOSAIC) (Fig. 3.3). The application of such a model has provided a good opportunity to predict the effects of the surrounding forest on the focal forest area, through the different tree levels in a tropical area. In addition, this model seems readily adaptable to non-tropical forest. The model predicts species richness, stand density and timber volume as exposed to management practices and abiotic and biotic factors that influence tree recruitment, growth and death. Four scales are used: land mosaic, focal forest area, grid cell and single tree. Originally authors called the "Landscape scale" the largest level, whereas I believe is more correct to call such scale a "land mosaic scale", although the difference from the "focal forest area" is mostly a subjective scale. The focal forest area scale could be called a patch forest scale in which the sub-unit, the grid cell, has no precise ecological meaning, only a procedural significance for the model. Without entering into details, there are a number of interesting proposals for managing tropical forests in a more sustainable way, such as having a species-rich seed zone around a focal forest area to compensate for negative harvesting impacts, or the harvesting of timber in optimal locations to encourage a new spread of seeds from the surroundings. Again the role of the surrounding mosaic appears fundamental to disturbance recovery.

3.5 Constraints in managing

The spatial and temporal scales at which it is possible to manage natural and human modified systems are more restricted than commonly believed. According to a general rule, when we try to manage a small piece of land the high variability of the events influencing such an area often frustrates our attempts and we have to invest a lot of energy to counterattack the external factors that "press" the selected area. Isolation of such areas based on a specific stewardship is the final result. A garden, or an orchard represent two good examples of how difficult it is to maintain such systems without direct human intervention. Both are fragile systems in which humus form, species composition and spatial arrangement are modified in short time.

On the other hand, managing large areas is not possible by direct stewardship. In this case, regulations are the most popular "procedures", based on the assumption: "This is allowed – this is not allowed", and so on. A compromise between the different views of the society components is the final result. Such a balanced result can be quite distant from the ecological (*sensu strictu*) minimum requirement of an area. This type of problem has recently received huge attention in the U.S. National Park Service which manages an impressive amount of natural areas (Soukup et al. 1999), areas that are increasingly surrounded and dominated by human activities. Changes of landscapes induced by natural events and human interventions, create dramatic challenges to the maintenance of some type of control inside and outside the

National parks. The recent policy of the U.S. National Park Service is devoted to maintaining "natural environments evolving through natural processes minimally influenced by human actions". Within this policy there emerges the concept of "natural regulation" that, theoretically, should ensure stability at population level even without predator control.

As reported by Peterson (1999), the recent decline of the wolf (*Canis lupus*) population (1980-1996) in Isle Royal National Park due to isolation and high rates of inbreeding, has produced a huge increase in moose (*Alces alces*) numbers without the activation of density dependence mechanisms. In 1996 the moose population suffered 80% of casualties mainly from starvation. This park may be considered a real laboratory in which intervention-non-intervention policies can be compared. Without a wolf reintroduction moose populations will also probably fluctuate greatly in the future and the "aesthetics" in park management should be reconsidered. Altough Isle Royale National Park is considered a relatively simple system (isolated and with little human pressure and only two predator-prey species) understanding how the ecosystem functions is still a long way from being adequately known for the development of efficient management actions. Again the context in which a park is considered assumes a great importance. In effect, in park management human intervention affects the whole of a wilderness area increasing it's isolation.

In the Yellowstone National Park, the debate on natural regulation is also quite open (Huff & Varley 1999) and it appears very hard to decide when a population of grazers is at the limit of the carrying capacity of a system. In fact, we have no reference system for comparison and long term studies require time investments that surpass the political and socio-economic decisions of local and central authorities. To interrupt the chain, more science is necessary to improve management approaches. I agree with the adaptive management approach in which management, experimentation and monitoring are common ingredients for increasing knowledge and park management (Wright 1999). In particular the management of large carnivores that are not habitat specific and require large areas pose great challenges to scientists and managers. The role of the large scale landscape has to be considered with more attention focusing on conservation not only in selected wilderness areas, but also considering the human dominated matrix that surrounds such "islands". As reported by Mladenoff et al. (1999), the wolf (*Canis lupus*) is a predator that is able to utilize roads for movements in wild areas yet the density of roads is a good indicator of unsuitable habitat for such a species. Wolves do not require wild areas to survive, in fact they are common in countries such as Spain and Italy where human distribution is quite diffuse, but require prey abundance and protection against human killing. For wolves, landscape structure at the human scale may be a good predictor of suitability.

3.6 Types of management

It is not possible to handle directly the complexity of the real world, hence a simplification of this realty is necessary although not entirely satisfactory. Species, patterns, processes and functional areas (bio-regions, watersheds and administrative units) are all popular approaches in ecosystem (landscape) management, but their full integration is necessary to improve management action (Fig. 3.4).

Habitat suitability population models consider the quality of a habitat; spatially explicit models consider both species-habitat relationships and the arrangement of habitats in space and time.

Such models consider variables such as: fragmentation, isolation, patch size and habitat shape, and so represent a useful tool for managers. For example, the effect of the 1988 fire in the Yellowstone National Park has been included in a model by Turner et al. (1994) to evaluate the effect of fire on wintering elk (*Cervus elaphus*) and bison (*Bison bison*). The integration of models for wildlife and economic goals is one of the more important possibilities offered by such procedures.

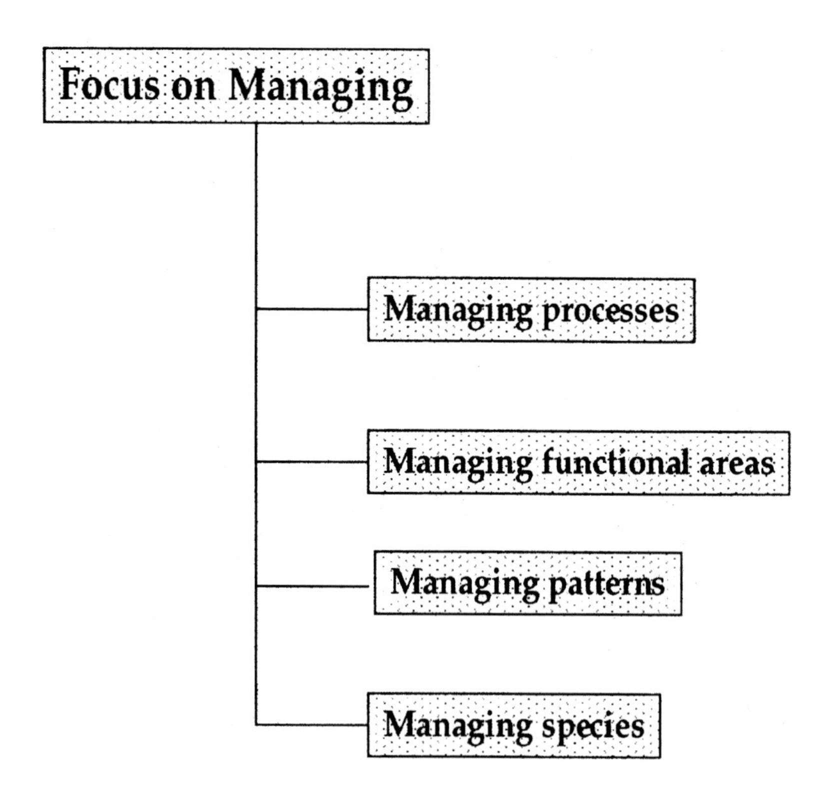

Fig. 3.4 - Classification of management typologies and their possible integration

3.6.1 Managing species: Land management and spatially explicit population models

In this direction must be considered ECOLECON, a model that incorporates timber production and the habitat requirements of Bachman's Sparrow (*Aimophila aestivalis*) in Southeastern US pine forests (Liu 1992). The limits of such models depend mainly on the possibility of extrapolating the predictions to other areas, and the possibilities for changing the scale of resolution.

The use of a GIS is essential to update the data and to process information. Another major difficulty is represented by the application of a model of several species contemporarily.

3.6.2 Managing species: Multi-species management

Management policies are complicated by the specific reaction of a species to environmental conditions and often it is not possible to preserve the complexity of an area based on the habitat requirements of a single species. It is necessary to implement the ecological knowledge of a large set of species before transferring the ecological requirements of a single species to activate a management procedure. The perception of a habitat mosaic may also be quite different for similar species and this can be of absolute importance when developing successful management strategies.

This is the case of the Florida scrub lizard (*Sceloporus woodi*) and six-lined racerunner (*Cnemidophorus six-lineatus*), two scrub sympatric lizards living in the Florida scrubs (a unique shrub-dominated community occupying relict sand dunes and forming strands long recent coastal sand dunes) (Hokit et al. 1999).

The two species have similar habitat and food requirements, but the six-lined racerunner has a higher dispersal capacity. The isolation of Florida scrub has a strong negative effect on the Florida scrub lizard but not on the other species.

An in-depth knowledge of the ecology of a species is an obligatory prerequisite to any design and management action. A multiscaled, multispecies approach is required to manage the Florida scrub based on differences in the dispersal abilities, area requirements and habitat preferences of the two species of lizard considered.

This example should be expanded as a general rule for any multispecies management and may be a good approach for formulating management scenarios based on selected key stone species.

3.6.3 Managing fragmented populations

As recently outlined by Opdam et al. (1994), fragmentation in many countries is a common process affecting the distribution and abundance of organisms such as birds.

The size of the fragment is an important parameter for predicting local extinctions. Fragmentation reduces the connectivity between different suitable habitats and creates isolation between populations. A huge literature has been produced recently on this subject relatively to the different types of impacts, such as an increase of predation (Andren et al. 1985, Burger et al. 1994, Leimgruber et al. 1994), extinction (Cutler 1991, Tscharntke 1992, Willson et al. 1994) and changes in the land mosaic (Hobbs & Hopkins 1986, Wilvove et al 1986, Kavanagh & Bamkin 1995, Nepstad et al. 1996).

We have much empirical evidence that birds utilize the metapopulation model to occupy the fragmented habitats in a matrix of cultivation. The minimum available area for the maintenance of a sub-population has been estimated by Opdam et al. (1994) as a threshold of 20 reproductive units (Fig. 3.5).

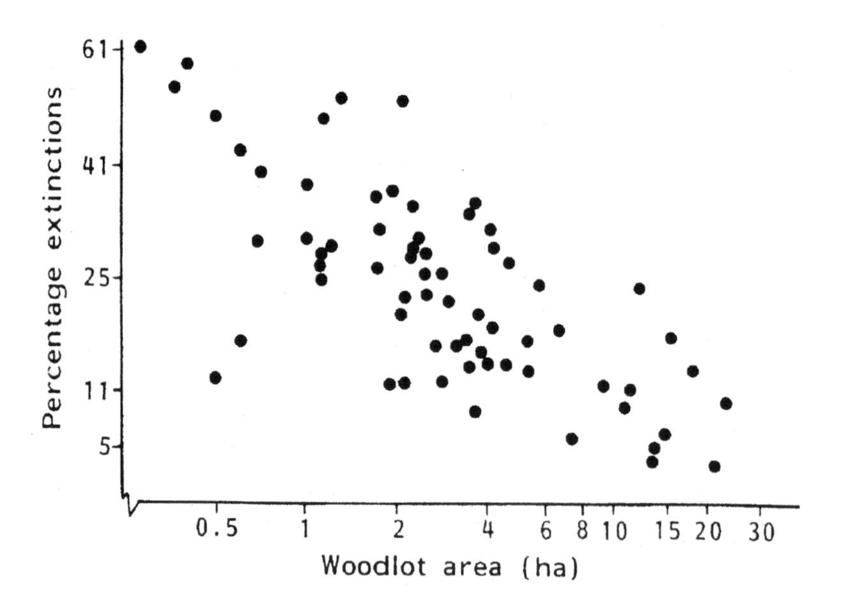

Fig. 3.5 - Relationship between local extinction of birds and the dimension of a relict woodlot area across a farmland mosaic (from Opdam et al. 1994).

Using such criteria for habitat sensitive species it is possible to build different scenarios which are useful for addressing planning solutions to the general decline of biodiversity. Such an approach must be carefully scaled according to the characteristics of the different areas of interest.

Conservation of small fragments is important in every biome, including tropical areas (see f.i. Turner & Corlett 1996). As stressed by Wiens (1994), it is important to recognize the value of remnant patches whilst also taking into consideration the surrounding context (the mosaic). The quality of the mosaic makes all the difference for fragment dynamics.

3.7 Managing patterns

3.7.1 Matrix biodiversity versus patch biodiversity

Managing landscapes means to recognize the importance of matrix biodiversity compared with patch biodiversity (Noss 1983). For many reasons, we often only consider the patch level, for instance a simpler location of a complex object, and we devote scant attention to the embedding matrix.

The matrix is important in maintaining biodiversity, especially across the Mediterranean countries. In this area patches are very small, ephemeral, and have a high inter-patch contrast. Per se, a patch has few possibilities to support the important components of biodiversity and to represent ecosystems (*s.l.*). The matrix is not a homogeneous medium but, if observed closely, appears as a structure with a fine grained mosaic of organisms and physical structures.

The sharp separation between the matrix and embedding patches is often the result of an impacting land use with illusory functioning. Within a matrix there are visible components (patches) and invisible patches composed of seed banks and millions of micro-organisms like bacteria, fungi, algae, protozoa, earthworms and insects. Dormant forms of vegetables such as bulbs and ephemeral plants, sprout creating temporary patches on the surface of the soil.

There is a reciprocal conditioning between the different organisms and the patterns they create for some life traits.

Every location has a unique, patchy distribution of organisms, and we can assume that the spatial overlap of coherent aggregations (patches) represents the "phenotype" of the land mosaic.

Patches can be classified according to the dominant land cover, but can also be classified according the length of their appearance. So, the spring growth of geophytes such as *Crocus* sp. lasts two weeks maximum as the blooming of most wild flowers is quite short, but for insects and other organisms this blooming is patchily distributed and is consequently perceived as a moasic (Fig. 3.6).

Fig. 3.6 - The spring blooming of *Crocus vernus* create a temporary patches relevant for many insects.

3.7.2 Managing ecotones

Where patches lie at the edge we can assume that in that location the functioning constraint is working. In this position many processes linked to organism traits are working as in a game in which each process tries to prevail. We call these edges 'ecotones'.

The location of ecotones assumes great importance in terms of predicting landscape dynamics across time. Ecotones are not stable, permanent borders but change position, function and interpatch constraint through time. Like patches, there is a broad range of ecotone permanency. For instance, in the same location ecotones can change in quality yet persist for a long time, or they can be more ephemeral forming new patterns at some distance from the original location. The first case pertains to ecotones created by secondary succession. The second case can be observed at a biome ecotone when climatic conditions change and the separation edge between biomes is forced to move in one direction or another. Management of ecotones may be an important strategy on the condition that we have full knowledge of the functions of such ecotones. Many studies have focused on the importance of riparian ecotones (see in another chapters) because such ecotones are quite stable, recognized worldwide and function in more or less the same way. The story of ecotones that are unique to a location is quite different, as in the California vernal ponds, the Mediterranean coastal sand dunes, or the Australian coral reefs.

The management of anadromous streams represents an important conservation hotspot, especially along the pacific coasts of North America. The relationship between grizzly abundance and salmon abundance is very strict. Although the relationships between terrestrial and aquatic systems are well known, little attention has been paid to the role aquatic organisms play on terrestrial systems like wildlife (predators) and birds (Willson et al. 1998). For example the riverine landscape is a fundamental source of insects for migratory birds throughout Mediterranean region (Farina 1993d) and the microclimate along rivers ensures a more precocious development of insect larvae and adult growth of aquatic larvae.

3.7.3 Managing linear habitats

Linear habitats like hedgerows, tree fences and riparian vegetation don't always form corridors but in these linear habitats the movement of organisms is more evident due to the effect of matrix contrast or concentration which, in some cases, represents a genuine ecological trap. The evaluation and conservation of these elements is important for understanding and conserving the centrifugal processes such as animal dispersion and energy and material flow.

Disturbance processes along such habitats assume a special emphasis for the fragile characteristics of such linear habitats. For instance, Naiman & Rogers (1997) have considered at least four levels of processes influencing such linear habitats.

The riverine landscape is characterized by a high variability in resource availability and in population dynamics. This landscape requires a different management approach based not on managing stability but on respecting disturbance regimes as a fundamental ecological pulse. In general large animals like beavers (*Castor canadensis*), moose (*Alces alces*), hippopotamus (*Hippopotamus amphibius*), elephants (*Loxodonta africana*), buffalo (*Bufalo cafrus*) and crocodyle (*Crocodylus niloticus*) are important factors of local disturbance enhancing the biodiversity of river systems.

Tab. 3.3 - The structure and dynamics of riparian corridors are conditioned by many factors according to a hierarchical sequence (Naiman & Rogers 1997).

Level of influence	**Factors**	**Actions**	**Consequences**	**Approximate scales**	
				Spatial (km2)	Temporal (years)
First level	Matter, energy, water, gravity, fire	Erosion, deposition, slope, aspect, altitude	Soil formation, macroclimate, geomorphology	10^6-10^8	10^6-10^8
Second level	Biophysical alterations: habitat modification	Dam building, wallows, herbivory, burrowing	Conversion of habitat from macropatches to mesopatches	10^{-1} - 10^3	10^1-10^4
Third level	Elemental distribution and cycling by biota	Metabolism, nutrient cycling, formation of specialized chemicals	Productivity, succession, biotic distribution, formation of meso- and micropatches	10^1-10^2	10^1-10^3
Fourth level	Biotic interactions (including disease)	Life history strategies, population and community processes, trophic pathways, epidemics	Competition, mutualism,abundance, micropatch distribution	10^{-2} –10^1	10^{-2} – 10^1

Naiman & Rogers (1997), distinguish at least three levels of human modification along a river, from the removal of large animals to the full persistence of animals (Tab. 3.3).

It is possible to observe that, in the presence of animals, the riverine dynamic allows a more obvious connection with the surrounding area and a more complicated mosaic of bare and vegetated soil. This pattern has an intrinsically higher diversity than the example shown in Fig 3.7 where only an edaphic gradient is operating. The presence of large animals along rivers ensures the combination of different processes that, "per se", act at specific scales. Animals have the capacity to transfer such processes to other process configurations.

Along rivers there are functional groups which, although taxonomically distant, operate in the same way. It is possible to synthesize such activity and the relative benefit to the entire scaling range of a riparian ecological system.

Animals increase the turnover of disturbed patches producing new ephemeral habitats for species. The productivity of soil is increased by manure and other biomass retention and recycling. The different parts of the river mosaic maintain their connection by pathways created by the movements of such large animals. The entire system assumes a more resilient characteristic from such dynamism. The impact of animal activity is produced by ponding water, digging soils, trampling, moving material, or the selected removal of riparian vegetation.

The escavation of deep ponds by large mammals allows the survival of amphibians and fishes during drought periods, and supplies drinking water for other animals like antelope.

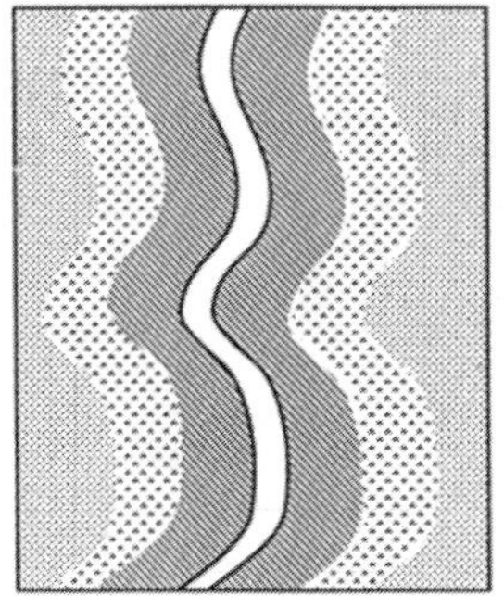
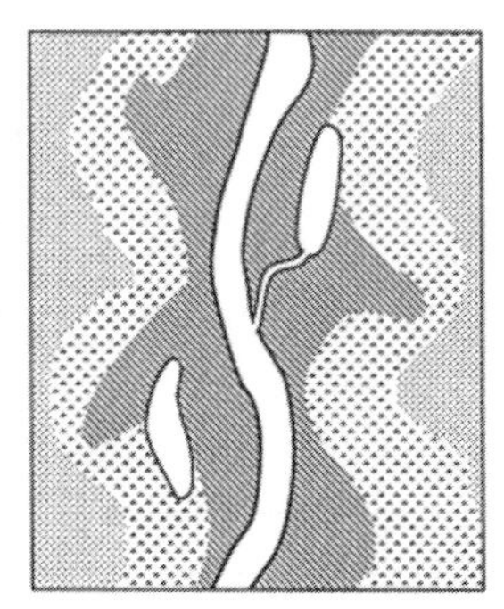
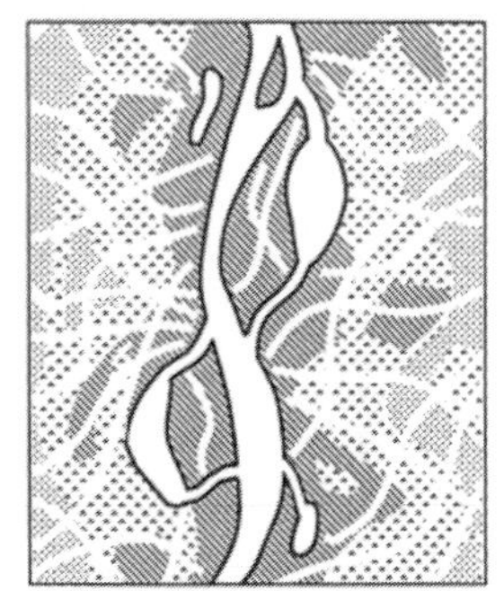
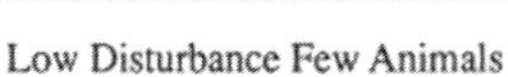

Moderate Hydrologic Disturbance

Hydrologic and Animal Disturbances

Fig. 3.7 - The complexity of riparian corridors is increased by an increase of disturbance regimes especially if physical disturbance is overlapped by animal disturbance (from Naiman & Rogers 1997).

Browsing activity modifies vegetation physiognomy and the competition, creating new vegetative communities that in turn affect other animal communities. Along rivers there can be found generalists, as in boreal forests, or specialists as in Africa. Often the effects of many specialists on a system equal the effect of few generalists.

For a more ecological, "ecosystemic" approach to land management it is important to recognize the role of such functional groups, which are true "engineers" of the landscape. These are key elements in the creation and maintenance of environmental heterogeneity and diversity, increasing the resilience of the system. We have to consider, in such a perspective, the concept that every ecological system changes with time due to the effect of a hierarchy of local, regional, inherent and external processes.

Previous concepts of "carrying capacity", which estimated the optimal number of animals per area, and "balance of nature" have to be converted into a more dynamic view of a system composed of the interactions of different organism-centered landscapes overlapping with a process landscape in which energy, nutrients and information move across complex scaled filters, like ecotones. "Balance of nature" should be changed to "flux of nature", from an homeostatic view to an homeorethic perspective. In a sound ecological management perspective, the role of large animals must be reconsidered because some landscapes are healthy on the condition that "disturbance" processes are maintained. This view is in contrast with the "siege" of new developments along rivers, deltas and sea coasts. But large animals, even if replaced by livestock, seems a promising management tool for diversity in fragmented landscapes. The case of the Maremma regional park in Italy is relevant here. Only the grazing of Maremmana cows coupled with horses, ensure the vegetation reduction needed by wintering birds foraging in open marshlands.

3.7.4 Managing remnant natural habitats

The increased fragmentation of natural habitats poses new challenges to managers. Remnants experience a decrease of biodiversity linked mostly to the small size of the area and to a reduction of natural disturbance regimes.

The prairie remnants of Wisconsin are currently in this condition (Leach & Givnish 1996). Despite 800,000 ha of prairies present in Wisconsin before European settlement, there are now less than 0.1% of the former distribution. These remnants are confined to small patches which suffer from the lack of a fire disturbance regime. Such a regime was frequent in the past and ensured open spaces whilst favoring short-stature, N-fixing, small-seeded species. Plants with N-fixing symbioses are favored only if frequent fires and open spaces concur to volatize N stocks. These plants can compete only in open, sunny, poor soil.

From the data collected Leach & Givnish (1996) argued that to stop the decline of remnant prairies and the extinction of many grass species it is necessary to prescribe burns, and to concentrate conservation towards short, small-seeded, N-fixing, regionally rare species.

3.7.5 Forest management

Forest management is a promising tool to retain biodiversity and ensuring the complexity. Experiments conducted by Chambers et al. (1999) in the Oregon coast range using three different silvocultural treatments on Douglas-fir forest mimicking low, medium and high disturbance regimes have indicated the necessity of creating a variety of stand types to meet needs of all species. This situation is quite common in southern Europe where woodlands and forests are managed according to a fine scale of resolution (small patches copying the fine grained distribution of ownership).

Evidence indicates that boreal swamp forests are important hotspots for biodiversity. These forests have a disturbance cycle of 400 ys. Few are prone to fire disturbance rather, in the past, they have been disturbed by human intrusions of cultivation, artificial flooding and prescribed fires (Hornberg et al. 1998). These forests are actually restricted to relict areas with a risk of extinction for many associated organisms. Using a conservation time lag of 300 years it is probable that a swamp forest can recover completely transforming into an old-growth forest. Historical data since the Pleistocene era indicate that swamp forests have been severely disturbed by natural as well as human induced events several times.

A policy devoted to reducing ecological contrast between managed dominant forests and pristine old-growth forest, seems extremely important for conserving biodiversity across the boreal biome. In such forests the dead biomass represents one of the major biodiversity supports, and is always rarer in managed or drier forests. Some years ago Zev Naveh recommended a multiple-use land as strategy to conserve semi-natural forests in the Middle East. The same suggestion has been pointed out by Hansen et al. (1991) in a completely different biome along Northwest Pacific Coast (USA).

In natural stands it is the disturbance regime which maintains diversity in mosaic, plant and animal assemblages. Managed stands generally present a very high contrast between cut and un-cut areas, so depressing biodiversity.

New attempts to mitigate the eradication of forests by logging consist of tree retention and maintaining fallen trees.

If, in the past, the old-growth forest was considered quite a unique requisite quality of forests, we are also aware that other successional stages are important for organisms and, in general, for ecological processes inside and outside forests. So a strategy to improve the quality of a forest is to have a full range of successional stages.

Another important point concerns the mosaic created by logged and unlogged stands. With regards to landscape design it is possible to find new designs for improving habitat suitability for species. We need a lot of knowledge to find real improvements for biodiversity at so large a scale. The retention, in managed forests, of natural tracts and not just old-growth stands, seems a very promising strategy towards the best conservation effort.

Other factors beyond the age and heterogeneity of a stand must be considered, one of which is the elevation. It is known that biodiversity decreases with elevation and that young stages in lowlands may have more species than old-growth forests on uplands.

Rarity in sere types is another important element to be considered. For instance, post-fire succession represented by grasses and small shrubs is relatively rare in managed forests where fires are under control.

In southern Europe, forestry is not as important as in North America or north Europe, but a good lesson can be learned from Mediterranean forestry where logging does not have such dramatic effects as in the boreal region. For example, tree retention is a common rule in logging across southern Europe and this, accompanied by the small size of logged stands, allows the persistence of a steady state mosaic of woodlands that ensure habitats for a broad range of living forms.

Landscape ecology offers a powerful tool for designing and managing complex mosaics. In fragmented forested areas, such as in north-central Wisconsin, Mladenoff et al. (1994) tried to produce a spatial model able to improve the mosaic quality of old-growth forest. The general assumption was based on the principle of connectivity between isolated patches, distinguishing edge zones of 100 m deep and a 300 m buffer zone in which logging activities were performed according to a precise protocol. A band of old-growth edge was differentiated from inner (core) old-growth remnant patch. The objective of these authors was to ensure the conservation (integrity) of old-growth patches. Two main types of analysis were made: interior analysis and adjacency analysis. A secondary zone (buffer), 300 m wide, of old-growth hedges was selected in which some disturbance regimes were considered. The function of this zone was to serve as a corridor between the old-growth patches.

Using the adjacency analysis it is possible to measure the degree of juxtaposition of different types of ecosystems. The application of such a model of landscape management considers the traditional land use (logging) with different levels of retention. Harvesting intensity increases with distance from the core area assuming different functions for secondary and outer zones and more integrated uses in which tree harvesting and landscape conservation are contemporarily considered.

3.8 Managing processes

3.8.1 Managing disturbance.

Disturbance is the basis process which enhances the diversity of living communities and land mosaics. Disturbance processes like flooding, insect outbreaks, weed invasion, animal grazing and trampling are considered either as potential tools for increasing biodiversity, or as events that spontaneously modify plant and animal composition of large areas.

For instance, the digging of grizzly bears (*Ursus arctos horribilis*) affects plant distribution and mineral availability in upland meadows. The effects on glacier lilies (*Erythronium grandiflorum*) by grizzly have been studied by Tardiff & Stanford (1998) in a subalpine meadow in the Glacier National Park. Like many other large animals, the digging activity of bears affects plant distribution, species richness and C:N ratio, modifying significantly the life traits of the glacier lily and producing changes in the land mosaic. These authors found a higher concentration of ammonium-N and nitrate-N in disturbed rather than in undisturbed adjacent sites (Fig. 3.8).

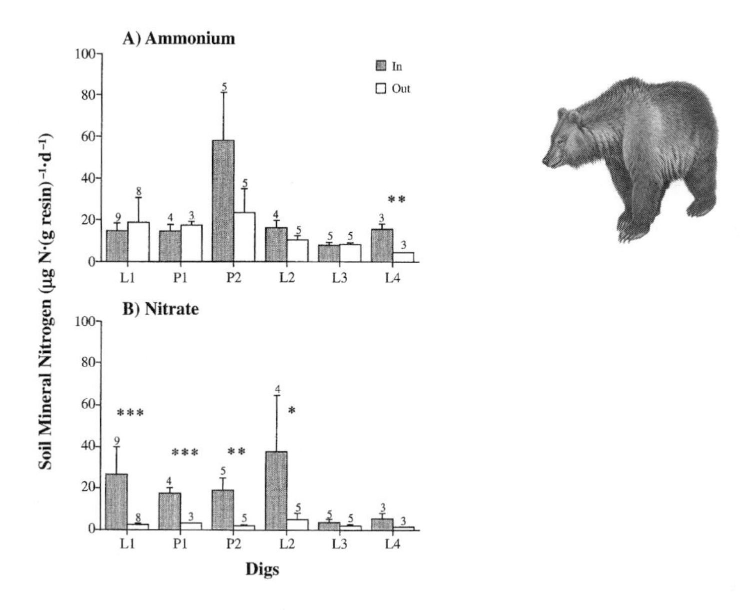

Fig. 3.8 - A comparison between the ammonium and nitrate contents of soil of 6 plots in Logan Pass (P) and Preston Park (P) relating to digging by grizzly bears (In) and adjacent undisturbed meadow (On) (Asterisks indicate significant differences within individual digs (*P<0.05, **P<0.01, *** P<0.001; t test) (from Tardiff & Stanford 1998).

The positive effects of digging have been evaluated based on the number of seeds produced by mature lilies which have escaped bear feeding. The large amount of bare soil in the digging patches favors seed establishment better than in adjacent meadows. In order to reduce biases due to patches rich in N "per se", or to avoid the effect of fertilization by bear manure deposits, experimental digging was conducted in adjacent areas. Again the same significant increase of N and a vigorous growth of disturbed lily were observed. Short and long-term effects on sub-alpine meadows by the digging activity of bears are important for determining the complexity of such montane mosaics. In the past, the large extent of grizzly bears across different landscapes probably had important consequences for plant distribution and abundance. Similar effects can be expected from the digs of wild boar (*Sus scrofa*) in Europe, but in this case the expansion of this species is expected to increase the disturbance of montane prairies and meadows. Most of these montane grasslands are the result of woodland clearing and, when abandoned by human stewardship rapidly evolve via secondary succession into shrublands (Farina 1991)(Figs. 3.9, 3.10). Actually, we don't know if the activity of wild boar will increase the speed of wood recovery or create new intermediate successional stages dominated more permanently by shrubs and forbs.

Fig. 3.9 - Effect of wild boar digging activity in montane prairies in the Northern Apennines.

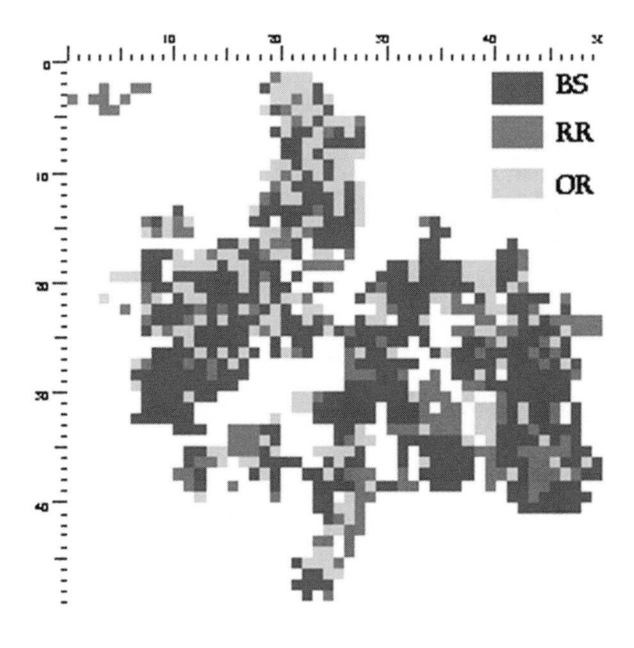

Fig. 3.10 - GIS of wild boar digging patches (grid size cm 25x25). BS) bare soil (present digging area), RR) area disturbed in the recent past (few months) and now recovered by vegetation, OR) area disturbed at least two years before and now with a well established vegetation. (from Guastalli 1998, unpl. dissertation).

3.8.2 Managing disturbance along river systems

Rivers are highly dynamic systems in which disturbances like flooding have a profound influence on the physical and biological components. Experiments were conducted by Wootton et al (1996), who changed the disturbance regime along a stream in California (Fig. 3.11). Through empirical observation during an extraordinary drought period (1990-1992 and 1994), the authors observed a marked increase in predatory resistant caddisfly, *Dicosmoecus gilvipes*. In the observed system at least two food chains existed; the first composed of algae and predators, grazers susceptible to predatory fish and insects, and the second comprising a range of algae to predator-resistant grazers. This multitrophic model responds in different ways to flood disturbance with the predator-resistant grazers reacting more negatively than the predator-susceptible grazers. A reduction of flood disturbance should increase predator-resistant grazers. This indirectly produces an increase of pressure on algae abundance and in turn a decrease of predator abundance because the food chain becomes more fragile.

However, predator-susceptible grazers are not affected by the change in abundance of their competitors. This model has been verified experimentally and the following output has been found: regulated rivers have a higher occurrence of predator-resistant grazers, a significantly lower occurrence of algae, and a lower though not significantly lower abundance of predator-susceptible grazers compared with rivers with natural flow.

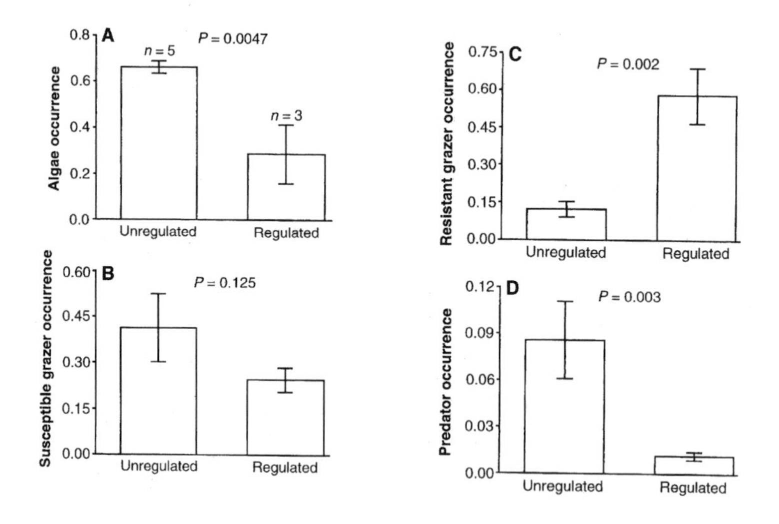

Fig. 3.11- Effect of unregulated and regulated regimes along a California river on predators, resistant grazers, susceptible grazers and algae (from Wooton et al. 1996).

These results are very important for conservation purposes, demonstrating that the autoecological approach is often not sufficient to solve the loss of diversity along rivers. Disturbance regimes are of more importance in management strategies than trying to solve the problems of each species separately. Disturbance regimes act, especially at the large scale which is in line with the regional landscape ecological approach.

3.8.3 Managing grazing regimes

In the past, grazing has been considered a disturbance, degrading vegetation cover with a negative influence on diversity. Actually, the recognition of its significance in controlling important cycles in the ecosystem is gaining popularity and the role of grazing in conservation management has been extensively reviewed (WallisDeVries et al. 1998). The importance of grazing in maintening plant and animal biodiversity. Most of the land mosaic is subjected to grazing but, according to the environmental context, it is important to graduate this grazing pressure in order to avoid negative effects such as soil erosion, soil trampling and nitrogen-dominant plants.

The grazing disturbance regime is characterized by a rapid cycling ecosystem and by a high level of spatial multiscalar heterogeneity. This system works very well with wild ungulates, but domestic ungulates are often true destroyers of such a system.

The removal of wild ungulates from rangeland generally produces negative effects on the natural vegetation which changes from a rapid cycling system into a detritivorous system with slow cycling. The natural grazing regime is based on nomadic or regular seasonal movement of grazers. Actually, wild rangelands are rare and becoming increasingly fragmented.

Grazing disturbance by livestock is a major concern in some areas where this regime has been imported and where ecological systems have no efficient defense for increasing their resilience. In these cases, grazing is the main cause of biodiversity degradation (see for example the grazing effects in tropical forests, or desertification around the Sahara).

Like many other disturbances, the effects depend on the ecological and geographical context in which the disturbance operates. The influence of the regional condition and the reaction of species are both relevant components in the process of grazing and related disturbances.

The impact of domestic grazers may seem controversial, but we have to distinguish the context in which this process is considered. In most of Europe, and especially in southern Europe, the grazing regime plays an important role in maintaining open spaces and highly diverse plant communities. Disturbance is of general importance for other processes too. There is common agreement regarding grazing management and the different experiences across Europe. For example, Bakker & Londo (1998) recommend that, for correct grazing management, it is important to practice grazing at the lowest possible stock rate for as long as possible during the year using breeds that need little care, and on as large an area as possible. These recommendations probably work well for The Netherlands situation as proposed by the authors.

A significant example of a grazing regime integrated with the maintenance of biodiversity can be find in the Maremma area (a geographical coastal region between Tuscany and Lazio, Italy). In this area reclaimed marshlands and Mediterranean maqui in the hills are grazed by "Maremmana" cattle. This species coupled with "Maremmani" horses maintain open spaces in poor sandy soil and under plantations of Italian stone pine (*Pinus pinea*). The Maremmana breed is not an indigenous breed (it was probably imported by central Europe centuries ago). This species does not need any care throughout the years if large wooded and open areas are available.

Grazing pressure modifies the structure and spatial arrangement of plants according to at least two orders of spatial scales: at an intermediate spatial scale (hundred meters), grazing pressure can alter the composition of vegetation communities; at a local micro-scale (few meters), it is the plant-to-plant selection which influences plant assemblages (Fig. 3.12) (Bakker 1998).

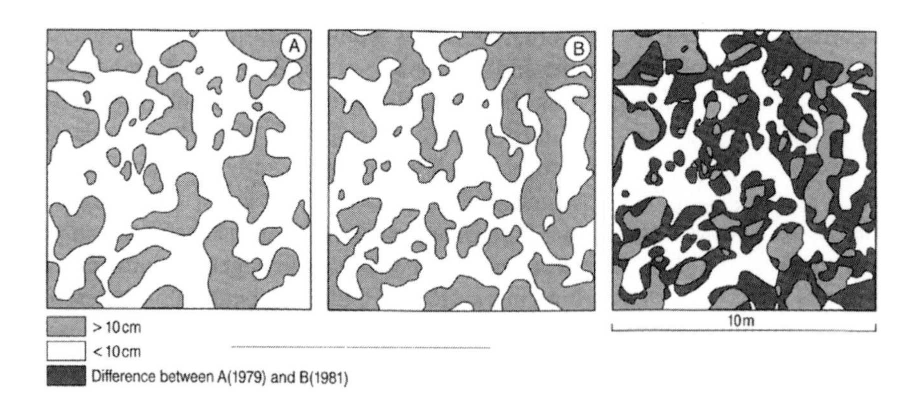

Fig. 3.12 - Comparison of the vegetation cover of a 10x10 m plot in Westerholt, The Netherlands in 1979 (a) and 1981(b), and the differences between the two years in which plant-to-plant selection by grazers has affected the distribution of plant assemblages (from Bakker 1998).

Prairies are endangered systems around the world. In particular, in the USA this environment has been dramatically reduced by overgrazing, agricultural transformation and urban development (Tab. 3.4). The decline of prairies is obvious, especially tallgrass prairies (Samson & Knopf 1994), and has caused modification of soil composition, surface and underground water circulation, and plant and animal distribution. This decline is accompanied by the loss of animal diversity, and by a general reduction in mosaic diversity. In fact bison, a key stone species in this system, produced fine scale modification of soil morphology by wallowing behavior, creating ephemeral micro-sites which are highly differentiated into the prairie mosaic.

Tab. 3.4 - Distribution of tallgrass, mixed grass and shortgrass prairies historically and contemporarily, and the % of decline and of protected areas across North America (from Samson & Knopf 1994).

		Historic (ha)	Current (ha)	Decline (%)	Current protected (%)
Tallgrass	Manitoba	600,000	300	99.9	N/A
	Illinois	8,900,000	930	99.9	<.01
	Indiana	2,800,000	404	99.9	<.01
	Iowa	12,500,000	12,140	99.9	<.01
	Kansas	6,900,000	1,200,000	82.6	N/A
	Minnesota	7,300,000	30,350	99.6	<1.0
	Missouri	5,700,000	30,350	99.5	<1.0
	Nebraska	6,100,000	123,000	98.0	<1.0
	North Dakota	1,200,000	1200	99.9	N/A
	Oklahoma	5,200,000	N/A	N/A	N/A
	South Dakota	3,000,000	449,000	85.0	N/A
	Texas	7,200,000	720,000	90.0	N/A
	Wisconsin	971,000	4000	99,9	N/A
Mixed grass	Alberta	8,700,000	3,400,000	61.0	<.01
	Manitoba	600,000	300	99.9	<.01
	Saskatchewan	13,400,000	2,500,000	81.3	<.01
	Nebraska	7,700,000	1,900,000	77.1	N/A
	North Dakota	13,900,000	3,900,000	71.9	N/A
	Oklahoma	2,500,000	N/A	N/A	N/A
	South Dakota	1,600,000	N/A	N/A	N/A
	Texas	14,100,000	9,800,000	30.0	N/A
Shortgrass	Saskatchewan	5,900,000	840,000	85.8	N/A
	Oklahoma	1,300,000	N/A	N/A	N/A
	South Dakota	179,000	N/A	N/A	N/A
	Texas	7,800,000	1,600,000	80.0	N/A
	Wyoming	3,000,000	2,400,000	20.0	N/A

Rangeland systems have similar patterns even if they are located in geographically different biomes, as demonstrated by a comparative study conducted by Frank et al. (1998) in Yellowstone National Park and the Serengeti Park. For example, the gradients affect the spatial heterogeneity of foraging. The spatial patterning of resources allows mobile animals to easily find the resources they need whilst, at the same time reducing local pressure due to their high vagility, assuring a natural rhythm to vegetation (Fig. 3.13). Grazers promote vegetation growth and a precise relationship explains the highly density of grazers in highly productive Serengeti rangeland compared with the less productive system of Yellowstone Park which is characterized by a modest number of ungulates.

Large parts of the Great Plain in the US for instance, have been negatively affected by cow overgrazing during the past century but the effect has not been the same for all parts. For instance, in the grasslands of New Mexico the overgrazing regime has transformed the grass dominated prairie into a *Larrea* and *Prosopis* shrubland, but in the northern shortgrass steppe this has not happened. Lauenroth et al. (1997) argued, this probably depends on the different species of grasses: *Bouteloua gracilis* lives in short grass areas and *B. eripoda* in the desert. *B. gracilis* is a long-lived species whilst *B. eripoda* is a short-lived species with frequent recruitment and so in more prone to the disturbance regime of grazers.

The same is true for the Mediterranean region where grazing regimes date back thousands of years. Uplands especially, have been affected by heavy grazing regimes with consequences also for soil stability. There is a common view about the effects of grazing in such a region. The first effect is that soil deprived of vegetation is exposed to erosion and in some parts of the Mediterranean this has been the case. The second effect on vegetation finds a consensus in the active role of grazing for the maintenance of high local diversity with an inversion of the paradigm that grazing depresses diversity (Perevolotsky & Seligman 1998). The main concern was about the role of goats which also influence shrub and tree growth and regeneration. The resilience capacity of Mediterranean vegetation is very high and plant communities have the capacity to react promptly to fire, grazing, trampling and manure deposition. These disturbances, which share the accumulated resources, increase the diversity and complexity of the system. An overgrazing regime maintains open spaces and reduces plant competition, developing highly diverse communities resistant to plant invasion. When the grazing regime is abandoned a thick impenetrable maqui develops. Such cover is very prone to fires and the control of this risk appears impossible.

Fig. 3.13 - Depending on the species, livestock have different effects on plant assemblages. Cows (National Park of Abruzzi, Central Apennines, Italy) (A), horses (Colorado, US)(B) goats (Orsaro Mt., Northern Apennines, Italy) (C), and sheep (Arizona shrubland, US) (D) create different levels of disturbance regimes.

The evaluation of the degradation status largely depends on the logic with which we observe the phenomenon. For a forester, the browsing of cows and goats of shrubs and trees is considered damaging, but for a shepherd this is an important available biomass. The reaction of shrubs and trees to grazing disturbance consists of an increase in thickness creating new important breeding niches for birds. Throughout the Mediterranean, the upland areas are the most affected by the grazing regime. These areas may be very deep or may have terraced slopes. In different historical periods cultivation was widespread (to the top of low mountains (1300 m) in the Northern Apennines, for instance) but such cultivations which have probably linked to a period of high demographic pressure have been substituted by periods of lower human pressure and abandoned. In all periods grazing has been the cause of "matrix disturbance" and plant adaptations are quite evident.

The alternation of periods of neglect with periods of intensive use has especially characterized the entire Mediterranean basin. The new view of the environment incorporating low human intervention has created difficult situations for management. The reinstatement of pine to open upland rangeland has been a real ecological drama.

Grazing is the main controller of vegetation structure. When the grazing regime is abandoned a very dense shrub cover develops, dominated mainly by a few species such as *Quercus* sp., is considered to be a "green desert" (Naveh 1971).

Erosion in grazed rangelands is often not so evident because vegetation has adapted so well that there is a larger root system in the soil of grazed areas than in that of ungrazed areas. In fact, in many mountains in the Northern Apennines (Italy) landslides in grazed regimes are less frequent than in ungrazed (abandoned). Grazed areas are hot spot sites for biodiversity. This is particular evident along the migratory routes of trans-Saharan migrants (Farina 1987, 1988). Grazed uplands attract many migratory birds either in Spring or in Fall time. These open spaces are rich in seeds which animals distribute on the soil surface or in grassy vegetation. Human stewardship coupled with moderate grazing disturbance by livestock, are the ingredients necessary to produce resources for insectivorous and granivorous birds.

A seasonality in bird assemblage is observed; in Spring time insectivorous (*Anthus trivialis, Anthus spinoletta, Anthus pratensis*) and frugivorous species (*Turdus philomelos, Turdus iliacus*) are the prevailing species whilst in fall time, granivourus species are dominant (*Alauda arvensis, Fringilla coelebs, Carduelis carduelis, Carduelis chloris*).

Grazing lead to selection within plant communities based, on the specific palatability of individual plants, reducing the dominance of the more aggressive species. The intermediate disturbance regime which can be considered important for ensuring high levels of richness in many biomes, seems less applicable to the Mediterranean where only a high level of disturbance can reduce the quick recovery of dense bush cover. The development of a dense topstory cover prevents the development of the ground layer. Considering that few species occupy the topstory layer, the loss of plant diversity due to the thickness of tree biomass can be considered to be a paradoxly negative effect. Potential productivity does not seem to be affected by overgrazing anywhere in the Mediterranean. This effect can be observed, for example, in the rapid closure of clearances in the beech forests of the Apennines. In just 50 years, these forests have recovered both the grazed clearings and the charcoal "piazze" used to transform wood charcoal. This great "vigor" of the Mediterranean region can be observed also in urban fringes, fragility and resilience are both important. Dominant plants are generally unpalatable and this create a strong resilience of the vegetation and contribute to the plasticity of the ecosystem, where plasticity is intended the capacity to maintain active functions. Land abandonment has created large areas prone to fires, and has caused serious problems to fire control.

In the Mediterranean basin, shrub encroachment can be controlled only by using a heavy grazing regime, although in other Mediterranean climate type ecosystems this practice must be lighter to avoid a depression in diversity. In Australia and California grazing regimes must be lighter to reduce the negative effects on fragile vegetation. This fragility largely depends on the recent story of grazing disturbance by large ungulates. In conclusion, undergrazing appears, today, to be a threat to the Mediterranean landscape when compared with grazed or in some cases overgrazed regimes.

3.9 Managing functional areas

3.9.1 Managing human dominated landscapes

To managing landscapes as perceived by humans, it is necessary to understand the importance of the processes that drive their dynamics (Baker 1992). This is not an easy task because it is difficult to separate the emerging processes responsible for land structure and dynamics of the land. For this reason, it seems important to sort relevant processes according to their effects on the considered landscape. It is a general rule that a process is distinguishable from a pattern due to the dynamism that is a property of the process. To distinguish a process from a pattern it is necessary to scan the phenomenon into a space/temporal window. From this perspective, studies on landscape (land-use) changes assume great importance. The study of modifications occurring in a mosaic helps us to understand the key processes responsible for the modification and allows us to build predictive models that can be usefully employed for management purposes. A landscape at the human scale can be considered a dynamic mosaic of patches created by human as well as natural disturbances. Within such landscapes, human processes such as the economy can play a more relevant role than natural ones. It is important for a general evaluation of human landscapes to understand the role of the main processes acting on the area in question. Generally in most human shaped landscapes, natural and anthropogenic processes interact and overlap in different ways. Land use such as forest or cultivation is decided in a human landscape by a hierarchy of natural and human constraints. Slope and aspect are important as well as the geological composition of soil, but distance from roads or from market centers is important too.

In Europe the story of the human landscape is more strictly related with a long term evolution of populations which in turn has affected the landscape according to temporal needs. If we exclude large scale reclamation activities, such as along the Maremma (Italy), land reclamation has not been a short term, intensive activity, but a long term, diffuse, weak, and continuous stewardship of individual farmers. This reality, common to many countries, is rare on the new continent (North and South America) where large scale planning has produced dramatic changes in the land mosaic in just a short time (from tropical areas to boreal forest on the north-western Pacific).

The landscape approach must consider all these factors which deeply influence the structure and functioning of human landscapes. In north America, private and public land are both important components of the land mosaic whereas in Europe, though most of the land is private, it is governed extensively by public rules. These differences are fundamental for distinguishing the effect of natural and human processes.

It is not hard to assume that human rules have effects on the landscape, as in the ecological conditions of different eco-regions (Fig. 3.14). I believe that the value of a land may be the first indicator of land use and land change. This value is the result of the attractiveness of an area for agriculture, urban settlement, industrial settlement and recreation. The economic value is the main force that today drives the fate of a landscape. Proximity to urban centers, or wildlife attractiveness both have the capacity to modify the land mosaic. Also the presence of road and railway infrastructures are powerful modifiers of a land mosaic. The evolution of a landscape is strictly liked to local situations (microclimate, morphology, local history and medium scale economy), but this evolution has two general, distinct patterns common to human dominated landscapes overall: fragmentation and aggregation (defragmentation). For the local case (see f.i. Turner et al. 1996, Wear et al. 1998) we can consider elevation, slope, distance to road, distance from market and population as indicators of change from forest to grass, forest to unvegetated, grass to forest, and grass to unvegetated, etc.). A deep relationship between the economic and social structure of the territory and the natural components exists. Human presence forces short term changes, on the land mosaic more than in the past, and these changes profoundly influence the composing ecosystems.

3.9.2 The ecosystem service concept

The functions of ecosystems can be approached according to a human oriented, utilitarian view and expressed as a service. The ecosystem service is a process by which nature produces food and incorporates waste disposal and multi-scale disturbances (Costanza et al. 1997). This service is of fundamental importance for life, particularly for humanity. Ecosystem services, range from gas regulation (CO_2/O_2 balance) to cultural services in which ecosystems (landscapes) are fountains of aesthetic, artistic, educational, spiritual and scientific values. Each of these services act at a specific family of spatial and temporal scales, ranging from few centimeters to the ecosphere overall.

Tab. 3.5 - List of ecosystem services (from Costanza et al. 1997).

Ecosystem service	Ecosystem functions	Examples
Gas regulation	Regulation of atmospheric chemical composition.	CO_2/O_2 balance,O_3 for UVB protection, and Sox levels.
Climate regulation	Regulation of global temperature, precipitation, and other biologically mediated climatic processes at global or local levels.	Greenhouse gas regulation, DMS production affecting cloud formation.
Disturbance regulation	Capacitance, damping and integrity of ecosystem response to environmental fluctuations.	Storm protection, flood control, drought recovery and other aspects of habitat response to environmental variability mainly controlled by vegetation structure.
Water regulation	Regulation of hydrological flows.	Provisioning of water for agricultural (such as irrigation) or industrial (such as milling) processes or transportation.
Water supply	Storage and retention of water.	Provisioning of water by watersheds, reservoirs and aquifers.
Erosion control and sediment retention	Retention of soil within an ecosystem.	Prevention of loss of soil by wind, runoff, or other removal processes, storage of stilt in lakes and wetlands.
Soil formation	Soil formation processes.	Weathering of rock and the accumulation of organic material.
Nutrient cycling	Storage, internal cycling, processing and acquisition of nutrients.	Nitrogen fixation, N,P and other elemental or nutrient cycles.
Waste treatment	Recovery of mobile nutrients and removal or breakdown of excess or xenic nutrients and compounds.	Waste treatment, pollution control, detoxification.
Pollination	Movement of floral gametes.	Provisioning of pollinators for the reproduction of plant populations.
Biological control	Trophic-dynamic regulations of populations.	Keystone predator control of prey species, reduction of herbivory by top predators.
Refugia	Habitat for resident and transient populations.	Nurseries, habitat for migratory species, regional habitats for locally harvested species, or overwintering grounds.
Food production	That portion of gross primary production extractable as food.	Production of fish, game, crops, nuts, fruits by huntic gathering, subsistence farming or fishing.
Raw materials	That portion of gross primary production extractable as raw materials.	The production of lumber, fuel or fodder.
Genetic resources	Sources of unique biological materials and products.	Medicine, products for materials science, genes for resistance to plant pathogens and crop pests, ornamental species (pets and horticultural varieties of plants).
Recreation	Providing opportunities for recreational activities.	Eco-tourism, sport fishing, and other outdoor recreational activities.
Cultural	Providing opportunities for non-commercial uses.	Aesthetic, artistic, educational,spiritual, and/or scientific values of ecosystems

The concept of capital is generally linked to economic issues but it is also reasonable to use this term in a more broad ecological meaning. The ecosystem service consists of a flow of material, energy and information from the natural capital, combined with the human capital, and to support human welfare (Tab. 3.5). In such types of "service" we can consider the role of fymbos vegetation in South Africa. Fymbos vegetation, a typical Mediterranean dense shrubby vegetation, plays a fundamental role at the watershed scale allowing water recharge (van Wilgen et al. 1996) (Fig. 3.15). The storage of water is of great importance for human settlements positioned in lowlands. This vegetation is susceptible to alien plant invasion, especially when it is heavily disturbed. Trees like Austrian pine (*Pinus nigra*) are common invaders; their distribution affects the availability of water in the soil and they are easily burned by natural as as well as human induced events. The increase of tree biomass has direct effects on streamflow .

3.9.3 Constraints in human dominated landscape management

Landscape management (ecosystem management is also a popular term) requires much information on the social and economic structure of the land to add to physical and biological data. This scenario is complicated by the presence of different collections of landowners. When we approach a broad scale area we have to consider a high level of inherent uncertainty of ecological responses to human influence. It seems quite difficult to implement ecosystem management, not only due to the lack of knowledge of ecological processes, but also due to the optimization of individual, private firms and public agencies (Wear et al. 1996). Landscape management is often a compromise between different social expectations, especially in areas dominated by human intervention. In addition public lands don't have the capacity to maintain biodiversity at adequate levels. In fact, public lands are often marginal areas with low soil fertility and hard climatic conditions. It is not the case that most natural reserves and parks are located in wild, remote, mountainous areas where productivity is strongly limited seasonally. The capacity of such areas to preserve a full range of biodiversity seems extremely poor, despite popular assumption that such wild, remote areas are important.

As recently pointed out by an IUCN workshop (Halladay & Gilmour 1995), "*significant elements of biodiversity are found outside the protected area system. Traditional agro-ecosystems are particularly rich sources of both biodiversity and indigenous knowledge about its management*". The value of traditional agro-ecosystems has been recognized as a source of biodiversity although such a system is often undervalued when economic and social worth are considered. Often, government policies favor the degradation of such agro-ecosystems, providing "perverse" subsidies to prevent the input of energy essential for stewardship maintenance. In such a way the management debt increases with devastating effects on landscape complexity.

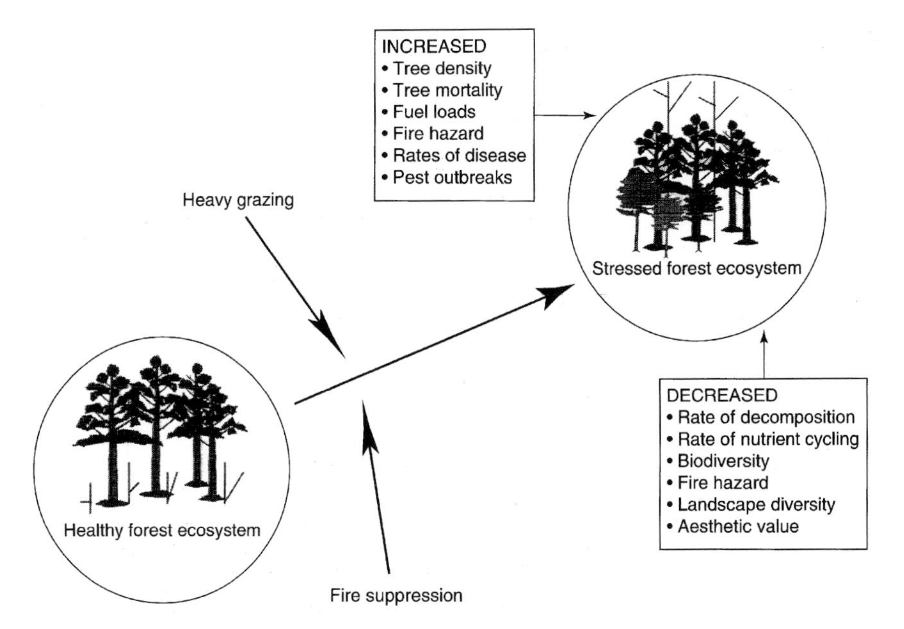

Fig. 3.14 - Example of changes in the Ponderosa pine ecosystem according to changes in the disturbance regime (Rapport et al. 1998).

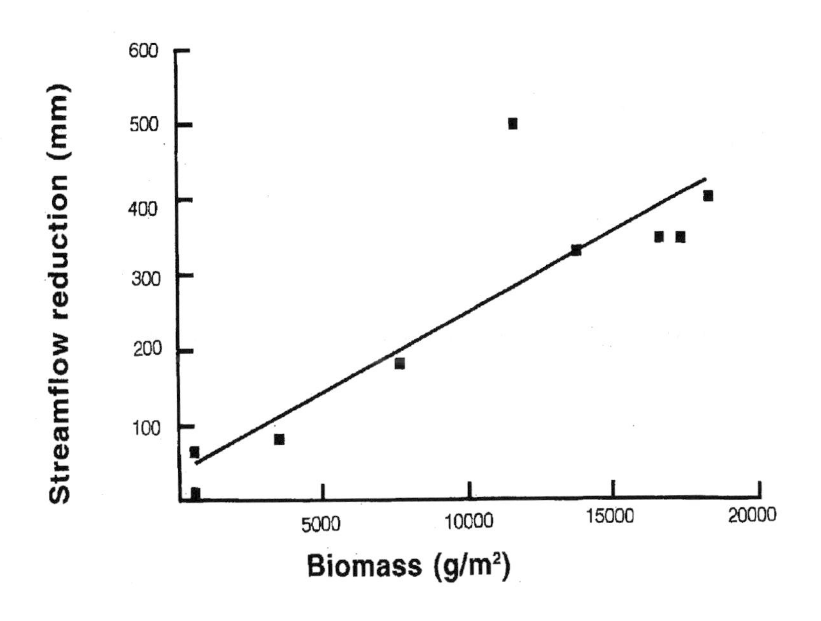

Fig. 3.15 - Effects of alien plant invasion, on streamflow within catchments (Western Cape, South Africa) (from van Wilgen 1996).

3.9.4 Ecosystem-based management (EM)

Within the conservation agencies, a new more integrated (holistic) approach based on the conservation of the ecosystem's functions is growing popularity. IUCN (1980) and WWF (1998), two major NGO agencies working at a world scale for nature conservation, have recognized the necessity to implement such an approach. Though, as stressed by McNeely (1999), there is some degree of confusion regarding the terminology used to indicate such an approach (e.g. ecosystem management, bioregional planning, ecoregion-based conservation, ecosystem-based approach, bioregional approach, integrated conservation and development projects, biosphere reserves, watershed management, etc.). However, despite this "plethora" of terms there are more common points than differences.

I'll try to clarify and implement the ecosystem-based management approach because it seems to be a strategy that utilizes the paradigms of landscape ecology more extensively and that doesn't have obvious contradictions with the hierarchical, multiscalar view of complexity. Ecosystem management has been conceptualized by Grumbine (1994) as follows: "Ecosystem management integrates scientific knowledge of ecological relationships within a complex sociopolitical and value framework, towards the general goal of protecting native ecosystem integrity over the long term".

Ten main themes of ecosystem management are considered by Grumbine (1994): Hierarchical context, ecological boundaries, ecological integrity, data collection, monitoring, adaptive management, interagency cooperation, organizational change, human's and their interactions with nature and consequent values. The goals of EM are: to maintain viable populations of all native species, to represent within protected areas all native ecosystem types, to achieve a sustainable land use, to maintain a balance between development and conservation, to maintain ecosystem functioning, to encourage integration at different scales of target processes, to manage for enough time enough to preserve the evolutionary potential of species and ecosystems, and to steer human use and occupancy according to ecosystem management guide lines.

Although the term ecosystem in this case is used as an extension of Tansley's original meaning (Tansley 1935), in effect the conceptual framework is respected. With such an approach, functions are considered more than individual species to be conservation and management priorities. The inclusion of a human sphere in such processes has been recognized to be a fundamental and integral part of ecosystems (UNEP 1998). The recent introduction of the concept of Ecosystem Service (Costanza et al. 1997) has increased the capacity of this approach to interact with the real world and to open a communication window with society and, in particular, with the stakeholders and the decision-makers.

Nutrient cycles, energy flow, water flow, disturbances, and processes connected with species and their aggregations are all considered. Adaptive management is recognized as a fundamental action for maintaining information control on the targeted system and monitoring fulfills a major role of verifing the feedback between processes and actions.

In this action human processes are important and their manipulation can be achieved through political procedures, such as a stakeholders consultancy, which decide goods and services that must receive priority, and in which way the costs and benefits have to be supported by the different components of society. The partnership of these different societal components involved in the decision process is fundamental to it's implementation. Cooperation and negotiation are two key stones in this strategy, in which the multi-cultural aspects of natural and man-made objects must be considered as one cause of controversial and delayed decisions.

Ecosystem-based management has to face some real constraints and difficulties due to the different temporal perspectives of decision-makers and ecosystem processes. The former acts over the short- term only whereas the latter requires long-tern action. In this conflict, the role of science is fundamental for the identification of indicators which are appropriate for describing both components.

Ecosystem-based management deals with biodiversity conservation and the ecological integrity of systems whilst focusing also on ecological boundaries. Biodiversity conservation assures the buffering of perturbations by the presence of the redundant actions of a rich species assemblage. In addition, biodiversity may be considered as a part of the "engine" that ensures the main ecosystem services. The definition of boundary in such an approach assumes a strategic priority. Often, there is inconsistency between human processes and natural borders, and the definition of functional units can categorize the real importance of these concurrent processes in different way. However, if we localize the functional units on the basis of the problems which we are facing and not on an "a priori" defined map, some of the proposed difficulties can vanish.

The ecological integrity of a system is recognized as the main goal for ensuring sustainable resource use; it should be respected and advanced via an integrated approach, though it is difficult to directly assess this goal since it requires indicators such as biodiversity which can be monitored more easily.

To transfer this approach into practice, landscape ecology provides fundamental tools under the umbrella of investigation and assessment procedures, and in terms of geographical definition. Ecosystem-based management can be utilized to solve problems from the small scale (e.g. stream catchment, parish) to the large scale (e.g. an entire watershed, provinces).

There are many hierarchical levels at which this approach can be applied, and for each level, boundaries are selected primarily according to an ecological criterion. The definition of a bio-region (sensu McNeely 1999, see also Miller 1999) assumes an important role only if connected to real problems that must be solved. This bio-region is defined in the same way as we define an ecotope. In fact there is no contradiction between these two terms when we compare the definitions of the procedures.

Integrity across a landscape is a well used concept in conservation policies and can mean different things according to different human perspectives and cultures. Ecological integrity is, again, a concept that can be applied better in areas where human presence is recent or in which human impact has been secondary compared with natural processes. Integrity means the presence of a set of viable populations, the maintenance of ecological and evolutionary processes, a low human impact, and the capacity of a system to recover from such an impact. However, this concept is strongly conditioned by the cultural model that we are using.

The new ecosystem-based management considers human presence as an important force that, when implementing an appropriate stewardship, reconciles the psychological separation of humans from the environment.

This approach moves from an exploitation-based to a conservation-based land use in an extreme attempt to fill the gaps of industrialized societies that are "peripherally informed" on ecological matters (Grumbine 1990).

Large scale conservation must change it's goals and perspectives when changing from a restricted area of management where conservation strategies target some biotopes or species only. It seems important to shift from focussing on a few species to focussing on all species, and from short-term, ephemeral conservation plan to long-term conservation strategies. At the large scale, large landscapes incorporate many natural as well as human processes, and it is their integration and their "sustanaibility" that must be achieved. We have to recognize the resource inter-dependence and we have to provide a more holistic decision-making framework.

The last part of the 20th century has been characterized by a growing concern about the depletion of natural resources with an evident shift in human behavior. The activity of local populations and their responsibility for environmental care at a large scale is a very hard task, and is often frustrated by concurrent goals like local social and economic welfare.

Direct policies on the environment are demanded from large-scale administrative, bureaucratic entities. The complexity of the environment requires actions across many scales and the local scale is as important as the large scale. Small scale policy is short-term scaled and can be controlled by direct feed-backs whereas large-scale policiy requires a long-time perspective and must be organized by highly ranked political entities moving beyond national borders. We have to recognize that the complexity of the natural environment and our limited knowledge of the controlling mechanisms maintains uncertainty around ecological issues, but new approaches like landscape ecology can forward and improve our capacity to interpret and act in the management realm.

3.9.5 Managing parks

Parks can be considered as biodiversity hot spots but also arenas in which ecological management (ecosystem-based management) can experiment. Most park management strategies focus on finding targets inside the administrative borders. Concepts like biodiversity, ecological integrity and ecosystem health are becoming popular in the recent development of new management strategies.

The initiative of "Parks Canada" (Canadian-Heritage 1996) of opening new perspectives to integrate natural processes with human processes by adopting procedures that are in effect the cultural basis of many European countries seems particularly relevant.

The integration of parks and protected areas into the surrounding landscape is required in order to offer and receive benefits from the exchange of different policies. The view of parks as sources of naturalness sounds good but only in some contexts. Again, strong differences emerge when we compare the environmental policies of Europe and North America. In Europe, the long history of human interaction has created an ecologically permanently human-disturbed matrix from which biodiversity hot spots spring by diffuse autopoietic mechanisms, and in which nature is an integral part of the human landscape.

3.9.6 Watershed management and water quality

As recently stressed by Wear et al. (1998), the use of the land by humans has profound and immediate effects on the quality of water. It seems important to understand the mechanisms that organize human behaviour and to relate these at a proper scale with the natural processes that interfer with them. In particular, water quality seems an important component of the human environment and the conservation of such quality assumes a special importance for environmental policy. By applying landscape metrics, such as dominance and contagion, it is possible to relate water quality with the characteristics of a landscape. The spatial organization of land cover as measured by dominance and contagion seems important for water quality.

In general non-forest cover has severe effects on water quality. Wear et al. (1998), suggest a non-forest index (*NFI*) where Yij is a binary equal to 1 when cell i,j lies in non-forest cover (and 0 when forested), dij is the distance from cell i,j to the stream, and $Ymax$ is the maximum possible value of the sum in brackets (if all cells are non-forest). This index ranges from 0 when the effects of non-forest cover are nil, to 1 for maximum effects. Water quality should be inversely related to this index.

$$NFI = \left| \sum_{i=1}^{n} \sum_{j=1}^{n} \left(\frac{Yij}{dij + 1} \right) \right| / Y \max$$

A great sensitivity has been found in water quality relating to different "landscape signatures". Just minimum changes in land cover quality (moving toward urban cover, increased paved roads, etc.) can greatly affect water quality.

In areas that experience land cover changes, as in the area studied by Wear et al. (1998) in the southern Appalachians, the impact of urban-rural change is apparently not based on a simple distance gradient from the urban center but shows at least two sensitive points (ecotones): one positioned very close to the urban area (0-100 m) and another at 500 - 400 m where land use is remote. This behaviour is very common in human dominated landscapes in which the major effects of land changes are localized in strategic positions according to physical as well as biological and human constraints.

In a hypothetical human landscape, the interactions between human and natural processes are placed, not along a gradient but in unique locations determined by local characteristics and events. Changes can be patterned in a theoretical way like circular waves where each wave is the result of interference between human and natural processes. For instance, in a gradient between urban and forest areas, an intermediate rural area is the general rule. However, the density and shape of the waves (ecotones in action) are determined by local constraints. Models that can be considered in theory isodiametric, modify their density and shape according to local conditions.

The location of ecotones, the areas in which land change is at a maximum, have a particular importance for water quality. In fact forestry, intensification of agriculture and new urban developments are stress factors which reduce nutrient cycling, modify temperatures and increase acid deposition in water bodies.

3.10 Conserving the landscape

3.10.1 General principles

Conservation is an explicit attitude of developed societies, but must assume a more important role also in countries where the development of human communities is not enough to guarantee food, shelter and a healthy life. Today, conservation should be scaled according to different human priorities; it is not possible to conserve nature or cultural landscapes *per se*, especially in regions in which the human status is critical for famine and diseases.

The multipurpose of conservation should be integrated into the concept of sustainable development. This is a fundamental rule that must be respected, but it is not the only one.

Landscape conservation assumes a central role in the modern ecology of nature conservation (Brandt & Agger 1984, Haber 1990, Antrop 1997, Farina 1997b, Phillips 1997, Naveh 1998). If the landscape is a scaled area in which we can observe the processes in which we are interested with the best of information then consequently this "scale" should also fit the needs of conservation.

The goal of landscape conservation is primarily to :

Protect processes that are the driving forces affecting organisms, their habitats and the landscape.
Protect the landscape for organisms and habitats.
Protect habitats.
Protect organisms

Each protecting action must be scaled according to the subject selected, producing an ambivalent result: high efficiency achieved towards the target subject, low efficiency towards the development of benefits for subjects outside the selected target.

Among relevant processes that we have to conserve, disturbance assumes a primarly importance. The abiotic and biotic disturbance regimes that dominate all aquatic and terrestrial ecosystems are now considered to be not only human-related processes but normal processes of the environment and basic drivers which ensure diversity and a shifting mosaic in landscapes.

For instance, the plant community richest in species is the one in which disturbance is high and resources are scarce and patchily distributed.

Competitive species spread when nutrients are abundant, and stressed species are present when nutrients are scarce. It is important in any conservation strategy to first consider the context in which processes operate before any action is taken. For instance in disturbance regimes, if the disturbance is man-made then the remnant undisturbed patches are generally rich in species. In contrast, in a naturally disturbed matrix the undisturbed patches have a lower diversity.

Every conservation practice, as a general rule, must be arranged according to a very broad spectrum of conditions and goals. Conserving for conservation's sake is generally a nonsense, we have to conserve for production, for health, for biodiversity, for amenity, etc.

Every conservation action requires a differently scaled landscape. Today conservation seems to be the way to correct the direction in which social and economic "developments" are moving. In the past, landscape quality was often incidentally assured after food production and the custodial role was processed by people living in the countryside.

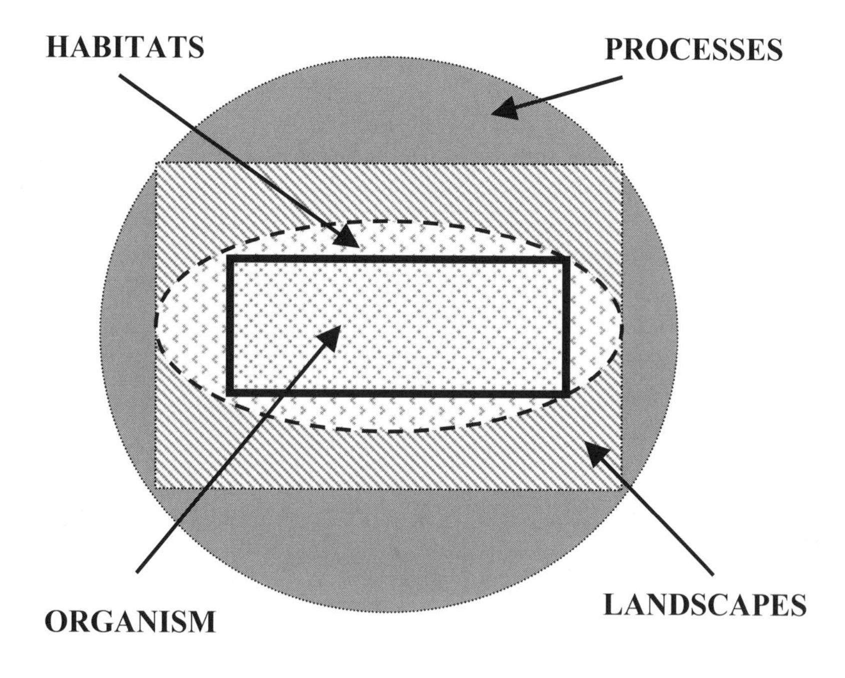

Fig. 3.16 - Hierarchical sequence to nature protection. Processes should represent the primarily objective in protection activity, followed by landscapes, habitats, and lastly by organisms. This sequence can be changed when one or more of these elements is not efficiently localized. Dashed lines indicate a weaker relationship between the different levels.

The principles that must be used for conservation are quite general and I'll try to balance the different philosophies and methodological approaches. The geographical component of a landscape is relevant in that different areas of the earth need different conservation regimes.

To be efficient and of long-term effect, these regimes should be completely incorporated into the social and economic perspectives, balancing the different actions that ensure human welfare and social and economic development. There are exceptions to this general statement as in remote, wild areas where human intrusion may be limited in space and time by precise, strict rules or by an efficient educational process. Conservation is the opposite of destruction, or of the destructive use of resources. Many resources are not renewable, for example like fossil oil and gold, but most of the resources necessary for our life are renewable, as in water, heat, vegetable and animal biomass.

Conservation action from a landscape perspective needs a well oriented approach that ensures the maintenance of some relevant characteristics as key elements of life. Again, conservation action should be detailed according to the four landscape perspectives (processes, geo-botanical, animal and human). Often, conservation is spoken of by use of the second and third perspective, creating conflicting interaction within human dynamics. These dynamics are often considered a developing process because the results are often unique and new compared with those last achieved. In the future it should be the fourth approach (human) which fits the best conservation efforts, but this is still a long story that needs further serious developments to achieve an integrated result.

Conservation means the *status quo* of a well defined situation for which we recognize an important role and/or value. It may be used to reverse a situation or to progress towards a new situation. Static conservation is very difficult to accomplish and is an ecological nonsense, but in some cases such as in the agricultural landscapes, this can be done via intense stewardship passing through crop rotation, tree pruning, and forest logging, that repeats the distribution of patterns and constantly maintains the effects of some successional processes

Some guidelines can be described for conserving biodiversity across a spatial scale but there are no fixed rules which encompass all purposes (see e.g. Norton& Ulanowicz 1992). According to the context in which we are operating approaches and actions must be accurately refined and regularly tested.

Large patches of natural vegetation are necessary to protect species richness for interior as well as for large home ranges (Forman & Collinge 1996). The total number of species in a patch is a good indicator of patch-suitability, but it is not enough for a conservation action for which other information is required.

3.10.2 Nature conservation: Criteria which fit the landscape paradigm

The multifunctional space represents the landscape *sensu strictu*. Landscape ecology should be one of the keys to open the door of environmental complexity and to penetrate deep into the mechanisms that create the conditions of a persistent life.

Landscape evaluation is quite a difficult procedure (see part II). I'll try to explain such an approach according to the different view allowed by landscape ecology.

Size (Extent)
Large areas have more heterogeneity, more organismal diversity, a lower edge effect and more intrinsically controlled processes.
In large areas, the disturbance mechanisms needed to ensure diversity and healthy conditions can work without the effects that appear when a small area is considered. A shifting mosaic seems a reasonable model to ensure species turnover, nutrient cycle efficiency and resource renewal.

Spatial arrangement
The position of reserves compared with neighboring protected areas seem important. When large reserves are not available a cluster of small protected areas could partially surrogate the fragmented dimension. In this case, the landscape seems an important criterion to use in the real world and corridors, core areas and meta-population models can be used extensively.
In the choice of conservation criterion, the distance from human dominated landscapes occupies a lead position. Often this is more a dogmatic than an ecological criterion. Remnant patches of lowland vegetation often have more species than remote patches, which are generally relegated in less productive uplands.

(Mosaic) Diversity
The diversity criterion is not popular in reserve creation decision procedures because stability and diversity are not easily related concepts. Nevertheless, a diverse mosaic can support a richer assemblage of plants and animals. In many regions such a mosaic pertains to the cultural landscape and, especially in the past, this condition was not in line with conservation criterion.

Rarity
The rarity criterion is especially important when considering the fact that rare species represent uniqueness and contribute greatly to the diversity of a region. Habitat rarity, patch rarity, mosaic rarity and landscape rarity have equal

importance in conservation measures. Rarity may represent a chance for diversity to be maintained when large scale or global changes occur, affecting the survivorship of common species. The role of rare species is important as a "source" of genotypes.

Typicalness

Moving to a different condition, the typicalness criterion seems as important as the rarity criterion. In fact according to this criterion the conservation of typical habitats, mosaics and landscapes that are often digressing far from "classical" typicalness, is of great importance.

For instance, the Tuscany landscape has typical patterns that can be found only in a few places after land abandonment or changes in land use.

Fragility

Fragility is a weak criterion because the conservation of an unstable system is merely a contradiction. Fragility is the attitude to change and a highly dynamic system is fragile. So, attempting to interrupt such process is an ecological nonsense.

3.10.3 Conserving hotspot biodiversity

The term "biodiversity hotspot" was coined by Meyers in the late 1980's. This term is used to indicate a geographical space that contains a particularly rich assemblage of species, or that preserves rare and/or endemic and threatened species.

Hotspots may be used either for localized mega areas in which species richness is very high, as in Brazil, Colombia, and Indonesia, or in which the number of endemic species is very high, as in Madagascar and Australia.

The use of taxonomic groups to identify hotspots is not often successful and little correspondence can be found. Of course, at a fine scale the hotspots of a taxonomic group do not overlap with other groups. In this case, it is important to integrate the geographic and taxonomic scales of resolution in order to properly use the hotspot paradigm. Higher-taxon diversity can be usefully employed as a surrogate for species diversity at a coarse as well as at a fine scale; 99% of bird richness variation across North America can be described by genus richness and 91% by family richness.

Richness at the continental scale can be assessed with similar results either by counting the number of families, by proportional family richness or by proportional family richness weighted for total species of each group.

At the coarse scale (continental), the hotspot strategy seems an efficient way to discriminate between non-overlapping taxonomic groups with a global conservation priority.

At the fine scale, the hotspot strategy seems more challenging because different taxonomic groups very seldom share the same geographical area. In this case a decision as to which group we must focus on needs to be considered. Following this procedure, rare species are often excluded by hotspot recognition because areas rich in species do not necessarily include those rare species living in restricted and peculiar areas.

Data for the United Kingdom and South Africa are concordant on this effect as the highest ranked hotspots share no more than 60% of the rarest species. This effect is scale dependent and so an enlargement of the geographical scale can capture such rarity.

The utility of hotspots is strongly linked to their capacity to localize the critical areas to conserve organism diversity. In the UK, hot spots covering just 5% of the entire territory contain 98% of British breeding birds. Similar percentages are shared by dragonflies, butterflies, and liverworts. In the US, 2% of hot spots contain 50% of endangered species of plants, birds, fish and mollusc.

Hotspots represent further evidence of ecological heterogeneity, in this case taxonomic heterogeneity. Such heterogeneity is due to the unequal distribution of habitats and their aggregations, land mosaics, and large scale landscapes. The hotspot procedure can be used in a landscape context considering the patchy distribution of richness at the same level as the vegetation patches in a geobotanical landscape.

This approach allows us to reconsider that species distribution is affected by functional heterogeneity and is not simply the effect of the geobotanical mould. By using this approach, it will be possible in the future to investigate species aggregation and the rules that create, maintain and modify such patterns. The concept of hotspots can be enlarged to also describe functions and landscape configurations. The recent recognition of cultural landscapes seems a very promising way to change the direction of environmental evaluation. This point will be expanded later in the section devoted to the cultural landscape but it is important to begin to consider that ecosystem functioning is a related, integrated and multi-process driven entity. Rare, threatened or unique landscapes, from human perceived complexity to remotely sensed complexity, are valuable components of the environment which need to be considered in any sustainable planning and ecologically related conservation plans (Reid 1998).

Landscape hotspots are often coincident with animal hotspots though this is not a universal rule (Fig. 3.17). In fact, landscape hotspots are often selected on the basis of scenic attributes that don't fit the biological attributes. Many natural protected areas have received attention mainly on the basis of their physical beauty, for example the Dolomites (Italy).

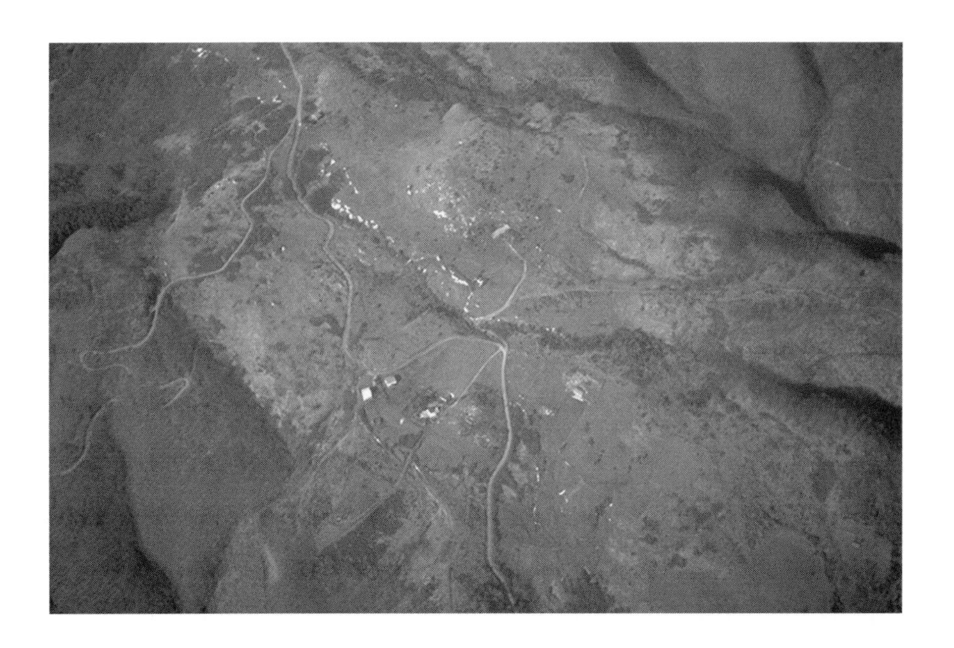

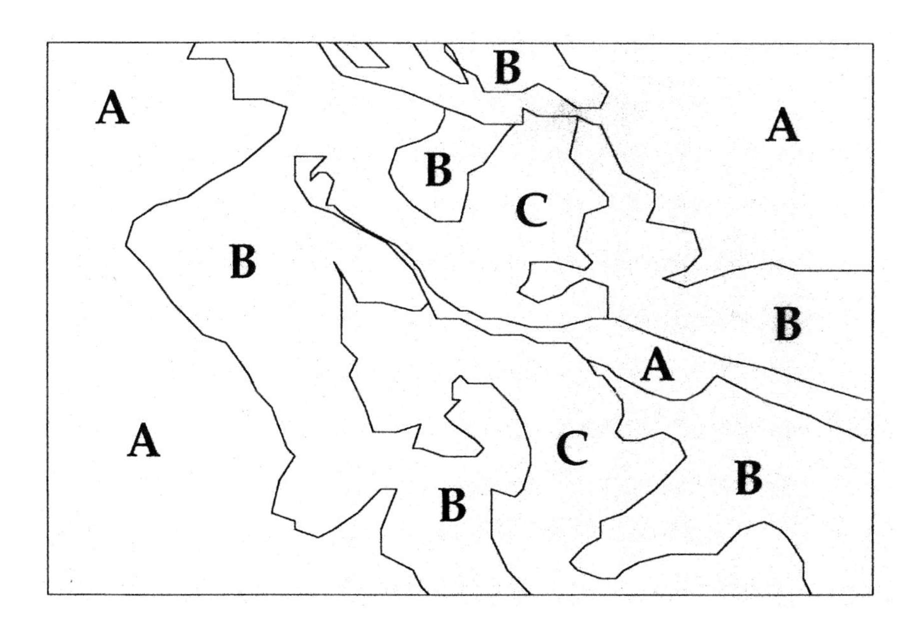

Fig. 3.17 - Sub-montane prairies are hot-spots for bird diversity in spring and summer time, when a complex mosaic of woodland (A), shrubland (B) and meadows (C), offers a great variety of resources (Logarghena, Northern Apennines, Italy).

3.10.4 Conserving riparian landscapes

One important element of a broad scale landscape is represented by riparian landscape rich in ecological processes that interact with the surrounding landscape (Naiman et al. 1993). This landscape has common characteristics shared with many biomes. First of all, it is a narrow strip of vegetation occupying different edaphic conditions compared with the neighboring area, and experiencing a different disturbance regime. High dynamism is a common attribute from streams to large rivers and such dynamism is responsible for the major changes in physical and biotic transformations. Coupled to this dynamism we find a very high diversity of organisms that, in different way's, receive benefits from the ephemeral conditions created by such a system. For instance, the riparian vegetation of western North America occupies less than 1% of all other landscape types, but has a higher bird richness than all other types of habitats combined (Knopf et al. 1988).

Actions to preserve this natural capital are necessary, especially at the large scale. Knopf et al. (1988) criticize local action when the importance of such landscapes is of nation wide dimension.

In many developed countries the riparian landscape is often the only remnant of a natural asset in a deeply modified broad landscape. In Padanian valley (Italy), river landscapes are the only natural relicts of past lowland forested landscapes. In many cases, river landscapes play a fundamental role as stop-over foraging sites for most trans-Saharan migratory birds in spring (Farina 1993b). In Autumn and Winter, riparian landscapes offer highly diversified resources from seeds to animal items for a very complex foraging guild of birds (Fig. 3.18).

The presence of vegetated linear habitats along rivers represents a good possibility to preserve biodiversity but also erosive processes and the nutrient cycle. In fact, riverine vegetation ensures the control of stream-bank erosion and reduces the loss of nutrients from soils to water. Roots and dead branches ensure protection and represent a real habitat for many species of fish. These linear habitats also ensure free movement for many species of organisms. The presence of vegetation shading the water reduces the temperature fluctuation ensuring more cold water suitable for organisms.

Rivers are systems with high levels of spatial and temporal complexity. Their human induced disturbance is the main concern in most countries, and a high priority is given to the conservation and restoration of such systems (Schimdt et al. 1998). The retention of the water flux by dams is the main perturbation along rivers. Managing this modification is a real challenge, especially because on the one hand there is a loss of diversity linked to a more restricted flood dynamic, and on the other hand, new species colonize banks and regulated river beds. One consequence is the spread of a different type of riparian vegetation more adapted to long periods of soil drought.

One important consequence of regulated rivers is the reduction and simplification of sandbars. Sandbars are created by the dynamics of eddies downstream to debris fans.

Fig. 3.18 - The riparian landscape offers important resources all year round to a broad assemblage of resident and migratory organims. In this case ringed plover (*Charadrius hiaticula*).

Dams have dramatic effects on the morphology of a river and on it's biological components. Downstream turbulence, turbidity, transported debris, temperature and nutrients all change. This has been described in detail by Schimdt et al. (1998) along the Grand Canyon where Glen Canyon dam has profoundly modified the river system. For instance, the macro-invertebrate composition changed after the dam construction. Released waters are colder, clearer and have little organic material, the consequence of which is a reduction in variety and biomass of fish. These effects are evident close to the dam, and the situation returns toward a more normal pattern due to resupply from unregulated tributaries. Profound changes occurr in riparian vegetation with more complexity and permanence of different types of vegetation from marshes to willow shrublands.

In Tab. 3.6, the management resources and related processes of the Colorado River in the Grand Canyon are illustrated. The restoration approach discussed by Schmidt et al (1998) considers at least five strategies:

a. Apply traditional techniques to ensure power production and water transfer. The ecological option is secondary.
b. Preserve the actual situation and some naturalized elements.

c. Simulate a natural ecosystem, in which the dam release should mimic pre-dam hydrography.

d. Substantial restoration of natural processes using an engineering approach to solve the problems of sediment concentration in the dam with their redistribution using a by-pass.

e. Full restoration. This approach considers the removal of dams along the river to return to pre-dam conditions. This action is full of problems due to the new condition in which the river now functions. The massive imput of sediments and the presence of alien species are some of the problems.

Flooding is the most spectacular process of a river system and, despite the severe damage that can be inflicted on settlements, properties and crop production, it is essential for river health. Floods increase the diversity of riparian zones with beneficial effects on agriculture and wildlife. They are considered a paradox for humanity which has an ambivalent response as there are always problems in managing this extreme and frequent events. On the other hand, regulated rivers experience a decrease of riparian ecosystem productivity. A policy window, as proposed by Kingdon (1984) and recently discussed by Haeuber & Michener (1998), represents an opportunity for action. Such a window is a critical event; it is short lived and often the furtuitous combination of circumstances. These windows allow to combine recognized problems, policy alternatives and politics (Fig. 3.19).

Tab. 3.6 - Comparison of processes and resources along the Colorado River as relicts of the pre-dam river and the post-dam river (from Schmidt et al. 1998).

Management resources and related processes of the Colorado River in the Grand Canyon
Processes and resources that are relicts of the pre-dam river: Seasonally fluctuating discharge, sediment transport, turbidity Seasonally changing temperature Large, unvegetaded sandbars that are emergent at low discharge Rapids dominated by large boulders Native fish assemblage, including species that are now endangered or extirpated Native upper riparian zone vegetation, native terrestrial species richness Archeological and historical sites
Processes and resources that are artifacts of the post-dam river: Low variability in annual discharge Substantial hourly variation of discharge in some years Low sediment transport and turbidity Constant low temperature Constricted rapids Blue-ribbon non-native trout fishery Biologically diverse marshes Dense lower riparian zone vegetation Endangered snail and bird species and other regionally significant populations occupying non-native riparian vegetation Hydroelectric power

Flood management can be considered a landscape procedure for maintaining a critical process for the entire large scale ecosystem. In this case, the choice of the watershed scale fits the dimension of the flood process which, though considered a disturbance, may be reconsidered as an energy pulse (Fig. 3.20).

Flooding is necessary to provide habitats for fish and for a large core of small animals. For instance, flood disturbance can prevent or reduce the settlement of invading plants like *Robinia pseudoacacia*, *Platanus orientalis*, and *Ailanthus* spp., along mediterranean rivers. The flood process reconnects the floodplain with the river bed distributing organic material and collecting nutrients for aquatic life.

Productivity and biodiversity are enhanced by the seasonal flooding regime, increasing local heterogeneity and creating ephemeral habitats (ponds) for breeding amphibians. The flood process influences secondary succession preventing the formation of dense and aged riparian woodlands.

In recent years, the regulatory role of floods has been recognized but it is very hard to apply a policy which is able to combine safety levels for populations living along rivers with the need to preserve river heterogeneity and organism diversity.

Large and small rivers have differently scaled mechanisms progressing from source to mouth. Capturing the dynamics along the river course is fundamental. At each lag along a river processes change in intensity and frequency. An adaptive policy must be used, of course, to fit the scale of such relevant processes. As outlined in Fig. 3.20 it is important to progress from flood control with only an hydraulic perspective to flood management in which people management (rules for local communities) and ecosystem management (the management of the surrounding mosaic) are considered. In many countries the human pressure along rivers is so high that it is not possible to adopt a policy of passive management and engineering defenses and dikes are required. On the other hand, it is possible to employ the tremendous energy developed by flowing waters to create new sand islands favoring new meanders and the development of riparian woodlands. The local heterogeneity can be easily restored by positioning stones or other natural debris to resurrect the "creative" action of flowing waters.

This action can be effective with explicit benefits for populations only if local and regional policy has established social and economic mitigation and compensation actions for short and medium term, and educational programs for a long term, changes of behavior.

Disturbance appears a very promising approach towards handling biodiversity (Tilman 1996) and, by definition, ecodiversity (*sensu* Naveh 1994).

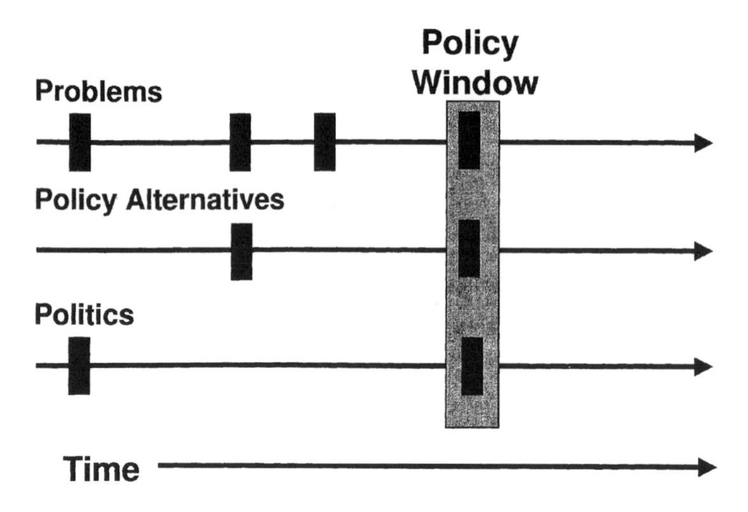

Fig. 3.19 - A policy window opens when problems, policy alternatives and politics overlap in time (from Haeuber & Michner 1998).

3.10.5 The importance of preserving oases

Recently the debate on the role of corridors has demonstrated the more subtle evidence that despite the dogmatic announcements corridors are not so evident in nature.

On the other hand, it has been quite well documented that in a context of natural reduction of tall and dense vegetation (for example, across the arid South-West of the United States) small, isolated oases and riparian vegetation both play a relevant role as temporary foraging habitat of migratory birds (Skagen et al 1998).

Connectivity does not seem to be an important component in the resource use of land birds during migratory time. Birds use connected and isolated patches indifferently, but in oases bird concentration is higher. The use of sparse vegetated patches embedded in a semi-desert matrix is of a great importance for migratory birds. If fragmentation has been evaluated in a context of large intact forested areas successively opened for agriculture or urban use, then in semidesert regions with a naturally poor vegetation cover, like in Arizona, fragmentation is not the process responsible for isolated oases or narrow riparian habitats. In this biome, it is the water deficit which creates the environmental constraint. The criterion of patch size for conserving biodiversity must be reconsidered carefully because this criterion works well only in fragmented forests. In naturally open areas like semideserts or drylands (for example, the Mediterranean) connectivity is not a workable process and so small isolated areas assume an important role in conserving biodiversity.

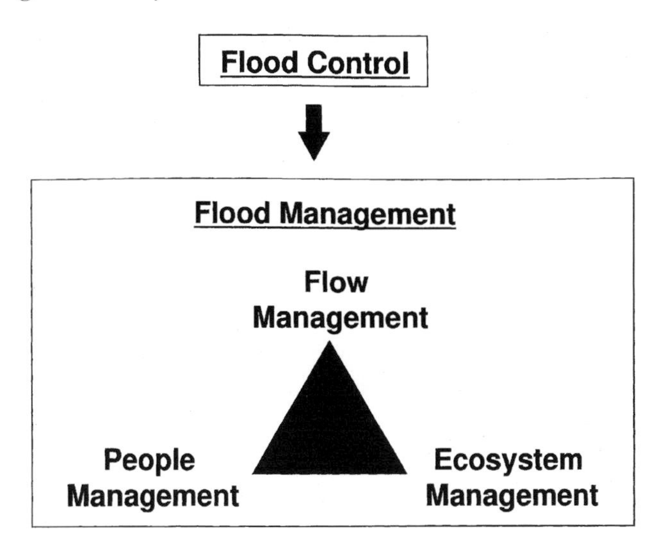

Fig. 3.20 - Schematic representation of the evolution from "flood control" policy to "flood management" (from Haeuber & Michener 1998).

In forestry today the practice of preserving residual patches instead of a complete clearcut appears a good conservation strategy for forest birds.

These residual patches apparently don't have a low environmental quality as the reduced number of species found seems to depend more on the small size than on the size "per se" (Merrill et al. 1997).

3.10.6 From protected areas to diffuse conservation

It is evident that isolated protected areas are not enough for long term conservation of biodiviersity. This is especially true if the matrix containing the protected patches is particularly hostile to organisms, for instance, urban fringes, intensive industrial agriculture and highly structured land.

A reduction of matrix hostility would improve the capacity to conserve the diversity of life. For this, landscape paradigms assume a special importance. In this case the landscape is considered a large heterogeneous area scaled against diversity as an indicator of the life-support mechanism process. Of course, the best option should be the conservation of the whole planet but, taking a more realistic perspective, it is reasonable to delimit geographic areas of such dimensions for which topography, vegetation cover, land use and settlements are delimiting the functional coherence.

From this perspective it is possible to accept the scientific validity of the landscape approach on the condition that the selection of a geographical area "per se" is recognized as an insufficient criterion.

The ecological functioning of large areas is not easily understood and in general the weakness of the relationship between entities, which are recognized more from a cultural filter than a functional filter, has caused embarrassment in ecosystemic ecology.

3.10.7 Countryside (cultural landscape) heritage conservation

In some human-dominated landscapes, natural dynamics are reduced or extirpated whilst in other cases human exploitation resembles natural disturbance such as flood, storm, fire and avalanches.

Often, historical settlements are located at deltas, on lake shores, and at river confluences, all of which are dynamic systems. Such dynamism is the major attraction because these systems are hotspots for nutrients, organisms and biomass.

Human settlement has, throughout time, reduced and damaged this dynamism and consequently natural events in such places find resistant structures that increase the catastrophe risk and reduce ecosystem function, thus progressing towards a permanent pathology in ecosystem functioning.

A new perspective, on the conservation of biodiversity can be achieved when we consider the cultural landscapes (countryside heritage) that are often outside any conservation planning strategy. These cultural landscapes were created for food production and it is only by coincidence that plant and animal diversity were high in such human dominated systems. The value of cultural landscapes has been recognized recently (van Droste et al. 1995), especially in Europe where the resilience capacity of such landscapes is more evident than in other regions such as the Tropics.

Countryside conservation is an elusive concept and many professionals such as foresters, farmers, landscape architects, geographers, ecologists, land agents and politicians are all engaged in conservation strategies. Countryside conservation is also a compromise between conflicting interests. In such conditions it is the people's welfare and needs that are the priority and the main driving force. Guide lines for conserving the countryside can be found in the conservation of amenity, wildlife, historical features and scenic beauty (Green 1996).

Actually, the demand for amenity, scenic beauty and wildlife is very strong and the over-capacity of European agriculture has created an unprecedented opportunity to conjugate food production, amenity, wildlife conservation and recreation. For this farmers have to play a new role in the stewardship of these landscapes.

Conserving tropical cultural landscape

Today, it is increasingly evident that some landscapes which were formerly considered pristine, such as the Amazon forest, have to be reconsidered in many cases as a special type of cultural landscape as, in fact, the presence of local populations over time has had a profound influence on the distribution and abundance of plants and trees. Native people have a perfect knowledge of hundred's of plants, with specific uses for construction, food, medicine, ornamental and veterinary use.

More than 90 of the tree species around Amazon villages and more than 96% of individual trees sampled are used (Bennett 1992). Many species are used at a low rate so that there is no risk of local or regional extinction, and often plants are voluntarily seeded, or removed. The total impact of native people on tropical forests has probably been less intense than in temperate regions and this, in large part, has been linked with the abundance of varieties of plants and animals in such forests. Hence human pressure has been shared across a large number of species so reducing the impact on a few species.

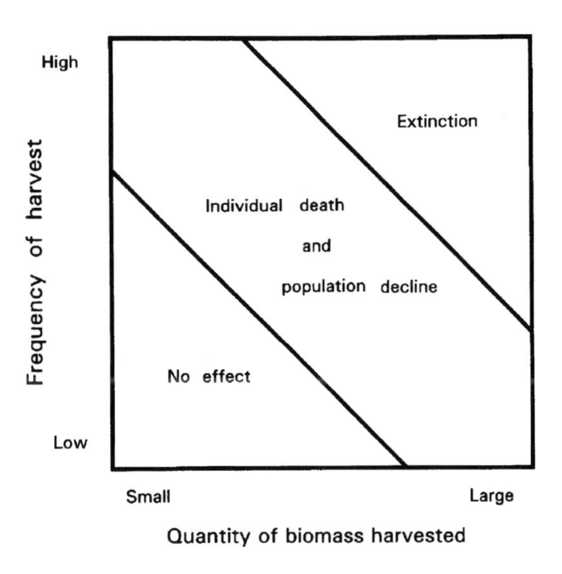

Fig. 3.21 - Harvest frequency and biomass harvested are the main factors controlling organism maintenance in a disturbed (tropical) landscape (from Bennett 1992).

The nomadic tendency of native forest peoples has produced a shifting mosaic in which the difference between cultivated and wild species is not as evident as in temperate regions. In this system recommendations for sustainable conservation do not seem necessary due to the long history of connections between indigenous people and their life style.

These people originally practiced agro-forestry which has now been rediscovered by ecologists and conservationists. As in other cultural landscapes, the feedback between natural and human processes is very strong and preserved in the "memory" of the forest, thus shaping it's structure and composition, and in human tradition. According to Bennett (1992), the relationship between individuals, species and plant populations and human pressure depends on the quantity of biomass harvested and the frequency of harvest (Tab. 3.7). It is possible to observe in fig 3.21 three plot's areas resulting from a combination of these two factors: no effect, individual death and population decline, and extinction. A shifting mosaic with a small biomass harvested at low frequency seems a more effective way for preserving resources. In this case, the living conditions of indigenous people can't be compared with developed areas so this is the weak link in of the chain of sustainability.

3.10.8 Soil conservation

It seems clear that erosion is a natural process often accelerated by human intrusion. This intrusion destroys the fertile soil at an increasing rate due to a combination of different factors of natural and human induced origin. As reported by Pimentel et al. (1995) soil erosion is a natural phenomenon but has been accelerated worldwide by agricultural practices. The increase of fertilizers, pesticides and irrigation compensates for soil loss but creates new problems for ecosystem resilience and the final carrying capacity.

Despite the loss of fertile soil the formation of new soil by transformation of parent material is very low - 1 ton per year compared with 17 tons of lost soil per ha per year in Europe and North America. Asia, Africa and South America have an erosion rate that depletes, on average, 30-40 tons per ha per year. Erosion is created mainly by rain drops launching soil particles into the air when a drop hits the soil, but also by water movement on the soil surface. Soil particles can also be transported for long distances by wind action. Erosion increases dramatically on steep slopes, especially when such slopes are fully deforested.

Soil structure can henance erosion if it is of medium or fine structure. Erosion depletes the fertile layer especially and the removal of N, and P can greatly reduce the crop yield. The soil removed by erosion contains 1.3 to 5 times more organic matter than the soil left behind.

Tab. 3.7 - Number of trees and lianas utilized by the Quijos Quichua population in one hectare at the Jatrun Sacha Biological Station in Amazonian Ecuador (from Bennett 1992).

Use category	Species		Individuals	
	Number	Percentage	Number	Percentage
Construction	191	77.0	637	87.4
Craft and dye/paint	7	2.8	29	4.0
Fiber	1	0.4	1	0.1
Fishing and hunting	3	1.2	11	1.5
Food	56	22.6	278	38.1
Food processing	2	0.8	10	1.4
Fuel	59	23.7	270	37.0
Medicine	17	6.9	215	29.5
Ornamental	0	0	0	0
Personal	1	0.4	14	1.9
Ritual/mythical	5	2.0	13	1.8
Veterinary	0	0	0	0
Miscellaneous	12	4.8	51	7.0

Soil biota are extremely important for the life cycle of crops and for the maintenance of soil functional efficiency. Technologies to control erosion are different according to the different scales. Generally they are based on terracing steep slopes, to planting trees along stream borders or on exposed slopes and ridges.

Different spatial architectures of the crop system encourage heterogeneous mosaics, for example growing grass strips, using mulches, planting in contours and creating windbreaks.

3.10.9 Conserving biodiversity in managed forests

Biological diversity cannot be conserved in parks and natural reserves only. In fact, protected areas are not enough to guarantee the conservation of processes and organisms.

Parks and protected areas are important hotspots for conservation, special areas in which education processes can be carried out, and heritage places for preserving some unique, relevant components of the environment.
Often we associate areas of great scenic value with parks and protected areas, but conservation efforts rarely covers the entire biodiversity potentiality.

Much effort must be devoted to maintaining a sufficiently conserved matrix around such areas (Harris 1984). These surrounding will have a semi-natural structure. In many forested areas, such as Pacific North America, new knowledge on ecological processes offers a great deal towards managing large areas which support timber income and biodiversity conservation (Hansen et al. 1991) (Fig. 3.22).

Previously large disturbance regimes were the most important driver of forest development. It has been estimated that there is a 435 years turnover in Mountain Rainier National Park (Hemstron & Franklin 1982) and a 150-276 year turnover in the central Oregon Cascades (Morrison & Swanson 1990).

Fire regimes have created a mosaic with patches of uneven size ranging from 0.2 to 100 ha. The replication of fire events throughout the history of these forests has created a mosaic of secondary successions with stands of different ages and size. The patchiness of this condition is increased at a local scale by small disturbances like tree death, low-severity fires, animal disturbance and diseases. The disturbance regimes generally do not cause an extirpation of all plants, some plants survive at an average rate of 30%. The contemporary presence of young and old stages have probably conditioned many of the organisms that can be found across the full forest sere.

Actually, natural disturbance in the American Northwest is replaced by the human disturbance of logging. This activity produces the main changes in the landscape mosaic. Logged areas are replanted by monospecific tree plantations and different practices are used to increase the survival rate of young planted trees.

There are big differences between natural disturbance and managed forests. The last practice is more constant, intense, and more frequent than the natural ones. The mosaic created by logging activity is more evenly distributed in the landscape compared with the natural mosaic. The edges of managed stands are more simple than the edges in natural mosaics, and so have a higher contrast. The design of such managed forests creates unfavorable conditions for interior species thus encouraging edge opportunistic species.

However, there are some elements of irregularity in these managed forests. For example, different ownerships and different strategies of stand use. This unpredictability is very important for the maintenance of an heterogeneous forest.

This pattern is very important in many regions such as Europe where small ownership prevents logging of large areas. In particular, if we look at the cadastrial maps of areas like the Apennines (Italy), it is possible to predict an uneven cutting regime simply by following the ownership edges. Again, the logging strategy is unevenly distributed in this area.

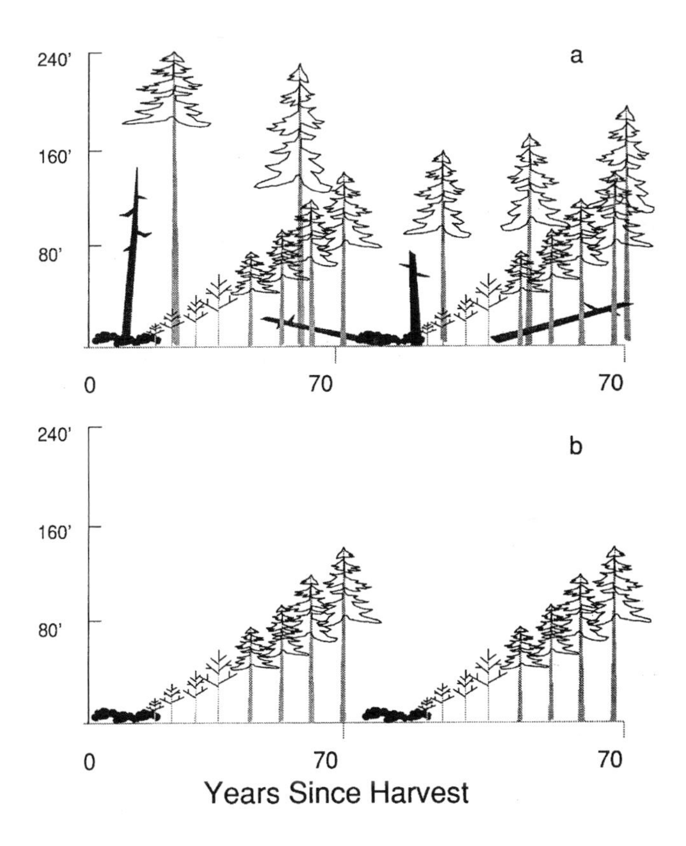

Fig. 3.22 - Schematic representation of an idealized stand for ensuring structural diversity (a) and maximum production(b) (from Hansen et al. 1991).

In the American Northwest, old-growth forests provide more old trees, more fallen trees and a more complete successional sere than managed forests. This has a profound influence on the diversity of organisms, especially for species that require large intact old-growth forest patches.

3.10.10 Conserving habitat fragments

A habitat patch may be defined as a discrete area used by a species for different uses (breeding, foraging, roosting, basking, etc.) and, for many species, a mosaic of habitat patches composes the overall habitat. In some habitats patches it is common for a population to go extinct. However, at a regional level this negative effect may be accommodated, providing it is not repeated throughout the region. After local extinction, colonization occurs largely depending on the availability of dispersing individuals and the possibility of crossing a hostile inter-patch matrix.

The spatial arrangement of habitat patches is a determinant for population dynamics. Size, shape and quality of patches are important for the permanence of populations.

In a world in which large human undisturbed areas are very rare, small fragments of natural environments acquire a growing importance (Fahrig & Merriam 1994, Shafer 1995). These fragments are comprised within a hostile matrix at the large landscape scale. The strategy of enlarging the borders of natural reserves and national parks is destined to fail due to socio-economic and political constraints. In conclusion, parks are too small to comprise a full set of organisms and processes. The application of the criterion of the "umbrella" species seems contrasting with the reality of a world in which nature conservation represents an "option" for developed countries, and rarely a true "sustainable" option for undeveloped regions.

The minimum viable area for large mammals during a time span of 1,000 years exceeds, for instance, 100 times the area of Yellowstone National Park (Shafer 1995). The conservation of small areas seems a very weak attempt to surrogate the loss of large areas. However, if small areas are not sufficient to preserve large animals and plants, they can play an important role for small organisms like amphibians, reptiles, micromammals and insects. Small reserves can also support wide-ranging species which need less hostile habitats during displacement, migration and dispersion. In addition, small reserves can be actively managed to maintain a dynamism that has disappeared in nature.

3.11 Perspectives in management and conservation

Humans depend on ecosystem services and need the maintenance of the integrity of ecosystems (biodiversity). It must be recognized that the complexity of an ecosystem can't be moved geographically (biological communities are place-based) and that ecosystem integrity can't be ensured inside protected areas only. This has to be coupled with the fact that these protected areas seem insufficient to guarantee the services required by a more demanding society. Climate change, fragmentation, small size and inappropriate shapes of valuable patches, changing demands in land use, and the application of new biotechnologies increase the marginal role of protected areas towards achieving human needs, even in a framework of sustainable development. Pressure on the scaled landscape is increasing due to over-harvesting, physical restructuring, introduction of exotic species and by modification of natural perturbations.

Protected areas could serve as core elements in a matrix of low quality ecosystem services, but in my opinion this is not enough to guarantee ecosystem functioning, although buffer zones and ecological networks concur to mitigate fragmentation, isolation and biodiversity degradation. From this perspective, any change in land use should be "biodiversity friendly".

Managing landscapes requires many conceptual and practical tools. It is important to transfer scientific knowledge into the arena of the real world, to integrate socio-economics into natural processes, to understand ecosystem functioning, to preview and adapt institutional mechanisms, and to build bridges between the different disciplines.

Managing complex areas requires a common aim between the different components of human society. This means to adopt partnership guidelines, to encourage communication strategies, to adopt a common vision, to focus on the target interests, to define issues and boundaries, and to ensure an intergenerational equity.

A new type of expert is required to handle such complex relationships. This expert should have a curriculum which is able to incorporate the key points of multiple disciplines, and the ability to open dialogue with stakeholders, policy makers and professionals. In other words, he should introduce into the ecological model variables which can assess the present efficiently and predict future scenarios, whilst integrating many of the ecological as well human related processes. To manage this complex commitment he should evaluate ecological, social and economic conditions, establish clear goals, select appropriate indicators, monitor results, introduce a decision-making procedure, and capitalize on the feedback obtained from previous actions.

Moving from principles and strategies to practice, it appears clear that the land mosaic represents the physical arena in which we operate. In such an arena, in which the matrix is often represented by a highly disturbed landscape, the improvement of the quality of core areas (by favoring "aggregation" processes inside the core area), the creation of linear vegetated belts able to intercept the dispersal and migratory movements of organisms, and the attempts to improve species-specific connectivity, acquire a particular importance. Several projects have been organized in this direction based on the reconstruction of ecological networks or "reserve networks". An ecological network can be considered a coherent system of natural and semi-natural land or sea-scape, that is configured and managed with the objective of maintaining or restoring ecological functions. At the same time, it provides opportunities for a sustainable use of natural and human resources. This model is under critical review by the government of The Netherlands in collaboration with IUCN (The World Conservation Union) using a global inventory and analysis of the different experiences gained in the sector of environmental management and design, and incorporating other initiatives such as bioregional planning, ecoregion-based conservation, greenways (Ahern 1994), wildlife corridors and Man and Biosphere reserves.

Some considerations must be pointed out regarding the importance of the ecological network approach. First of all in many regions (countries) there are no other possibilities for modelling the land mosaic in order to retain, restore and improve biodiversity. This situation is common in many regions (e.g. the Padania valley in Italy) and especially in small countries (Belgium, The Netherlands, Denmark, etc.). In these areas the matrix is represented by a dominant urban cover, infrastructures and industrialized agriculture. In such a matrix, only small core areas, buffer zones around these core areas, and ecological networks can be maintained.

But in other regions and countries where the matrix is represented by the prevailing natural system, this model assumes scant importance and application. In such conditions, the matrix is the valuable component of the land mosaic and, in order to ensure the functioning of system, it is necessary to preserve the autopoietic and autocatalytic capacity of this matrix. A diffuse naturalness appears in these conditions and management should be oriented towards maintaining the processes rather than focusing on the existing patterns.

As already discussed in other parts of this book, the Mediterranean region offers a good example of the importance of a matrix composed of ephemeral patches that result from different disturbance regimes (e.g. grazing, fire, trampling, plowing) combined with the seasonal variation of climatic conditions. The management of such landscapes represents new challenges for landscape ecology since it has based most of it's theories and models on patches. New theories will probably be required to correctly and efficiently approaching the conditions of fine grained mosaics that, at the last analysis, seem more resilient than coarse-grained mosaics. This may well be a matter of scale; human interactions have an increasing impact when their intrinsic scale is close to the ecological scale of natural processes.

With regards to the Ecosystem-based approach to management, new cultural innovations are required for the landscape ecologist; new areas of interest are disclosed and unexplored fields of human behavior may represent a real challenge. For instance, if considering human behavior, it seems quite clear that actions like raising people's awareness (by education, training, confrontation), analyzing the context in which people work, improving working habits (dissemination of good practice), and collecting external stimuli (new ideas, new markets) are important goals.

However, conflicting objectives must be faced in such an approach. Within our society short-term planning is required and it is not an easy task to translate this into long term action, which fit a sustainable approach. This is especially true when long-term planning try to operates within the private sector where perverse subsidies can invalidate any action.

In many cases, the procedures of the ecosystem-based approach to management are not understandable/efficient enough, and the need to clarify concepts, define benefits, promote user alliances, ensuring fairer apportionment of costs and benefits, and to produce real examples (learning from experience) acquires an urgent priority.

It is no less important to integrate action across different administrative levels, to encourage a more creative use of tools and mechanisms, and to understand other interests and values.The creation of a political framework for the environment in which procedures such as detection, diagnosis, and prediction, along with the permanent uncertainty of the systems evolution, should be the central focus.

Acting in landscape ecology: Concluding remarks

•

Acting in landscape ecology: Concluding remarks

Different approaches are registered in landscape ecology studies and each of these approaches seems completely different and dissonant. One approach considers processes as actors affecting, and responding to landscape constraints. Another approach is to consider the landscape as a large area. At this scale of resolution the spatial patterns of patches are compared and studied. Finally, another approach is to consider organisms in heterogeneous systems and to study their behavioural response. This approach also includes the human being. In this case landscape is considered as a large area in which remnants of original vegetation persist in a human dominated landscape. This approach is very familiar to geographers, human ecologists, planners, and landscape architects. The analysis of a constructed landscape is carried out, i.e., the landscape is considered the vital space of humanity, countries, regions, cities, villages, countryside, farms. It is at this level that landscape ecology has to be applied to improve the quality of human life, and to protect the diversity of life, assuring a healthy system.

The historical, economical and social context are to be considered in this landscape ecology human oriented approach, and a patterned landscape is overlapped by a virtual landscape in which economic, political and social decisions are relevant processes. Often the human approach is more popular than the others and this has created conflict between different discipline-oriented practitioners. One relevant goal of this book is to unify these different approaches so as to create a common framework in which to place the pletora of different concepts and applications.

Another relevant approach concern the study of the land mosaic. This approach considers the spatial heterogeneity as a relevant pattern at any scale of resolution and extension. Often this approach may be confused with the human oriented approach, but differs in that it has the possibility of changing scales, moving from micro to macro and mega mosaic. In this case again the mosaic is considered as a level of functioning, although the organization of the mosaic is evaluated. This approach is particularly useful in evaluating processes, in the studies of disease spread, in pest outbreaks and in disturbance regime reactions. The choice of the right scale depends mostly on the geographical perspective than on the organism-centered perspective. It could be considered a sub-level of the human perceived landscape but in reality differs from the human-related approach because the subject of interest is the mosaic like as a super-organism.

Where landscape ecology may be useful and what might be the degree of affordability in the real world is not easy to detect. But it makes a good sense to recognize that this discipline has the capacity to create a permanent forum worldwide concerning the most urgent environmental issues challenging the existence of large human populations.

At least in two main fields of the human realm, landscape ecology shows it's potentiality as an ecological problem-solving science. The first is the mitigation of human impact in developed countries in which a jungle of roads, urban and industrial settlements are reducing the distribution and the movement of organisms. The second "niche" is located in undeveloped countries in which the depletion of natural resources seems the prime goal and in which human populations still live in degraded lands dominated by diseases and food scarcity.

In the first case landscape ecology seems more a cosmetic, amenity-caring science and in the second case acts more as a primary tool to improve the life style of billions of people. In developed countries landscape ecology can play a fundamental role as a restoration and mitigation science combining a highly technologically developed and informed society with a more diffuse naturalness, reducing the risk of local extinctions and assuring a reasonable level of biodiversity. For this landscape ecology acts primarily in the locating and structuring of residual natural areas contributing to protected zones and parks.

Due to the dynamic scale of resolution at which this discipline can move, it's role is determinant in increasing amenity, wildlife protection, and water quality maintenance in the countryside and in urban fringes.

The role of landscape ecology in developing countries seems extremely important because it's holistic, intra-disciplinary approach can act in a balanced way in areas where the priorities are food harvesting, soil erosion protection, infrastructure development, and pollutants control.

In countries where the cultural heritage seems completely lost in the constraint between extreme poverty and super-technological development, this approach seems a reasonable strategy, with which to discover now past cultural heritage could be combined with modern procedures to produce food, to control diseases and to guarantee a social and economic development.

Planning in developing countries through the use of landscape ecology should avoid the errors that developed societies have made in the domestic arena during recent decades.

References

Acuna, J.A. & Yortsos, Y.C. (1995) Application of fractal geometry to the study of networks of fractures and their pressure transient. *Water Resources Research* **31**: 527-540.

Ahern, J. (1994) Greenways as ecological networks in rural areas. *In:* Cook, E.A. and H.N. van Lier (eds.), *Landscape planning and ecological networks.* Elsevier. Pp.159-177.

Alados, C.L., Escos, J.M., Emlen, J.M. (1996) Fractal structure of sequential behaviour patterns: an indicator of stress. *Anim. Behav.* **51**: 437-443.

Allen, M.F., Hipps, L.E., Wooldridge, G.L. (1989). Wind dispersal and subsequent establishment of VA mycorrhizal fungi across a successional arid landscape. *Landscape Ecology* **2**: 165-171.

Allen, R.B., Bellingham, P.J., Wiser, S.K. (1999) Immediate damage by an earthquake to a temperate montane forest. *Ecology* **80(2)**: 708-714.

Allen, T.F.H. & Starr, T.B. (1982) *Hierarchy, perspectives for ecological complexity.* The University of Chicago Press, Chicago.

Allen, T.F.H. & Hoekstra, T.W. (1992) *Toward a unified ecology.* Columbia University Press, New York.

Anderson, J.E. (1991) A conceptual framework for evaluating and quantifying naturalness. *Conservation Biology* **5**: 347-352.

Anderson, A.N. & McBratney, A.B. (1995) Soil aggregates as mass fractals. *Aust. J. Soil. Res.* **33**: 757-72.

Andreassen, H.P., Halle, S., Ims, R.A. (1996) Optimal width of movement corridors for root voles: not too narrow and not too wide. *Journal of Applied Ecology* **33**: 63-70.

Andren, H., Angelstam, P., Lindtstrom, E., Widen, P. (1985) Differences in predation pressure in relation to habitat fragmentation: an experiment. *Oikos* **45**: 273-277.

Andrews, J. & Rebane, M. (1994) *Farming & Wildlife.* RSPB, The Lodge, Sandy, UK.

Angermeier, P.L. & Karr, J.R. (1994) Biological integrity versus biological diversity as policy directives. *Bioscience* **44**: 690-697.

Antrop, M. (1997) *Landscape, planning and landscape ecology.* Lecture at the International PH.D. Course on Landscape Ecology, Roskilde University Center, Denmark (unpl.).

Arthur, W.B. (1999) Complexity and the economy. *Science* **284**: 107-109.

Austad, I., Hauge, L., Helle (1993) *Maintenance and conservation of the cultural landscape in Sogn og Fjordane, Norway.* , T. Department of Landscape Ecology, Sogn og Fjordane College, Norway .

Awimbo, J.A., Norton, D.A., Overmars, F.B. (1996) An evaluation of representativeness for nature conservation, Hokitika ecological district, New Zealand. *Biological Conservation* **75**: 177-186.

Bailey, R.G. (1995) *Description of the ecoregions of the United States.* 2d ed.Misc.publ.No. 1391, Washigton, D.C. USDA Forest Service.

Baker, B.D. (1996) Landscape pattern, spatial behavior, and a dynamic state variable model. *Ecological Modelling* **89**: 147-160.

Baker, W.L. (1992) The landscape ecology of large disturbances in the design and management of nature reserves. *Landscape Ecology* **7**: 181-194.

Baker, W.L. & Cai, Y (1992) The r.le programs for multiscale analysis of landscape structure using the GRASS geographical information system. *Landscape Ecology* **7**: 291-302.

Bakker, J.P. (1998) The impact of grazing on plant communities. Pp. 137-184. *In:* WallisDeVries, M.F., Bakker, J.P., Van Wieren, S.E. (eds.) 1998 *Grazing and conservation management.* Kluwer Academic Publishers, Dordrecht.

Bakker, J.P. & Londo, G. (1998) Grazing for conservation management in historical perspective. Pp. 23-54. *In:* WallisDeVries, M.F., Bakker, J.P., Van Wieren, S.E. (eds.) 1998 *Grazing and conservation management.* Kluwer Academic Publishers, Dordrecht.

Bancroft, G.T., Strong, A.M., Carrington, M. (1995) Deforestation and its effects on forest-nesting birds in the Florida Keys. *Conservation Biology* **9**: 835-844.

Barak, P. , Seybold, C.A., McSweeney, K. (1996) Self-similitude and fractal dimension of sand grain. *Soil Sci. Soc. Am. J.* **60**: 72-76.

Barker, W. (1987) *Favorite animals of North America.* Portland House, New York.

Baudry, J. (1984) Effects of landscape structures on biological communities: the case of hedgerow network landscapes. *In:* Brandt, J., Agger, P. (eds.), *Methodologies in landscape ecological research and planning.* Vol. 1 Proceedings of the first international seminar of the International Association of Landscape Ecology. Roskilde, Denmark, October 15-19. Pp. 55-65.

Baudry, J. & Merriam, G. (1987) Connectivity and connectedness: functional versus structural patterns in landscapes. *In:* Schreiber, K.F. (ed.) *Connectivity in landscape ecology.* Proceedings of the 2nd International Seminar of the "International Association for Landscape Ecology". Munstersche Geographische Arbeiten 29, 1988. Pp. 23-28.

Beerling, D.J. (1998) Biological flora of the British Isles. *Salix erbacea. Journal of Ecology* **86**: 872-895.

Beier, P. (1993) Determining minimum habitat areas and habitat corridors for cougars. *Conservation Biology* **7**: 94-108.

Beier, P. & Noss, R.F. (1998) Do habitat corridors provide connectivity? *Conservation Biology* **12 (6)**: 1241-1252.

Bennett, B.C. (1992) Plants and people of the Amazonian rainforests. *Bioscience* **42(8)**: 599-607.

Bennett, K.D. (1998) The power of movement in plant. *TREE* **13**:339-340.

Benvenuti S. & Joalé P. (1982) Seasonal and diurnal variation of weight in four passeriformes in Autumn and Winter. *Avocetta* **6**: 63-74.

Berthold, P. (1993) *Bird migration. A general survey.* Oxford University Press, Oxford.

Bierregaard, R.O. Jr., Lovejoy, T.E., Kapos, V., dos Santos, A.A., Hutchings, R.W. (1992) The biological dynamics of tropical rainforest fragments. *Bioscience* **42**: 859-866.

Blackburn N. , Fenchel T., Mitchell, J. (1998) Microscale nutrient patches in planktonic habitat show b by chemotactic bacteria. *Science* **282**: 2254-2256.

Blake, J.G. (1986) Species-are relationship of migrants in isolated woodlot in east-central Illinois. *Wilson Bull.* **98**: 291-296.

Blake, J.G. & Hoppes, W.G. (1986) Influence of resource abundance on use of tree-fall gaps by birds in an isolated woodlot. *Auk* **103**: 328-340.

Blake, J.G. & Karr, J.R. (1987) Breeding birds of isolated woodlot: area and habitat relationships. *Ecology* **68**: 1724-1734.

Blondel, J. & Cuvillier, R. (1977) Une methode simple et rapide pour decrire les habitats d'oiseaux: le stratiscope. *Oikos* **29**: 326-331.

Blondel, J., Perret, P., Maistre, M., Dias, P.C. (1992) Do arlequin Mediterranean environments function as source-sink for Blue Tits (*Parus caeruleus* L.)? *Landscape Ecology* **6**:212-219.

Bogaert, J. & Impens, I. (1998) An improvement on area-perimeter ratios for interior-edge evaluation of habitats. *In:* . F. Muge, R.C. Pinto, M. Piedade (eds.) *Proceedings 10th Portuguese Conference on Pattern Recognition*, Lisbon, Portugal. Pp.55-61.

Bogaert, J., Van Hecke, P., Moermans, R., Impens , I. (in press) Twist number statistics as an additional measure of habitat perimeter irregularity*? Environmental and ecological statistics.*

Boyd I.L. (1996) Temporal scales of foraging in a marine predator. *Ecology* **77(2)**: 426-434.

Brandt, J. & Agger, P. (ed.) (1984) *Proceedings of the First International Seminar on methodology in Landscape Ecological Research and Planning.* Roskilde, University Center, DK.

Brittingham, M.C. & Temple, S.A. (1983) Have cowbirds caused forest songbirds to decline? *Bioscience* **33**: 31-35.

Brown, J.H., Mehlman, D.W., Stevens, G.C. (1995) Spatial variation in abundance. *Ecology* **76(7)**: 2028-2043.

Brothers, T.S. & Spingarn, A. (1992) Forest fragmentation and alien plant invasion of Central Indiana old-growth Forests. *Conservation Biology* **6**: 91-100.

Burel, F. (1996) Hedgerows and their role in agricultural landscapes. *Critical Reviews in Plant Sciences* **15**: 169-190.

Burel, F. & Baudry, J. (1990a) Hedgerows network pattern and processes in France.*In:* Zonneveld, I.S. & Forman R.T.T. (eds.) *Changing Landscapes: An Ecological Perspective*, Springer-Verlag. Pp. 99-120.

Burel, F. & Baudry, J. (1990b) Structural dynamic of a hedgerow network landscape in Brittany France. *Landscape Ecology* **4**: 197-210.

Burger, L.D., Burger, L.W., Faaborg, J. (1994) Effects of prairies fragmentation on predation on artificial nests. *J. Wildl. Manage.* **58**: 249-254.

Burkey, T.V. (1995) Extinction rates in archipelagoes: implications for population in fragmented habitats. *Conservation Biology* **9**: 527-541.

Burrough, P.A. (1986) *Principles of geographic information systems for land resources assessment.* Clarendon, Oxford.

Burrough, P.A. & McDonnell, R.A. (1998) *Principles of geographical information systems.* Oxford University Press, New York.

Butler, D.R. (1995) *Zoogeomorphology. Animals as geomorphic agents.* Cambridge University Press, New York.

Campbell, R.D. (1996) Describing the shapes of fern leaves: a fractal geometrical approach. *Acta Biotheorica* **44**: 119-142.

Canadian Heritage-Parks Canada (1996) *Principles and standards for ecosystem-based management for Parks Canada.* Canadian Heritage-Parks Canada, Natural Resources Branch, Hull, Québec.

Carlile, D.W., Skalski, J.R., Baker, J.E., Thomas, J.M., Cullinan, V.I. (1989) Determination of ecological scale. *Landscape Ecology* 2: 203-213.

Chambers, C.L., McComb, W.C., Tappeneir II, J.C.(1999) Breeding bird responses to three silvicultural treatments in the Oregon coast range. *Ecological Applications* **9(1)**: 171-185.

Chapin III, F.S., Sala, O.E., Burke, I.C., Grime, J.P., Hooper, D.U., Lauenroth, W.K., Lombard, A., Mooney, H.A., Mosier, A.R., Naeem, S., Pacala, S.W., Roy, J., Steffen, W.L., Tilman, D. (1998) Ecosystem consequences of changing biodiversity. *Bioscience* 48(1): 45-52.

Chen, S.G., Ceulman, R., Impen, I. (1992) A fractal-based *Populus* canopy structure model for the calculation of light interception. *Forest Ecology and Management* 69: 97-110.

Cheng, K. & Spetch, M.L. (1998) Mechanisms of landmark use in mammals and birds. *In:* Healy, S. (ed.) *Spatial representation in animals.*Oxford University Press, New York. Pp. 1-17.

Clarks, J.S., Fastie, C., Hurtt, G., Jackson, S.T., Johnson, C., King, G.A., Lewis, M., Lynch, J., Pacala, S., Prentice, C., Schupp, E.W., Webb III, T., Wyckoff, P. (1998) Reid's Paradox of rapid plant migration. *Bioscience* **48(1)**: 13-24.

Clements, F.C. (1905) *Research methods in ecology*. University Publishing Co., Lincoln, Nebraska, USA.

Colwell R.C. (1996) Global climate and infectiuous disease: The cholera paradigm. *Science* **274**: 2025-2031.

Correll D.L. (1997) Buffer zones and water quality protection: general principles. In Haycock N.E., Burt T.P., Goulding K W T, Pinay G. (eds.*) Buffer zones: Their processes and potential in water protection.*Pp. 7-20.

Correll, D.L., Jordan, T.E., Weller, D.E. (1992) Nutrient flux in a landscape: effects of coastal land use and terrestrial community mosaic on nutrient transport to coastal waters. *Estuaries* **15**: 431-442.

Costanza R., d'Arge, R., de Groot, R., Farber, S., Grasso, M., Hannon, B., Limburg, K., Naeem, S., O'Neill, R.V., Paruelo, J., Raskin, R.G., Sutton, P., van den Belt, M. (1997) The value of the world's ecosystem services and natural capital. *Nature* **387**: 253-260.

Coulson, R.N., Lovelady, C.N., Flamm, R.O., Spradling, S.L., Saunders, M.C. (1991) Intelligent Geographic Information Systems for natural resource management *In:* Turner, M.G. and R.H. Gardner (eds.), *Quantitative methods in landscape ecology*. Springer-Verlag New York. Pp.153-172.

Cracknell, A.P. & Hayes, L.W.B. (1993) *Introduction to remote sensing*. Taylor & Francis, London.

Cutler, A. (1991) Nested faunas and extinction in fragmented habitats. *Conservation Biology* **5**: 496-505.

Dale, V., Gardner, R.H., Turner, M.G. (1989) Predicting across scales: theory development and testing. *Landscape Ecology* **3**:147-252.

Daubenmire, R. (1968) *Plant communities*. Harper & Row, New York, USA.

Dayton, P.K. & Tegner, M.J. (1984) The importance of scale in community ecology: a kelp forest example with terrestrial analogs. Price, P.W., Slobodchinof, C.N., Gaud, W.S. (eds.), *A new ecology*. John Wiley and Sons, New York. Pp. 457-481.

Delcourt, P.A. & Delcourt, H.R. (1987) Long-term forest dynamics of temperate forests: applications of paleoecology to issues of global environmental change. *Quat. Sci. Rev.* **6**:129-146.

Delcourt, H.R., Delcourt, P.A. (1988) Quaternary landscape ecology: relevant scales in space and time. *Landscape Ecology* **2**:23-44.

Delcourt, P.A. & Delcourt, H.R. (1992) Ecotone dynamics in space and time. *In:* Hansen, A.J. & di Castri, F. (eds.) *Landscape boundaries. Consequences for biotic diversity and ecological flows.* Springer-Verlag, New York.Pp. 19-54.

Delcourt, H.R., Delcourt, P.A., Webb, T. III (1983) Dynamic plant ecology: the spectrum of vegetational change in space and time. *Quat. Sci. Rev.* **1**:153-175.

De Leo, G.A. & Levin, S. (1997) The multifacet aspects of ecosystem integrity. *Conservation Ecology* (on line), **1(1)**:3.

Desaigues, B. (1990) The socio-economic value of ecotones. *In:* Naiman, R.J. and H. Decamps(eds.), *The ecology and management of aquatic-terrestrial ecotones.* MAB, UNESCO, Paris. Pp. 263-293.

Dias, P.C. (1996) Sources and sinks in population biology. *TREE* **11**: 326-330.

DiCastri, F. , Hansen, A.J., Holland, M.M. (1988) A new look at ecotones: emerging international projects on landscape boundaries. *Biology International, Special Issue* **17**:1-163.

Diffendorfer, J.E., Gaines, M.S., Holt, R.D. (1995) Habitat fragmentation and movements of three small mammals (*Sigmodon, Microtus,* and *Peromyscus*). *Ecology* **76**: 827-839.

Diggle, P.J. (1983) *Statistical analysis of spatial point patterns.* Academic Press, London .

Donovan, T.M., Thompson III, F.R., Faaborg, J., Probst, J.R. (1995a) Reproductive success of migratory birds in habitat sources and sinks. *Conservation Biology* **9**: 1380-1395.

Donovan, T.M., Lamberson, R.H., Kimber, A., Thompson III, F.R., Faaborg, J. (1995b) Modeling the effects of habitat fragmentation on source and sink demography of neotropical migrant birds. *Conservation Biology* **9**: 1396-1407.

Eghball, B. & Power, J.F. (1995) Fractal description of temporal yeld variability of 10 crops in the United States. *Agronomy Journal* **87(2)**: 152-156.

Etzenhouser, M.J., Owens, M.K., Spalinger, D.E., Murden, S.B. (1998) Foraging behavior of browsing ruminants in a heterogeneous landscape. *Landscape Ecology* **13**: 55-64.

Fahrig, L. & Merriam, G. (1994) Conservation of fragmented populations. *Conservation Bio*logy **8**: 50-59.

Fahrig, L. & Freemark, K. (1995) Landscape-scale effects of toxic events for ecological risk assessment. *In:* Cairns, J.Jr and B.R. Niederlehner (eds.) *Ecological Toxicity Testing.* Lewis Publishers, Boca Raton. Pp. 193-208.

Farina, A. (1971) Osservazioni sull'avifauna dell'alta Lunigiana orientale. *Ricerche di Zool. appl. alla caccia. Lab. Zool.appl. Caccia*: 1-50pp, Tip. Compositori, Bologna.

Farina, A. (1978) Breeding biology of the crag martin *Hirundo rupestris. Avocetta (N.S.)* **2**: 35-46.

Farina, A. (1981) Contributo alla conoscenza dell'avifauna nidificante nella Lunigiana. *Boll. Mus. S. Nat. Lunigiana* **1(1)**: 21-70.

Farina, A. (1985) Habitat use and structure of a bird community? *S.It.E. Atti* **5**:679-686.

Farina, A. (1987) Autumn-winter structure of bird communities in selected habitats of central-north Italy. *Boll. Zool.* **54**:243-249.

Farina, A. (1988a) Birds communities structure and dynamism in spring migration in selected habitats of northern Italy. *Boll. Zool.* **55**:327-336.

Farina, A.(1988b) Observations on the swift's *Apus apus* (L.) social flights at the breeding sites. *Mon. Zool. It. (N.S.)* **22**:255-261.

Farina, A. (1991) Recent changes of the mosaic patterns in a montane landscape (North Italy) and consequences on vertebrate fauna. *Option Mediterraneenes - Serie seminaires* **15**: 121-134.

Farina, A. (1993a) *L'ecologia dei sistemi ambientali.* CLEUP, Padova.

Farina, A. (1993b) Birds in a riparian landscape. *In:* E.J.M. Hagemeijr and T.J. Vestrael (eds.), *Bird Numbers 1992*, Statistics Netherlands, Voorburg/Heerlen.Pp. 565-578.

Farina, A. (1993c) From global to regional landscape ecology. *Landscape Ecology* **8(3)**: 153.

Farina, A. (1994) Birds in a mountain landscape. *In*: Dower, J.W. (ed.), *Fragmentation in agricultural landscapes.* Proceedings of the third annual IALE(UK) conference, held at Myerscough College, Preston. Pp.153-160.

Farina, A. (1995a) Cultural landscapes and fauna. *In:* van Droste, B., Plachter, H., Rossler, M. (eds.) *Cultural landscapes of universal value.* Gustav Fischer, Jena.Pp. 60-77.

Farina, A. (1995b) *Ecotoni. Patterns e processi ai margini.* CLEUP, Padova.

Farina, A. (1996) The cultural landscape of Lunigiana. *Memorie Accademia Lunigianese Scienze "Giovanni Capellini"* 56: 83-90.

Farina, A. (1997a) Resources allocation and bird distribution in a sub-montane ecotone. *In*: Novak, M.M. and Dewey, T.G. (eds.), *Fractal Frontiers.* World Scientific, Singapore.

Farina, A. (1997b) Landscape ecology as a basis for protecting threatened landscapes. *In:* Nelson, G.J. and R. Serafin (eds.), *National parks and protected areas.* NATO ASI Series G: Ecological Sciences, vol. 40:24-30.

Farina, A. (1997c) Landscape structure and breeding bird distribution in a sub-Mediterranean agro-ecosystem. *Landscape Ecology* **12**: 265-378.

Farina, A. (1998) *Principles and methods in landscape ecology.* Chapman & Hall, London.

Feder, J. (1988) *Fractals.* Plenum, New York.

Foran, B. & Wardle, K. (1995) Transition in land use and the problems of planning: a case study from mountainlands of New Zealand. *Journal of Environmental Management* **43**: 97-127.

Forman, R.T.T. (1981) Interaction among landscape elements: A core of landscape ecology. *In*: Tjallingii, S.P. and A.A. de Veer (eds.), *Perspective in landscape ecology.* Proc. Int. Congr. Neth. Soc. Landscape Ecology, Vedhoven.Pudoc Wageningen. Pp. 35-48.

Forman, R.T.T. & Godron, M. (1986) *Landscape ecology.* Wiley & Sons, New York.

Forman, R.T.T. & Moore, P.N. (1992) Theoretical foundations for understanding boundaries in landscape mosaics. *In:* Hansen, A.J. and F. di Castri (eds.), *Landscape boundaries. Consequences for biotic diversity and ecological flows.* Springer-Verlag, New York, pp 236-258.

Forman, R.T.T. (1995) *Land mosaics. The ecology of landscapes and regions.* Cambridge Academic Press, Cambridge, UK.

Forman R.T.T & Collinge S.K. (1996) . The spatial solution to conserving biodiversity in landscapes and regions. *In:* R.M. DeGraaf and R.I. Miller. (eds.), *Conservation of faunal diversity in forested landscapes.* Pp. 537-568. Chapman & Hall, London

Foster, D.R., Motzkin, G., Slater, B. (1998) Land-use history as long-term broad-scale disturbance: regional forest dynamics in central New England. *Ecosystems* **1**: 96-119.

Fourcassié, V., Coughlin, D., Traniello, J.F.A. (1992) Fractal analysis of search behavior in ants. *Naturwissenschaften* **79**: 87-89.

Frank, D.A., McNaughton, S.J., Tracy B.F. (1998) The ecology of the earth's grazing ecosystems. *Bioscience* **48(7)**: 513-521.

Frontier, S.A. (1987) Fractals in marine ecology. *In*: Legendre, L. (ed.), *Development in numerical ecology* . Springer-Verlag, Berlin. Pp. 335-78.

Furness, R.W. & Greenwood, J.J.D. (1993) *Birds as monitors of environmental change.* Chapman & Hall, London.

Galli, A.E. , Leck, C.F., Forman, R.T.T. (1976) Avian distribution patterns in forest islands of different sizes in central New Jersey. *Auk* **93**: 356-364.

Gardner, R.H. & O'Neill, R.V. (1991) Pattern, process, and predictability: the use of neutral models for landscape analysis. *In:* Turner, M.G. and R.H. Gardner (eds.), *Quantitative methods in landscape ecology: the analysis and interpretation of landscape heterogeneity.* Springer-Verlag, New York, New York, USA. Pp. 289-307.

Gardner, R.H., Milne, B.T., Turner, M.G., O'Neill, R.V. (1987) Neutral models for the analysis of broad-scale landscape pattern. *Landscape Ecology* **1**: 19-28.

Gardner, R.H., Turner, M.G., Dale, V.H., O'Neill, R.V. (1992) A percolation model of ecological flows. *In:* Hansen, A.J. and F. di Castri (eds.), *Landscape boundaries. Consequences for biotic diversity and ecological flows.* Springer-Verlag, New York. Pp. 259-269.

Gates, J.E. & Gysel, L.W. (1978) Avian nest dispersion and fledging success in field forest ecotones. *Ecology* **59**: 871-883.

Gibbs, J.P. & Faarborg, J. (1990) Estimating the viability of ovenbird and Kentucky warbler populations in forest fragments. *Conservation Biology* **4**: 193-196.

Gilpin, M. & Hanski, I. (eds.) (1991) *Metapopulation dynamics: empirical and theoretical investigations.* Academic Press, London.

Goldenfeld, N. & Kadanoff, L.P. (1999) Simple lessons from complexity. *Science* **284**:87-89.

Golley, F.B. (1993) *A history of the ecosystem concept in ecology.* Yale University Press, New Haven.

Goossens, R., D'Haluin, E., Larnoe, G. (1991) Satellite image interpretation (SPOT) for the survey of the ecological infrastructure in a small scaled landscape (Kempenland, Belgium). *Landscape Ecology* **5**: 175-182.

Gosz, J.R. (1993) Ecotone hierarchies. *Ecological Applications* **3**:369-376.

Gotzmark, F. (1992) Naturalness as an evaluation criterion in nature conservation: a response to Anderson. *Conservation Biology* **6**: 455-460

Green B.H. (1996) *Countryside Conservation.* E & FN Spon, London (Third ed.).

Green, B.H., Simmons, E.A., Woltjer, I. (1996) *Landscape conservation. Some steps towards developing a new conservation dimension. A draft report of the IUCN-CESP landscape Conservation Working Group.* Department of Agriculture, Horticulture and Environment, Wye College, Ashford, Kent, UK.

Gregory, S.V., Swanson, F.J., McKee, W.A., Cummins, K.W. (1991) An ecosystem perspective of riparian zones. *Bioscience* **41(8)**: 540-551.

Groffman, P.M. & Turner, C.L. (1995) Plant productivity and nitrogen gas fluxes in a tallgrass prairie landscape. *Landscape Ecology* **10**: 255-266.

Gross, J.E., Zank, C., Hobbs, N.T., Spalinger, D.E. (1995) Movements rules for herbivores in spatially heterogeneous environments: responses to small scale pattern. *Landscape Ecology* **10**: 209-217.

Grumbine, R.E. (1990) Protecting biological diversity through the greater ecosystem concept. *Natural Areas Journal* **10(3)**: 114-120.

Grumbine, R.E. (1994) What is ecosystem management? *Conservation Biology* **8(1)**: 27-38.

Guastalli, G. (1998) *Modalità di utilizzazione da parte del cinghiale* (Sus scrofa) *di biotopi erbacei submontani della Toscana settentrionale.* Unpl. Thesis Dissertation, Pisa University.

Gustafson, E.J. & Parker, G.R. (1992) Relationships between landcover proportion and indices of landscape spatial pattern. *Landscape Ecology* **7**: 101-110.

Haber, W. (1990) Using landscape ecology in planning and management. *In:* Zonneveld, I.S. and R.T.T. Forman (eds.), *Changing landscapes: An ecological perspective.* Springer-Verlag, New York. Pp. 217-232.

Haeuber, R.A. & Michener, W.K. (1998) Policy implications of recent natural and managed floods. *Bioscience* **48(9)**: 765-772.

Haila, Y. (1988) Calculating and miscalculating density: the role of habitat geometry. *Ornis Scandinavica* **19**: 88-92.

Haila Y. (1998) Assessing ecosystem health across spatial scales. In. Rapport, D., Costanza, R., Epstein, P.R., Gaudet, C., Levins, R. (eds) 1998 *Ecosystem health.* Blackwell, USA.

Haines-Young, R.H. (1992) The use of remotely-sensed satellite imagery for landscape classification in Wales (UK). *Landscape Ecology* **7**: 253-274.

Halladay, D. & Gilmour, D.A. (eds.) (1995) *Conservation biodiversity outside protected areas. The role of traditional agro-ecosystems.* IUCN (Gland), Switzerland, and Cambridge, UK .

Hall, F.G., Botkin, D.B., Strebel, D.E., Woods, K.D., Goetz, S.J. (1991) Large scale patterns of forest succession as determined by remote sensing. *Ecology* **72**: 628-640.

Hansen, A.J. & di Castri, F. (eds.) (1992*) Landscape boundaries. Consequences for biotic diversity and ecological flows.* Springer-Verlag, New York.

Hansen, A.J., Spies, T.A., Swanson, F.J., Ohmann, J.L. (1991) Conserving biodiversity in managed forests. *Bioscience* **41(6)**: 382-392.

Hansen, A.J., di Castri, F. , Naiman, R.J. (1992a) Ecotones: what and why? *In:* Hansen, A.J. and F. di Castri (eds.), *Landscape boundaries. Consequences for biotic diversity and ecological flows.* Springer Verlag, New York. Pp. 9-46.

Hansen, A.J., Risser, P.G., di Castri, F. (1992b) Epilogue: biodiversity and ecological flows across ecotones. *In:* Hansen, A.J. and F. di Castri (eds.), *Landscape boundaries. Consequences for biotic diversity and ecological flows.* Springer Verlag, New York. Pp. 423-438.

Hanski, I.A. (1997) Metapopulation dynamics. From Concepts and observations to predictive models. *In:* Hanski, I. A. and M.E. Gilpin (eds.), *Metapopulation Biology.* Academic Press, San Diego, Ca.Pp. 69-91.

Hanski, I. A. & Gilpin, M. E. (1991) Metapopulation dynamics: brief history and conceptual domain. *Biological Journal of Linnean Society* **42**: 3-16.

Hanski, I. A. & Gilpin, M.E. (eds.) (1997) *Metapopulation Biology.* Academic Press, San Diego, Ca.

Hanski, I., Kuussaari, M., Nieminen, M. (1994) Metapopulation structure and migration in the butterfly *Melitaea cinxia*. *Ecology* **75**: 747-762.

Hansson, L. (1991) Dispersal and connectivity in metapopulation. *Biological Journal of Linnean Society* **42**: 89-103.

Hansson, L. & Angelstam, P. (1991) Landscape ecology as a theoretical basis for nature conservation. *Landscape Ecology* **5**: 191-201.

Hantush, M. M. & Marino, M.A. (1995) Continuous time stochastic analyses of groundwater flow in heterogeneous aquifers. *Water Resources Research* **31**: 565-575.

Haralick, R.M., Shanmugam, K., Dinstein, I. (1973) Textural features for image classification. IEEE Transactions on Systems. *Man, and Cybernetics SMC* **3**: 610-21.

Hargis C.C., Bissonette J.A., David J.L. (1997) Understanding measures of landscape pattern. *In:* Bissonette J.A. (ed.), *Wildlife and landscape ecology. Effects of pattern and scale.* Springer, New York.

Harris, L.D (1984) *The fragmented forest. Island biogeography theory and the preservation of biotic diversity.* The University of Chicago Press .

Harris, L.D. (1988) Edge effects and conservation of biotic diversity. *Conservation Biology* **2**: 330-332.

Harrison, R.L. (1992) Toward a theory of inter-refuge corridor design. *Conservation Biology* **6**: 293-295.

Harrison, S. & Taylor, A.D. (1997) Empirical evidence for metapopulation dynamics. *In:* Hanski, I. A. and M.E. Gilpin (eds.), *Metapopulation Biology.* Academic Press, San Diego, Ca.Pp. 27-42.

Hastings, A. (1990) Spatial heterogeneity and ecological models. *Ecology* **71**: 426-428.

Hastings, H.M. & Sugihara, G. (1993) *Fractals. A user's guide for the natural sciences.* Oxford University Press, Oxford, UK.

Healy, S. (ed.) (1998) *Spatial representation in animals.*Oxford Univeristy Press, New York.

Henein, K., and G. Merriam (1990) The elements of connectivity where corridor quality is variable. *Landscape Ecology* **4(2/3)**:157-170.

Herkert,J.R. (1994) The effect of habitat fragmentation on midwestern grassland bird communities. *Ecological Applications* **4**: 461-471.

Hill, G.E. (1995) Ornamental traits as indicators of environmental health. *Bioscience* **45(1)**: 25-31

Hinsley, S.A., Bellamy, P.E., Newton, I. (1995) Bird species turnover and stochastic extinction in woodland fragments. *Ecography* **18**: 41-50.

Hobbs, R.J. & Hopkins, A.J.M. (1990) From frontier to fragments: European impact on Australia's vegetation. *Proc. ecol. Soc. Aust.* **16**: 93-114.

Hoekstra, T.W., Allen, T.F.H., Flather, C.H. (1991) Implicit scaling in ecological research. *Bioscience* **41**: 148-154.

Hof, J. & Bevers, M. (1998) *Spatial optimization for managed ecosystems.* Columbia University Press, New York.

Hofmann-Wellenhof B. , Lichteneger H. , Collins J. (1993) *Global Positioning System, Theory and Practice,* Second Edition. Springer Verlag, Wien, New York.

Hokit, D.G., Stith, B.M., Branch, L.C. (1999) Effects of landscape structure in Florida scrub: A population perspective. *Ecological Applications* **9(1)**: 124-134.

Holland, M.M. (1988) SCOPE/MAB Technical consultations on landscape boundaries. report of a SCOPE/MAB workshop ecotones. *In:* Di Castri, F., Hansen, A.J.,

Holland, M.M. (eds.), *A new look at ecotones: emerging international projects on landscape boundaries.* Biology International, Special Issue 17.

Holland, L.D. & Hansen, A.J. (1988) Meeting reviews: ecotones. *Bulletin of the Ecological Society of America* **69**: 54-56.

Holland, M.M., Risser, P.G., Naiman, R.J. (1991) *Ecotone. The role of landscape boundaries in the management and restoration of changing environments.* (eds.), Chapman & Hall, London, UK.

Holling, C.S. (1992) Cross -scale morphology, geometry, and dynamics of ecosystems. *Ecological Monographs* **62**: 447-502.

Hornberg, G., Zackrisson, O., Segerstrom, U., Svensson, B.W., Ohlson, M., Bradshaw, R.H.W. (1998) Boreal swamp forests. *Bioscience* **48(10)**: 795-802.

Hubbell, S.P., Foster, R.B., O'Brien, S.T., Harms, K.E., Condit, R., Wechsler, B., Wright, S.J., Loo de Lao, S. (1999) Light-gap disturbances, recruitment limitation, and tree diversity in a neotropical forest. *Science* **283**: 554-557.

Huff, D.E. & Varley, J.D. (1999) Natural regulation in Yellowstone National Park's northern range. *Ecological Applications* **9(1)**: 17-29.

Huggett, R.J. (1975) Soil landscape systems: a model of soil genesis. *Geoderma* **13**: 1-22.

Huggett, R.J. (1995) *Geoecology. An evolutionary approach.* Routledge, London and New York.

Hughes, T.P., Baird, A.H., Dinsdale, E.A., Moltschaniwskyj, N.A., Pratchett, M.S., Tanner, J.E., Willis, B.L. (1999) Patterns of recruitment and abundance of coral along the Great Barrier Reef. *Nature* **397**: 59-62.

Hulse, D.W. & Larsen, K. (1989) *MacGIS 2.0, A geographic information system for Macintosh.* University of Oregon.

Hunter, M.L. Jr. (1990) *Wildlife, forests, and forestry.* Prentice Hall, Englewood Cliffs, New Jersey.

Hutchinson, G.E. (1957) Concluding remarks. Population Studies: Animal Ecology and Demography. *Cold Spring Harbor Symposia on Quantitative Biology* **22**: 415-27.

Iannaccone, P.M. & Khokha, M. (1995) Mosaic pattern in tissues from chimeras. *In:* Iannaccone, P.M., Khokha, M.(eds), *Fractal geometry in biological systems. An analytical approach.*CRC Press, Boca Raton, FL.

Ichoku, C., Karnieli, A., Verchovsky, I. (1996) Application of fractal techniques to the comparative evaluation of two methods of extracting channel networks from digital models. *Water Resources Research* **32**: 389-399.

Isaaks, E.H. & Srivastava, R.M. (1989) *An introduction to applied geostatistics.* Oxford University Press, New York.

Iturbe, I. R. & Rinaldo, A. (1997) *Fractal river basins. Change and self-organization.* Cambridge University Press, UK.

IUCN, (1980) *World Conservation Strategy: Living Resource Conservation for Sustainable Development.* IUCN-UNEP-WWF, Gland 44 pp.

Iverson, L.R., Graham, R.L., Cook, E.A. (1989) Applications of satellite remote sensing to forested ecosystems. *Landscape Ecology* **3**: 131-143.

Iverson, L.R., Cook, E.A., Graham, R.L. (1994) Regional forest cover estimation via remote sensing: the calibration center concept. *Landscape Ecology* **9**: 159-174.

Jeffrey, D.W. & Madden, B. (1994) *Bioindicators and environmental management.* Academic Press, London.

Joalé, P. & Benvenuti, S. (1982) Seasonal and diurnal variation of weight in four passeriformes in Autumn and Winter. *Avocetta* **6**: 63-74.

Johnson, G.D., Tempelman, A., Patil, G.P. (1995) Fractal based methods in ecology: a review for analysis at multiple spatial scale. *Coenoses* **10(2)**: 123-131.

Johnson, A.R., Wiens, J.A., Milne, B.T., Crist, T.O. (1992) Animal movements and population dynamics in heterogeneous landscapes. *Landscape Ecology* **7**: 63-75.

Johnston, C.A. & Naiman, R.J. (1987) Boundary dynamics at the aquatic-terrestrial interface: the influence of beaver and geomorphology. *Landscape Ecology* **1**:47-58.

Kareiva P (1987) Habitat fragmentation and the stability of predator-prey interactions. *Nature* **326**: 368-369 and 390.

Kareiva, P. (1994) Space: the final frontier for ecological theory. *Ecology* **95**: 1.

Karr, J.R. & Dudley, D.R. (1981) Ecological perspective on water quality goals. *Environ. Manage.* **5**: 55-68.

Kattan, G.H., Alvarez-Lopez, H., Giraldo, M. (1994) Forest fragmentation and bird extinctions: San Antonio eighty years later. *Conservation Biology* **8**: 138-146.

Kavanagh, R.P. & Bamkin, K.L. (1995) Distribution of nocturnal forest birds and mammals in relation to the logging mosaic in south-eastern New South Wales, Australia. *Biological Conservation* **71**: 41-53.

Kenkel, N.C. & Walker, D.J. (1996) Fractals in the biological sciences. *Coenoses* **11(2)**: 77-100.

Kesner, B.T. & Meentemeyer, V. (1989) A regional analysis of total nitrogen in an agricultural landscape. *Landscape Ecology* **2**: 151-163.

Kief, T. L., White, C.S., Loftin, S.R., Aguilar, R., Craig, J.A., Skaar, D.A. (1998) Temporal dynamics in soil carbon and nitrogen resources at a grassland-shrubland ecotone. *Ecology* **79(2)**: 671-683.

King , A.W. (1997) Hierarchy theory: A guide to system structure for wildlife biologists. *In:* A. Bissonette (ed.), *Wildlife and landscape ecology. Effects of pattern and scale.* Springer Verlag, New York. Pp. 185-212.

Klaassen, W. & Claussen, M (1994) Landscape variability and surface flux parameterization in climate models. *Agricultural and Forest Meteorology* **73**: 181-188.

Klein, B.C. (1989) Effects of forest fragmentation on dung and carrion beetle communities in central Amazonia. *Ecology* **70**: 1715-1725.

Knapp, A.K., Blair, J.M., Briggs, J.M., Collins, S.L., Hartnett, D.C., Johnson, L.C., Towne, E.G. (1999) The keystone role of bison in north American tallgrass prairie. *Bioscience* **49(1)**: 39-50.

Knick, S.T. & Rotenberry, J.T. (1995) Landscape characteristics of fragmented shrubsteppe habitats and breeding passerine birds. *Conservation Biology* **9**: 1059-10 - 71.

Knopf, F.L., Johnson, R.R., Rich, T., Samson, F.B., Szaro, R.C. (1988) Conservation of riparian ecosystems in the United States. *Wilson Bull.* 100(2): 272-284.

Koenig, W.D. (1997) Spatial autocorrelation in California land birds. *Conservation Biology* 12: 612-619.

Koenig, W.D. (1999) Spatial autocorrelation of ecological phenomena. *TREE* **14**: 22-26.

Kolasa, J. & Pickett, S.T.A (1991) *Ecological heterogeneity.* Springer-Verlag, New York.

Kolasa, J. & Rollo, CD. (1991) Introduction: the heterogeneity of heterogeneity: a glossary. *In:* Kolasa, J. and S.T.A. Pickett (eds.), *Ecological heterogeneity*. Springer-Verlag, New York. Pp. 1-23.

Kotliar, N.B. & Wiens, J.A. (1990) Multiple scales of patchiness and patch structure: a hierarchical framework for the study of heterogeneity. *Oikos* **59**: 253-260.

Kosko, B. (1993) *Fuzzy thinking. The new science of fuzzy logic.* Hyperion, New York.

Kozak, E., Pachepsky, Y.A., Sokolowski, S., Sokolowska, Z., Stepniewski, W. (1996) A modified number-based method for estimating fragmentation fractal dimensions of soils. *Soil Science* **60**: 1291-1297.

Kozakiewicz, M. & Szacki, J. (1995) Movements of small mammals in a landscape: Patch restriction or nomadisms? Pag; 78-94. *In:* Lidicker, W.Z. (ed.), *Landscape approaches in mammalian ecology and conservation.* University of Minnesota Press, Minneapolis, US.

Krajick, K. (1999) Scientists and climbers - discover cliff ecosystems. *Science* **283**: 1623-1625.

Krebs, J.R. & McCleery, R.H. (1984) Optimization in behavioural ecology. *In:* Krebs, J.R. and N.B. Davies (eds.), *Behavioural Ecology*. Blackwell Scientific Publications, Oxford (2nd edition). Pp. 91-121.

Krummel, J.R., Gardner, R.H., Sugihara, G., O'Neill, R.V., Coleman, P.R. (1987) Landscape patterns in a disturbed environment. *Oikos* **48**: 321-324.

Kurki, S., Nikula, A., Helle, P., Linden, H. (1998) Abundance of red fox and pine marten in relationship to the composition of boreal forest landscapes. *Journal of Animal Ecology* **67**: 874-886.

Lamberson, R.H., McKelvey, R., Noon, B.R., Voss, C. (1992) A dynamic analysis of northern spotted owl viability in a fragmented forest landscape. *Conservation Biology* **6(4)**: 505-512.

Laszlo, E. (1996) *The whispering pond.* Element Books, Inc., Rockport, US.

Lathrop, R. & Peterson, D.L. (1992) Identifying structural self-similarity in mountainous landscapes. *Landscape Ecology* **6**: 233-238.

Lauenroth, W.K., Coffin, D.P. , Burke, I.C., Virginia, R.A. (1997) Interactions between demographic and ecosystem processes in a semi-arid and arid grassland: a challenge for plant functional types. *In*: Smith T.M., Shugart, H.H., Woodward, F.I. (eds.), *Plant functional types: Their relevance to ecosystem properties and global change.* Cambridge University Press, Cambride, UK. Pp. 234-254.

Leach, M.K. & Givnish, T.J. (1996) Ecological determinants of species loss in remnant prairies. *Science* **273**: 1555-1558.

Leduc, A. Prairies, Y.T., Bergeron, Y. (1994) Fractal dimension estimates of a fragmented landscape: sources of variability. *Landscape Ecology* **9**: 279-286.

Leick A. (1990) *GPS Satellite Surveying* .Wiley and Sons, New York.

Leimgruber, P., McShea, W.J., Rappole, J.H. (1994) Predation on artificial nests in large forest blocks. *J. Wildl. Manage.* **58**: 254-260.

Leopold, A. (1933) *Game management.* Scriber, New York, USA.

Levin, S.A. (1992) The problem of pattern and scale in ecology. *Ecology* **7**: 1943-1967.

Levins, R. (1969) Some demographic and genetic consequences of environmental heterogeneity for biological control. *Bull. Entomol. Soc. Am.* **15**: 237-240.

Levins, R. (1970) Extinction. *In:* Gertenshaubert, M. (ed.), *Some Mathematical questions in biology. Lectures in mathematics in the life sciences.* American Mathematical Society, Providence, Rhode Island. Pp. 77-107.

Li, B.L., Loehle, C., Malon, D. (1996) Microbial transport through heterogeneous porous media: random walk, fractal, and percolation approaches. *Ecological Modelling* **85**: 285-302.

Li, H. & Reynolds, J.F. (1994) A simulation experiment to quantify spatial heterogeneity in categorical maps. *Ecology 75*: 2446-2455.

Lidicker, W.Z. Jr (1995) The landscape concept: something old, something new. *In:* W.Z.Lidicker Jr. (ed.), *Landscape approaches in mammalian ecology and conservation.* University of Minnesota Press, Minneapolis, US. Pp. 3-19.

Lillesand, T.M. & Kiefer, R.W. (1987) *Remote sensing and image interpretation.* Second ed. Wiley & Sons, New York.

Lima, S.L. & Zollner, P.A. (1996) Towards a behavioural ecology of ecological landscapes. *TREE* **11**:131--135.

Lindemayer, D.B. & Nix, H.A. (1993) Ecological principles for the design of wildlife corridors. *Conservation Biology* **7**: 627-630.

Liu, J. (1992) ECOLOLECON: A spatially-explicit model for ecological economics of species conservation in complex forest landscapes. *Ecological Modelling* **70**: 63-87.

Liu, J. & Ashton, P.S. (1999) Simulating effects of landscape context and timber harvest on tree species diversity. *Ecological Applications* **9(1)**: 186-201.

Loehle, C. (1990) Home range: A fractal approach. *Landscape Ecology* **5**: 39-52.

Loehle, C. & Li, B.L. (1996) Statistical properties of ecological and geologic fractals. *Ecological Modelling* **85**: 271-284.

Lord, J.M. & Norton, D.A. (1990) Scale and the spatial concept of fragmentation. *Conservation Biology* **4**: 197-202.

Lubchenco, J., Olson, A.M., Brudbaker, L.B., Carpenter, S.R., Holland, M.M., Hubbell, S.P., Levin, S.A., MacMahon, J.A., Matson, P.A., Melillo, J.M., Mooney, H.A., Peterson, C.H., Pulliam, H.R., Real, L.A., Regal, P.J., Risser, P.G. (1991) The sustainable biosphere initiative: an ecological research agenda. *Ecology* **72**: 371-412.

Lucas, O.W.R. (1991) *The design of forest landscapes.* Oxford University Press, Oxford, UK.

MacArthur, R.H. (1972) *Geographical ecology, patterns in the distribution of species.* Princeton University Press, Princeton.

MacArthur, R.H. & MacArthur, J.W. (1961) On bird species diversity. *Ecology* **42**: 594-598.

MacArthur, R.H., MacArthur, J.W. and Preer, J. (1962) On bird species diversity. II. Prediction of bird census from habitat measurements. *American Naturalist* **96**: 167-174.

MacArthur, R.H. & Wilson, E.O. (1967) *The theory of island biogeography.* Princeton University Press, Princeton.

Mack MC. & D'Antonio C.M. (1998) Impacts of biological invasions on disturbance regimes. *TREE* **13**: 195-198.

Magnuson, J.J. (1990) Long-term ecological research and the invisible present. *Bioscience* **40**:495-501.

Maguire, D.J., Goodchild, M.F., Rhind, D.W (1991) *Geographical information systems.* Longman Scientific & Technical., Harlow, England.

Maguire, D.J. (1991) An overview and definition of GIS. *In:*Maguire, D.J. , Goodchild, M.F., Rhind, D.W. (eds.), *Geographical information systems.* Longman Scientific & Technical., Harlow, England. Pp. 9-20.

Mandelbrot, B.B. (1975) *Les objects fractals: Forme, hasard et dimension.* Flammarion, Paris.

Mandelbrot, B. (1982) *The fractal geometry of nature.* Freeman, New York.

Mandelbrot, B. (1986) Self-affine fractal sets. *In:* Pietronero, L. and E. Tosatti (eds.), *Fractals in physics.* North-Holland, Amsterdam. Pp. 3-28.

Mann, C.C. & Plummer, M.L. (1993) The high cost of biodiversity. *Science* **260**: 1868-1871.

Margules, C.R., Milkovits, G.A., Smith, G.T. (1994) Contrasting effects of habitat fragmentation on the scorpion *Cercophonius squama* and an amphipod. *Ecology* **75**: 2033-2042.

McAuliffe, J. R. (1994) Landscape evolution, soil formation, and ecological patterns and processes in Sonoran desert Bajadas. *Ecological Monographs* **64**: 111-148.

McDowell, W.H. & Likens, G.E. (1988) Origin, composition, and flux of dissolved organic carbon in the Hubbard Brook valley. *Ecological Monographs* **58**: 177-195.

McGarigal, K. & Marks, B.J. (1995) *Fragstats: Spatial pattern analysis program for quantifying landscape structure.* USDA, Pacific Northwest Research Statio. General Technical Report PNW-GTR-351.

McNeely, J.A. (1999) *Bioregional planning and ecosystem-based management: Commonalties, contrasts, constraints, and convergences.* Proceedings of the Workshop on Integrated Planning at different scales: Policy and Practice. Scottish Natural heritage, Perth, Scotland, 7-9 April 1999, in press.

Meentemeyer, V. & Box, E.O. (1987) Scale effects in landscape studies. *In:* Turner, M.G.(ed.), *Landscape heterogeneity and disturbance.* Springer-Verlag, New York. Pp. 15-36.

Merriam, G. (1984) Connectivity: a fundamental ecological characteristic of landscape pattern. *In:* Brandt, J., Agger, P. (eds.), *Methodologies in landscape ecological research and planning.* Vol. 1 Proceedings of the first international seminar of the International Association of Landscape Ecology. Roskilde, Denmark, October 15-19. Pp. 5-15.

Merrill, S.B., Cuthbert, F.J., Oehlert, G. (1997) Residual patches and their contribution to forest-bird diversity on Northern Minnesota aspen clearcut. *Conservation Biology* 12(1): 190-199.

Metzger, J.P. & Muller, E. (1996) Characterizing the complexity of landscape boundaries by remote sensing. *Landscape Ecology* 11: 65-77.

Meyers, N. (1989)Threatened biotas: "Hotspots" in tropical forests. *Environmentalist* **8**: 1-20.

Meyers, N. (1990) The biodiversity challenge: expande hotspots analysis. *Environmentalist* **10:** 243-256.

Meyers, N. (1995) The environmental unknowns. *Science* **269**: 358-360

Miller, D.E. (1981) *Energy at the surface of the earth. An introduction to the energetics of ecosystems.* Academic Press, New York.

Miller, K.R. (1999) What is a bioregional-planning? *Proceedings of the Workshop on Integrated Planning at different scales: Policy and Practice.* Scottish Natural heritage, Perth, Scotland, 7-9 April 1999 (in press).

Mills, L.S., Soulé, M.E., Doak, D.F. (1993) The keystone-species concept in ecology and conservation. *Bioscience* **43**: 219-224.

Milne,B.T. (1991) Lessons from applying fractal models to landscape patterns. *In:* M.G. Turner and R.H. Gardner (eds.), *Quantitative methods in landscape ecology.* Springer- Verlag, New York.Pp.199-235.

Milne, B.T. (1997) Applications of fractal geometry in wildlife biology. *In:* Bissonette, J.A. (ed.), *Wildlife and landscape ecology. Effects of pattern and scale.* Springer Verlag, New York. Pp. 32-69.

Mladenoff, D.J., White, M.A., Crow, T.R., Pastor, J. (1994) Applying principles of landscape design and management to integrate old-growth forest enhancement an commodity use. *Conservation Biology* **8(3)**: 752-762.

Mladenoff, D.J., Sickley, T.A., Wydeven, A.P. (1999) Predicting gray wolf landscape recolonization: logistic regression models vs. new field data. *Ecological Applications* **9(1)**: 37-44.

Morreale, S.J., Standora, E.A., Spotila, J.R., Paladino, F.V. (1996) Migration corridor for sea turtles. *Nature* **384**: 319-320.

Mueller-Dumbois, D. & Ellemberg, H. (1974) *Aims and methods of vegetation ecology.* John Wiley and Sons, New York.

Musick, H.B. & Grover, H.D. (1991) Image textural measures as indices of landscape pattern. *In:* M.G. Turner and R.H. Gardner (eds.), *Quantitative methods in landscape ecology.* Springer-Verlag, New York. Pp. 77-103.

Naeem, S., Thompson, L.J., Lawler, S.P., Lawton, J.H., Woodfin, R.M. (1994) Declining biodiversity can alter the performance of ecosystems. *Nature* **368**: 734-737.

Naiman, R.J., Holland, M.M. , Decamps, H., Risser, P.G. (1988) A new UNESCO program: research and management of land:inland water ecotones. *Biology International, Special Issue* **17**: 107-136.

Naiman, R.J. & Decamps, H. (1990) *The ecology and management of aquatic-terrestrial ecotones.* The Parthenon Publishing Group, Park Ridge, NJ, USA.

Naiman, R.J., Rogers, K.H. (1997) Large animals and system-level characteristics in river corridors. *Bioscience* **47(8)**: 521-529.

Naiman, R.J., Decamps, H., Pollock, M. (1993) The role of riparian corridors in maintaining regional biodiversity. *Ecological Applications* **3(2)**: 209-212.

Nardelli, R. (1996) Distribuzione ed abbondanza del pettirosso (*Erithacus rubecula*) attraverso un gradiente ambientale del sistema appeninico tosco-emiliano. Unpl. Thesis Dissertation, Parma University 1995-96.

Naveh, Z. (1971) The conservation of ecological diversity of Mediterranean ecosystems through ecological management. *In*: Duffey E, Watt A.S. (eds). *The scientific management of animal and plant communities for conservation.* Oxford, Blackwell Science. Pp. 603-622.

Naveh, Z. (1992) A landscape ecological approach to urban systems as part of the total human ecosystem. *J.Nat.Hist.Mus., Inst. Chiba,* **2(1)**: 47-52.

Naveh, Z. (1998) Culture and landscape conservation: A landscape-ecological perspective. *In:* Gopal, B., Pathak, P.S., Saxena, KG. (eds.), *Ecology Today: An Anthology of Contemporary Ecological Research*: 19-48.

Naveh, Z. & Lieberman, A.S. (1984) *Landscape ecology. Theory and application.* Springer-Verlag, New York.

Naveh, Z. (1987) Biocybernetic and thermodynamic perspectives of landscape functions and land use patterns. *Landscape Ecology* **1**: 75-83.

Naveh, Z. (1994) From biodiversity to ecodiversity: A landscape-ecology approach to conservation and restoration. *Restoration Ecology* **2**: 180-189.

Nepstad, D.C., Moutinho, P.R., Uhl, C., Vieira, I.C., da Silva,C. J.M. (1996) The ecological importance of forest remnants in an eastern Amazonian frontier landscape. *In:* Schelhas, J. and R. Greenberg (eds.), *Forest patches in tropical landscapes.* Island Press, Washington, D.C. Pp. 133-149.

Newmark, W.D. (1990) Tropical fragmentation and the local extinction of understory birds in the eastern Usambara Mountains, Tanzania. *Conservation Biology* **5**: 67-78

Nilsson, C. & Grelsson, G. (1995) The fragility of ecosystems: a review. *Journal Appl. Ecol.* **32**: 677-692.

Norton, B.G. & Ulanowicz, R.E. (1992) Scale and biological policy: A hierarchical approach. *Ambio* **21**: 244-249.

Norton, D.A., Hobbs, R.J., Atkins, L. (1995) Fragmentation, disturbance and plant distribution: mistletoes in woodland remnants in the Western Australian wheatbelt. *Conservation Biology* **9**: 426-438.

Noss, R.F. (1983) A regional landscape approach to maintain diversity. *Bioscience* **33**:700-706.

Noss, R.F. (1990) Indicators for monitoring biodiversity: A hierarchical approach. *Conservation Biology* **4(4)**: 355-364.

Odum, E.P. (1959) *Fundamentals of ecology.* Second Edition W.B. Saunders Company, Philadelphia, Pennsylvania, USA.

Odum, W.E. (1990) Internal processes influencing the maintenance of ecotones: do they exists? *In:* Naiman, R.J. and H. Decamps (eds.), *The ecology and management of aquatic-terrestrial ecotones.* MAB, UNESCO, Paris. Pp. 91-102.

Olsen, E.R., Ramsey, R.D., Winn, D.S. (1991) A modified fractal dimension as a measure of landscape diversity. *Photogrammetric Engineering & Remote Sensing* **59(10)**: 1517-1520.

O'Neill, R.V., DeAngelis, D.L., Waide, J.B., Allen, T.F.H. (1986) *A hierarchical concept of ecosystems.* Princeton University Press, Princeton.

O'Neill, R.V., Krummel, J.R., Gardner, R.H., Sugihara, G., Jackson, B., De Angelis, D.L., Milne, B.T., Turner, M.G., Zygmunt, B., Christensen, S.W., Dale, V.H., Graham, R.L., (1988) Indices of landscape pattern. *Landscape Ecology* **1**: 153-162.

O'Neill, R.V., Johnson, A.R., King, A.W. (1989) A hierarchical framework for the analysis of scale. *Landscape Ecology* **3**:193-205.

O'Neill, R.V., Kahn, J.R., Russell, C.S. (1998) Economics and ecology: The need for detente in conservation ecology. *Conservation Ecology (online)* **2(1)**: 4.

Opdam, P., Foppen, R., Reijnen, R., Schotman, A. (1994) The landscape ecological approach in bird conservation: integrating the metapopulation concept into spatial planning. *Ibis* **137**: S139-S146.

Ostfeld, R.S. (1992) Small-mammal herbivores in a patchy environment: individual strategies and population responses. *In:* Hunter, M.D., Ohgushi, T., Price, P.W. (eds.), *Effects of resource distribution on animal-plant interaction.* Academic Press. Pp. 43-74.

Paine, R.T. (1966) Food web complexity and species diversity. *American Naturalist* **100**: 65-75.

Pachepsky, Y.A., Shcherbakov, R.A., Korsunskaya, L.P. (1995) Scaling of soil water retention using a fractal model. *Soil Science* **159(2)**: 99-104.

Paine, R.T. (1966) Food web complexity and species diversity. *Am. Nat.* **100**: 65-75.

Paine, R.T. (1969) A note on trophic complexity and community stability. *American Naturalist* **103**: 91-93.

Parrish, J.K., Edelstein-Keshet, L. (1999) Complexity, pattern, and evolutionary trade-off in animal aggregation. *Science* **284**: 99-101.

Pasitschniak, M. & Messier, F. (1995) Risk of predation on waterfowl nests in the Canadian prairies: effects of habitat edges and agricultural practices. *Oikos* **73**: 347-355.

Pearson, S.M. (1991) Food patches and the spacing of individual foragers. *Auk* **108**: 355-362.

Pearson, S.M. & Gardner, R.H. (1997) Neutral models: Useful tools for understanding landscape patterns. *In:* A. Bissonette (ed.), *Wildlife and landscape ecology. Effects of pattern and scale.* Springer Verlag, New York. Pp. 215-230.

Perevolotsky, A., Seligman, N.G. (1998) Role of grazing in Mediterranean rangeland ecosystems. *Bioscience* **48(12)**: 1007-1017.

Perfect, E. & Blevins, R.L. (1997) Fractal characterization of soil aggregation and fragmentation as influenced by tillage treatment. *Soil Science* **61(3)**: 896-900.

Perfect, E., Kay, B.D., Rasiah, V. (1993) Multifractal model for soil aggregate fragmentation. *Soil Science* **57(4)**: 896-900.

Perfect, E. & Kay, B.D. (1995) Applications of fractals in soil and tillage research: a review. *Soil & Tillage Research* **36**: 1-20.

Perfect, E., McLaughlin, N.B., Kay, B.D., Topp, G.C. (1996) An improved fractal equation for the soil water retention curve. *Water Resources Research* **32**: 281-287.

Perrier, E., Mullon, C., Rieu, M. (1995) Computer construction of fractal structures: Simulation of their hydraulic and shrinkage properties. *Water Resources Research* **31**: 2927-2943.

Peterjohn, W.T., Correll, D.L. (1984) Nutrient dynamics in an agricultural watershed: observations on the role of a riparian forest. *Ecology* **65**: 1466-1475.

Peterson, R.O. (1999) Wolf-moose interaction on Isle Royale: the end of natural regulation? *Ecological Applications* **9(1)**: 10-16.

Peterson, G., Allen, C.R., Holling, C.S. (1998) Ecological resilience, biodiversity, and scale. *Ecosystem* **1**: 6-18.

Phillips, A. (1997) Landscape approaches to national parks and protected areas. *In:* Nelson, G.J. and R. Serafin (eds.), *National parks and protected areas.* NATO ASI Series G: Ecological Sciences, vol. 40:31-42.

Pickett, S.T.A. & White, P.S. (1985) *The ecology of natural disturbance and patch dynamics.* Academic Press, London.

Pickett, S.T.A. & Cadenasso, M.L. (1995) Landscape ecology: spatial heterogeneity in ecological systems. *Science* **269**:331-334.

Pielke, R.A. & Avissar, R. (1990) Influence of landscape structure on local and regional climate. *Landscape Ecology* **4**: 133-155.

Pickett S.T.A. & Rogers, K.H. (1997) Patch dynamics: The transformation of landscape structure and function. *In:* A. Bissonette (ed.), *Wildlife and landscape ecology. Effects of pattern and scale*, Springer Verlag, New York. Pp. 101-127.

Pimentel, D., Harvey, C., Resosudarmo, K., Sinclair, K., Kurz, D., MnNair, M., Crist, S., Shpritz, L., Fitton, L., Saffouri, R., Blair, R. (1995) Environmental and economic costs of soil erosion and conservation benefits. *Science* **267**: 1117-1123.

Plachter, H. & Rossler, M. (1995) Cultural landscapes: reconnecting culture and nature. *In:* van Droste, B., Plachter, H., Rossler, M. (eds.), *Cultural landscapes of universal value.* Gustav Fischer, Jena.

Plotnick, R.E., Gardner, R.H., O'Neil, R.V. (1993) Lacunarity indices as measures of landscape texture. *Landscape Ecology* **8**: 201-211.

Plowright, R.C. & Galen, C. (1985) Landmarks or obstacles: the effect of spatial heterogeneity on bumble bee foraging behavior. *Oikos* **44**: 459-464.

Pollard, E. & Yates, T.J. (1993) *Monitoring butterflies for ecology and conservation.* Chapman & Hall, London.

Powell, D.M. & Ashworth, P.J. (1995) Spatial pattern of flow competence and bed load transport in a divided gravel bed river. *Water Resource Research* **31**: 741-752.

Power, M.E., Tilman, D., Estes, J.A., Menge, B.A., Bond, W.J., Mills, B.L., Daily, G., Castilla, J.C., Lubchenco, J., Paine, R.T. (1996) Challenges in the quest for keystones. *Bioscience* **46(8)**: 609-620.

Probst, J.R. & Weinrich, J. (1993) Relating Kirtland's warbler population to changing landscape composition and structure. *Landscape Ecology* **8(4)**: 257-271.

Pulliam, R. (1988) Sources-sinks, and population regulation. *American Naturalist* **132**: 652-661.

Pulliam, H.R. (1996) Sources and sinks: empirical evidence and population consequences. *In:* Rhodes, O.E., Chesser, R.K., Smith, M.H. (eds.), *Population dynamics in ecological space and time.* The University of Chicago Press, Chicago.Pp. 45-69.

Ranney, J.W., Bruner, M.C., Levenson, J.B. (1981) The importance of edge in the structure and dynamics of forest islands. *In*: Burgess, R.L., Sharpe, D.M. (eds.), *Forest island dynamics in man-dominated landscapes.* Springer-Verlag, New York. Pp. 67-95.

Ranta, E. Lindstrom, J., Kaitala, V., Kokko, H., Linden, H., Helle, E. (1997) Solar activity and hare dynamics: a cross-continental comparison. *American Naturalist* **149**: 765-775.

Rapport D.J., Costanza, R., McMichael, A.J. (1998) Assessing ecosystem health. *TREE* **13**: 397-402.

Rapport, D., Costanza, R., Epstein, P.R., Gaudet, C., Levins, R. (1998) *Ecosystem health.* Blackwell Science, Malden, USA.

Rapport, D.J. & Whitford, W.G. (1999) How ecosystems respond to stress. *Bioscience* **49(3)**: 193-203.

Rapport, D.J., Regier, H.A., Hutchinson, T.C. (1985) Ecosystem behavior under stress. *American Naturalist* **125**: 617-640.

Rasiah, V. (1995) Fractal dimension of surface-connected macrpore count-size distributions. *Soil Science* **159(2)**: 105-108.

Redpath, S.M. (1995) Impact of habitat fragmentation on activity and hunting behaviour in the tawny owl, *Strix aluco. Behavioural Ecology* **6**: 410-415.

Reid, W.V. (1998) Biodiversity hotspots. *TREE* **13**:275-280.

Rex, K.D. & Malanson, G.P. (1990) The fractal shape of riparian forest patches. *Landscape Ecology* **4:** 249-258.

Ricklefs, R.E. (1973) *Ecology.* Chiron Press, Inc.

Riitters K.H., O'Neill R.V., Jones K.B. (1997) Assessing habitat suitability at multiple scales: A landscape-level approach. *Biological Conservation* **81**: 191-202.

Ripley, B.D., (1981) *Spatial statistics.* John Wiley & Sons. New York.

Risser, P.G., Karr, J.R., Forman, R.T.T. (1984) *Landscape ecology. Directions and approaches.* Illinois Natural History Survey Special Publication number 2, Champaign, Illinois.

Risser, P.G. (1987) Landscape ecology: state-of-art. *In:* M.G. Turner (ed.), *Landscape Heterogeneity and Disturbance.* Springer-Verlag, New York. Pp. 3-14.

Risser, P.G. (1989) The movement of nutrients across heterogeneous landscapes. *In:* M. Clarholm and L. Bergstrom (eds.), *Ecology of arable land.* Kluwer Academic Publisher.Pp. 247-251.

Risser, P.G. (1993) Ecotones. Ecotones at local to regional scales from around the world. *Ecological Applications* **3(3)**:367-368.

Risser, P.G. (1995) Recommendations for managing ecotones. *Ecology International* **22**: 95-102.

Risser, P.G. (ed.) (1995) *Understanding and managing ecotones.* Ecology International 22.

Roland, J. (1993) Large-scale forest fragmentation increases the duration of tent caterpillar outbreak. *Oecologia* **93**: 25-30.

Root, R.B. & Kareiva, P.M. (1984) The search for resources by cabbage butterflies (*Pieris rapae*): ecological consequences and adaptive significance of markovian movements in a patchy environment. *Ecology* **65**: 147-165.

Rosenberg, D.K, Noon, B.R., Meslow, E.C. (1997) Biological corridors: Form, function, and efficay. *Bioscience* **47(10)**: 677-687.

Rudis, V.A. (1995) Regional forest fragmentation effects on bottomland hardwood community types and resource values. *Landscape Ecology* **10**: 291-307.

Rusek, J. (1992) Distribution and dynamics of soil organisms across ecotones. *In:* Hansen, A.J. and F. di Castri (eds.), *Landscape boundaries. Consequences for biotic diversity and ecological flows.* Springer Verlag, New York. Pp. 196-214.

Russell, R.W., Hunt, G.L. Jr., Coyle, K.O., Cooney, R.T. (1992) Foraging in a fractal environment: Spatial patterns in marine predator-prey system. *Landscape Ecology* **7**: 195-209.

Saab, V. (1999) Importance of spatial scale to habitat use by breeding birds in riparian forests: A hierarchical analysis. *Ecological Applications* **9(1)**: 135-151.

Samson, F. & Knopf, F. (1994) Prairie conservation in North America. *Bioscience* **44 (6)**: 418-421.

Saunders, D.A., Hobbs, R.J., Margules, C.R. (1991) Biological consequences of ecosystem fragmentation: a review. *Conservation Biology* **5**: 18-32.

Schieck, J., Lertzman, K., Nyberg, B., Page, R. (1994) Effects of patch size on birds in old-growth montane forests. *Conservation Biology* **9**: 1072-1084.

Schimel, D., Stillwell, M.A., Woodmansee, R.G. (1985) Biogeochemistry of C, N, P in a soil catena in the short-grass steppe. *Ecology* **66**: 276-282.

Schippers, P., Verboom, J., Knaapen, J.P., van Apeldoorn, R.C. (1996) Dispersal and habitat connectivity in complex heterogeneous landscapes: an analysis with a GIS-based random walk model. *Ecography* **19**: 97-106.

Schlesinger, W.H., Raikes, J.A., Hartley, A.E., Cross, A.F. (1995) On the spatial pattern of soil nutrients in desert ecosystems. *Ecology* **77**: 364-374.

Schlosser, I.J. 1991 Stream fish ecology: A landscape perspective. *Bioscience* **41(10)**: 704-712.

Schmidt, J.C., Webb, R.H., Valdez, R.A., Marzolf, G.R., Stevens, L.E. (1998) Science and values in river restoration in the Grand Canyon. *Bioscience* **48(9)**: 735-747.

Schreiber, , K-F (1990) The history of landscape ecology in Europe. *In:* Zonneveld, I.S. & Forman, R.T.T. (eds.), *Changing landscapes: An ecological perspective.* Springer Verlag.Pp. 21-33.

Schumaker, N.H. (1996) Using landscape indices to predict habitat connectivity. *Ecology* **77(4)**:1210-1225.

Schwarz, W.L., Malanson, G.P., Weirich, F.H. (1996) Effect of landscape position on the sediment chemistry of abandoned-channel wetlands. *Landscape Ecology* **11**: 27-38.

Scoones, I. (1995) Exploiting heterogeneity: habitat use by cattle in dryland Zimbabwe. *Journal of Arid Environment* **29**: 221-237.

Scott, J.M., Davis, F., Csuti, B., Noss, R., Butterfield, B., Groves, C., Anderson, H., Caicco, S., D'Erchia, F., Edwards, T.C., Ulliman, J.Jr, Whright, R.G. (1993) Gap analysis: A geographical approach to protection of biological diversity. *Wildlife Monographs* **123**: 1-41.

Shafer, C.L. (1995) - Values and shortcomings of small reserves. *Bioscience* **45**: 80-88.

Shannon, C. & Weaver, W. (1949) *The mathematical theory of communication.* Urbana: University of Illinois Press, 117 p.

Shea K and NCEAS Working Group on Population Management (1998) Management of populations in conservation, harvesting and control. *TREE* **13**: 371-375.

Shelford, V.E. (1913) *Animal communities in temperate America.* The University of Chicago Press, Chicago, Illinois, USA.

Shorrocks B. & Swingland I.R. (1990) *Living in a patchy environment.* Oxford University Press, New York.

Simberloff, D., Farr, J.A., Cox, J., Mehlman, D.W. (1992) Movement corridors: conservation bargains or poor investment? *Conservation Biology* **6**: 493-504.

Simmons, M.A., Culliman, V.I., Thomas, J.M. (1992) Satellite imagery as a tool to evaluate ecological scale. *Landscape Ecology* **7**: 77-85.

Simpson, J.E. (1967) Swifts in a sea-breeze fronts. *British Birds* **60 (6)**: 225-239.

Sinclair, A.R.E. (1984) The function of distance movement in vertebrates. *In:* Swingland, I.R., Greenwood, P.J. (eds.), *The ecology of animal movement.* Oxford University Press, Oxford, UK. Pp. 240-258.

Skagen, S.K., Melcher, C.P., Howe, W.H., Knopf, F.L. (1998) Comparative use of riparian corridors and oases by migrating birds in Southeast Arizona. *Conservation Biology* **12(4)**: 896-909.

Skole, D.L., Chomentowski, W.H., Salas, W.A., Nobre, A.D. (1994) Physical and human dimensions of deforestation in Amazonia. *Bioscience* **44(5)**: 314-322.

Smith, H., Feber, R.E., Johnson, P.J. , McCallum, K., Jensen, S.P., Younes, M., Macdonald, D.W. (1993) *The conservation management of arable field margins.* English Nature Science, No. 18.

Smith, T.G. Jr., Lange, G.D., Marks, W.B. (1995) Fractal methods and results in cellular morphology dimensions, lacunarity and multifractals. *Journal of Neuroscience Methods* **69**: 123-136.

Soukup, M., Foley, M.K., Hiebert, R., Huff, D.E. (1999) Wildlife management in U.S. National Parks: Natural Regulation revisited. *Ecological Applications* **9(1)**: 1-2.

Southwick, E.E. & Buchmann, S.L. (1995) - Effects of horizon landmarks on homing success in honey bees. *American Naturalist* **146**:748-764.

Stamp, J.A., Buechner, M., Krishnan, V.V. (1987) The effects of edge permeability and habitat geometry on emigration from patches of habitat. *American Naturalist* **129**:533-552.

Stauffer, D. (1985) *Introduction of percolation theory*. Taylor & Francis, London.

Stouffer, P.C. & Bierregaard, R.O. (1995) Use of Amazonian forest fragments by understory insectivorous birds. *Ecology* **76**: 2429-2445.

Sugihara, G. & May, R.M. (1990) Applications of fractals in ecology. *TREE* **5**:79-86.

Swanson, F.J. & Sparks, R.E. (1990) Long-term ecological research and the invisible place. *Bioscience* **40**:502-508.

Swingland, I.R., Greenwood, P.J. (1984) *The ecology of animal movement.* Oxford University Press, Oxford, UK.

Tansley, A.G. (1935) The use and abuse of vegetation concepts and terms. *Ecology* **16**: 284-307.

Tardiff, S.E. & Stanford, J.A. (1998) Grizzly bear digging: effects on subalpine meadow plants in relation to mineral nitrogen availability. *Ecology* **79(7)**: 2219-2228.

Thomas, J.W., Maser, C., Rodiek, J.E. (1979). Edges. *In:* Thomas, J.W. (ed.*), Wildlife habitats in managed forests: the Blue Mountains of Oregon and Washington.* USDA, Forest service Agricultural Handbook, Washington D.C., n. 553. Pp.48-59.

Tilman, D. (1996) The benefits of natural disasters. *Science* **273**: 1518.

Tilman, D., May, R.M., Lehman, C.L., Nowak, M.A. (1994) Habitat destruction and the extinction debt. *Nature* **371**: 65-66.

Tomlin, C.D. (1990) *Geographic information systems and cartographic modeling.* Prentice-Hall, Englewood Cliffs, New Jersey.

Trimble Navigation (1994) *GPS Mapping Systems, General reference.* Trimble Navigation, Ltd. Sunnyvale, Ca.

Troll, C. (1968) Landschaftsokologie. In: Tuxen R. (ed.), *Pflanzensoziologie und Landschaftsokologie.* Berichte das Internalen Symposiums der Internationalen Vereinigung fur Vegetationskunde. Stolzenau/Weser. The Hague.Pp. 1-21.

Tscharntke, T. (1992) Fragmentation of *Phragmites* habitats, minimum viable population size, habitat suitability, and local extinction of moths, midges, flies, aphids, and birds. *Conservation Biology* **6**: 530-536.

Turner, I.M. & Corlett, R.T. (1996) The conservation value of small, isolated fragments of lowland tropical rain forest. *TREE* **11**:330-333.

Turner, M.G. (1987a) Spatial simulation of landscape changes in Georgia: A comparison of 3 transition models. *Landscape Ecology* **1**: 29-36.

Turner M.G.(1987b) *Landscape heterogeneity and disturbance.* Springer-Verlag .

Turner, M.G. (1989) Landscape ecology: the effect of pattern on process. *Annu. Rev. Ecol. Syst.* **20**:171-197.

Turner, M.G. (1990) Spatial and temporal analysis of landscape patterns. *Landscape Ecology* **4**: 21-30.

Turner, M.G. (1998) Landscape ecology. Living in a mosaic. *In:* Dodson, S.I., Allen, T.F.H., Carpenter, S.R., Ives, A.R., Jeanne, R.L., Kitchell, J.F., Langston, N.E., Turner, M.G. (eds.) *Ecology.* Oxford University Press, New York. Pp. 77-121.

Turner, M.G. & Gardner, R.H. (eds.) (1991) *Quantitative methods in landscape ecology: the analysis and interpretation of landscape heterogeneity.* Springer-Verlag, New York, NY,USA.

Turner, M.G, Dale, V.H., Gardner, R.H. (1989a) Predicting across scales: theory development and testing. *Landscape Ecology* **3**:245-252.

Turner, M.G., O'Neill, R.V., Gardner, R.H., Milne, B. (1989b) Effect of changing spatial scale on the analysis of landscape pattern. *Landscape Ecology* **3**: 153-162.

Turner, S.J., O'Neill, R.V., Conley, W., Conley, M., Humphries, H.(1991) Pattern and scale: statistics for landscape ecology. *In:* Turner, M.G. and R.H. Gardner (eds.), *Quantitative methods in landscape ecology: the analysis and interpretation of landscape heterogeneity.* Springer-Verlag, New York, NY,USA.Pp. 17-49.

Turner, M.G., Wu, Y., Romme, W.H., Wallace, L.L., Brenkert, A. (1994) Simulating winter interactions between ungulates, vegetation and fire in northern Yellowstone Park. *Ecological Applications* **4**: 472-496.

Turner, M.G., Arthaud, G.J., Engstrom, R.T., Hejl, S.J., Liu, J., Loeb, S., McKelvey, K. (1995) Usefulness of spatially explicit population models in land management. *Ecological Applications* **5**: 12-16.

Turner, M.G., Wear, D.N., Flamm, R.O. (1996) Land ownership and land-cover change in the southern Appalachian highlands and the Olympic peninsula. *Ecological Applications* **6(4)**: 1150-1172.

Turner, M.G., Pearson, S.M. , Romme, W.H., Wallace, L.L. (1997) Landscape heterogeneity and ungulate dynamics: What spatial scales are important? *In:* A. Bissonette (ed.), *Wildlife and landscape ecology. Effects of pattern and scale.* Springer Verlag, New York. Pp. 331-348.

Ulanowicz, R.E. (1997) *Ecology, the ascendent perspective.* Columbia University Press, New York.

Underwood, A.J. & Chapman, M.G. (1996) Scales of spatial patterns of distribution of intertidal invertebrates. *Oecologia* **107**: 212-224.

UNEP (1998) *Report of the workshop on the Ecosystem Approach.* United Nations Environment Program, Nairobi.

Urban, D.L., O'Neill, R.V., Shugart, H.H., Jr (1987) Landscape ecology. A hierarchical perspective can help scientists understand spatial patterns. *Bioscience* **37 (2)**:119-127.

van der Heijden M.G.A., Klironomos J.N., Ursic M, Moutoglis P, Streitwolf-Engel R, Boller T, Wiemken A, Sandres I.R. (1998) Mycorrhizal fungal diversity determines plant biodiversity, ecosystem variability and productivity. *Nature* **396**: 69-72.

van Dorp, D. & Opdam, P.F.M. (1987) Effects of patch size, isolation and regional abundance of forest bird communities. *Landscape Ecology* **1**: 59-73.

van Droste, B., Plachter, H., Rossler, M. (1995) *Cultural landscapes of universal value.* Gustav Fischer, Jena.

van Hees, W.W.S. (1994) A fractal model of vegetation complexity in Alaska. *Landscape Ecology* **9**: 271-278.

van Wilgen, B.W., Cowlig, R.M., Burgers, C.J. (1996) Valuation of ecosystem services. A case study from South African fymbos ecosystems. *Bioscience* **46**: 184-189.

Villard, M.-A. & Taylor, P.D. (1994) Tolerance to habitat fragmentation influences the colonization of new habitat by forest birds. *Oecologia* **98**: 393-401.

Villard, M.-A., Merriam, G., Maurer, B.A. (1995) Dynamics in subdivided populations of neotropical migratory birds in a fragmented temperate forest. *Ecology* **76**: 27-40.

Vos, W. & Stortelder, A. (1992) *Vanishing Tuscan landscapes. landscape ecology of a sub-Mediterranean-Montane area (Solano Basin, Tuscany,Italy).*Pudoc Scientific Publishers, Wageningen, NL.

Wagener, S.M., Oswood, M.W., Schimel, J.P. (1998) Rivers and soils: Parallels in carbon and nutrient processing. *Bioscience* **48**: 104-108.

WallisDeVries, M.F., Bakker, J.P., Van Wieren, S.E. (eds.) (1998*) Grazing and conservation management.* Kluwer Academic Publishers, Dordrecht.

Walker, D.A., Halfpenny, J.C., Walker, M.D., Wessman, C.A. (1993) Long-term studies of snow-vegetation interactions. *Bioscience* **43**:287-301.

Ward, D. & Saltz, D. (1994) Foraging at different spatial scales: Dorcas gazelles foraging lilies in the Negev desert. *Ecology* **75**: 48-58.

Watkinson, A.R. & Sutherland, W.J. (1995) Source, sinks and pseudo-sinks. *Journal of Animal Ecology* **64**: 126-30.

Wear, D.N., Turner, M.G., Flamm, R.O. (1996) Ecosystem management with multiple owners: landscape dynamics in a Southern Appalachian watershed. *Ecological Applications* **6(4)**: 1173-1188.

Wear, D.N., Turner, M.G., Naiman, R.J. (1998) Land cover along an urban-rural gradient: implications for water quality. *Ecological Applications* **8(3)**: 619-630.

Weaver, J.E. & Clements, F.C. (1928) *Plant ecology*. McGraw, New York,USA.

Wegner, J.F. & Merriam, G. (1979) Movements by birds and small mammals between a wood and adjoining farmland habitats. *Journal of Applied Ecology* **16**: 349-358.

Weng, G., Bhalla, U.S., Iyengar, R. (1999) Complexity in biological signaling systems. *Science* **284**: 92-96.

Werner B.T. (1999) Complexity in natural land patterns. *Science* **284**: 102-104.

White, P.S. & Pickett, S.T.A. (1985) Natural disturbance and patch dynamics, an introduction. *In*: Pickett, S.T.A. and White, P.S. (eds.), *The ecology of natural disturbance and patch dynamics.* Academic Press. New York, New York, USA.

Whitesides, G. & Ismagilov, R.F. (1999) Complexity in chemistry. *Science* **284**:89-92.

With, K.A. & King, A.W. (1997) The use and misuse of neutral landscape models in ecology. *Oikos* **79**:219-22.

Whittaker, R.H. (1975) *Community and ecosystems.* McMillan Publishing, New York (2nd ed.).

Whittaker, R.H. (1977) Evolution of species diversity in land communities. *Evolutionary Biology* **10**: 1-67.

Wiens, J.A., Crawford, C.S., Gosz, R. (1985) Boundary dynamics: a conceptual framework for studying landscape ecosystems. *Oikos* **45**: 421-427.

Wiens, J.A. (1986) Spatial scale and temporal variation in studies of shrubsteppe birds. *In:* Diamond, J. and T.J. Case (eds.), *Community Ecology*. Harper & Row Publishers, New York. Pp. 154-172.

Wiens, J.A. & Milne, B.T. (1989) Scaling of "landscape" in landscape ecology, or, landscape ecology from a beetle's perspective. *Landscape Ecology* **3**:87-96.

Wiens, J. (1989) *The ecology of bird communities*. Cambridge University Press, Cambridge, UK.

Wiens, J.A. (1992) Ecological flows across landscape boundaries: a conceptual overwiev. *In:* Hansen, A.J. and F. di Castri (eds.), *Landscape boundaries. Consequences for biotic diversity and ecological flows*. Springer-Verlag, New York. Pp. 217-235.

Wiens, J.A. (1992) Ecology 2000: An essay on future directions in ecology. *Ecological Society of America Bulletin* **73(3)**: 165-170.

Wiens, J., Crist, T.O., Milne, B.T. (1993) On quantifying insect movements. *Environmental Entomology* **22**: 709715.

Wiens, J.A. (1997) Metapopulation dynamics and landscape ecology. *In:* Hanski, I. A. and M.E. Gilpin (eds.), *Metapopulation Biology*. Academic Press, San Diego, Ca.Pp. 43-62.

Wiens, J.A., Crist, T.O., With, K.A., Milne, B.T. (1995) Fractal patterns of insect movement in microlandscape mosaics. *Ecology* **76**: 663-666.

Wilcove, D.S. (1985) Nest predation in forest tracts and the decline of migratory songbirds. *Ecology* **66**: 1211-1214.

Wilcove, D.S., McLellan, C.H., Dobson, A.P. (1986) Habitat fragmentation in the temperate. *In:* Soulé, M.E. (ed.), *Conservation biology*. Sinauer Associates, Inc. Pp. 237-256.

Wilcove, D.S., Rothstein, D., Dubow, J., Philipps, A., Losos, E. (1998) Quantifying threats to imperiled species in the United States. *Bioscience* **48(8)**: 607-615.

Willson, M.E., Gende, S.M., Marston, B.H. (1998) Fishes and the forest. *Bioscience* **48(6)**: 455-462.

Willson, M.F., De Santo, T., Sabag, C., Armesto, J.J. (1994) Avian communities of fragmented south temperate rainforests in Chile. *Conservation Biology* **8**: 508-520.

With, K.A. (1994) Using fractal analysis to assess how species perceive landscape structure. *Landscape Ecology* **9**: 25-36.

Wooton, J.T., Parker, M.S., Power, M.E. (1996) Effects of disturbance on river food webs. *Science* **273**: 1558-1561.

Wright, R.G. (1999) Wildlife management in the national parks: Questions in search of answers. *Ecological Applications* **9(1)**: 30-36.

Woodmansee, R.G. (1990) Biogeochemical cycles and ecological hierarchies. *In:* Zonneveld, I.S. and R.T.T. Forman (eds.), *Changing landscapes: An ecological perspective*. Springer-Verlag.Pp. 57-71.

WWF (1998) *Proceedings: Ecoregion-based Conservation Workshop*. WWF, Washington D.C.

Zhang, H., Brandle, J.R., Meyer, G.E., Hodges, L. (1995) The relationship between open windspeed and windspeed reduction in shelter. *Agroforestry Systems* **32**: 297-311.

Ziff, R. (1986) Test of scaling exponents for percolation-cluster perimeters. *Phys. Rev. Lett.* **56**: 545-548.

Zonneveld, I.S. (1995) *Land ecology*. SPB Academic Publishing, Amsterdam.

Glossary and Acronyms

Aggregation - The process opposite to fragmentation, occurring when a fragmented object (land cover, forested range, etc.) gains coverage thus filling gaps.
Alfa diversity - The number of species in a collection.
Animal landscape - A land mosaic structured by animal activity.
Autocatalytic patch - Relatively distinct patches embedded into an active mosaic and in which communities find a good place to interact.
Autopoietic capacity - The capacity of a system to be self-organizing and to maintain a «creative» attitude across homeostatic and homeorhetic responses to changing conditions.
Beta Diversity - The rate of change in species along a gradient from one habitat to another.
Biodiversity - The number of species present at a site, the variety of living organisms.
Biodiversity debt - The decrease of the number of species, mostly masked and not easily visible.
Bit - The measure of information inherent in a single binary decision.
Bit map - A sequence of bits (i.e. 0/1) on a grid.
Buffer - Refers to a transitional area acting as a filter or a mitigator of disturbance processes.
Cadastral maps - Maps at the scale of 1:2.000 that describe the bounds of properties , roads and idrographic nets.
Centrifugal process - A process dominated by fluxes of energy, nutrients , and organisms.
Centripetal process - A process that creates internal structural complexity and edge contrast.
Chemotactic bacteria - Bacteria that react to the release of chemical signals like food availability.
Chorological dimension - The dimension in which the spatial arrangement of objects is relevant.
Cluster - A group of cells or pixels that are connected to each other.
Coarse grained - When a pattern or a mosaic has a large component.
Complexity (ecological) - Combination and integration of functions and structures in the real world.
Connectedness - The physical distance between elements of the same type used generally for forest patches.

Connectivity - Functional attribute of connectedness related to a species-specific scaling property.

Contagion - A measure of the degree of clumping of land cover or vegetation types.

Contrast (between patches) - Difference in attributes of patches.

Core patch - The interior part of a patch according to a reference organism.

Core habitat - The central part of a habitat with very predictive (typical) conditions.

Corridor - A narrow strip of habitat surrounded by habitats of different types, with connection functions.

Corridor patch - A patch or a habitat that has the functions of a corridor.

Cnidaria - Phylum comprising the sea anemones, jellyfish and corals.

Cultural landscape - A landscape deeply changed by long history of feedback between ecological processes and human activities such as agriculture, forestry and pastoralism.

Dehesa - A belt of a mosaic of pastures and scattered trees of central Spain.

Discontinuity - The abrupt change of some characteristics of a system.

Dispersion - The capacity from individuals to populations to move to new habitats or to new parts of a landscape.

Disturbance debt - The rate of disturbance reduction in a disturbed-prone ecological system.

Ecodiversity - The diversity of land cover type or forest type. May be used also in cultural landscape to describe the diversity of land use and human culture integrated with ecological processes.

Eco-field - The interactions between an organism and it's surroundings.

Ecological debt - The progressive simplification of ecological complexity.

Ecological niche - The functional and structural dimensions occupied by an organism.

Ecosphere - Portions of the universe favorable for living organisms and in which all the ecological processes are contained.

Ecotone - A transition site between different habitats; a tension zone between systems of different maturity and where energy exchange and matter are highest.

Ecotope - The elementary unit of a landscape, with an homogeneous pattern or function.

Edge effect - The presence of a higher concentration of organisms at the edges.

Ephemeral - Attribute of a phenomenon or an organism lasting for only a short time (a few days or hours).

Evaporation - The process of transformation of liquid water into vapor.

Evenness - The distribution of abundance between a collection of organisms or patches of a landscape.

Foraging patches - Portion of a land mosaic used by an animal to find food.

Forested mosaic - The aggregation of different patches in a forest.

Fractal - An object that has a fractional dimension and that, by changing the scale of resolution, shows a self-similarity.

Fragility - An attribute of ecological systems; a system is fragile when, under a perturbation regime, a change of biological diversity occurs.

Fragmentation - A process by which forest cover is opened and stands become disjointed as isolated woodlots.

Functional heterogeneity - The heterogeneity in the spatial distribution of ecological entities (individual, populations, species, communities).
Functional landscape - Spatial dimension occupied by an organism.
Functional patch - A patch that has homogeneous characteristics for a particular function.
Fymbos - Shrub cover in South Africa (Cape Town Region) similar in shape to Mediterranean maqui.
Gamma Diversity - The diversity of species in different habitats along a geographical area.
Gap analysis - The search for concordances between different map layers.
Gap-phase - The process that follows tree fall in a forest producing clearings which permit the growth , by secondary succession, of new trees.
Geobotanical landscape - The spatial arrangement of plants and their physical and biological substrate.
Geo-statistic - The mathematical procedure of testing the significance of the spatial arrangement of data.
GIS - Geographical Information Systems.
GPS - Global Positioning System , a satellite-based positioning system.
Grain - The resolution of an image or the minimum area perceived as distinct by an organism.
Guild - A group of animals with similar characteristics associated with functions (foraging guild, breeding guilds, etc.).
Gully - Incision in bed rocks or in sediments by water run-off.
Habitat patch - A patch selected by individuals of the same species.
Hierarchical neutral model - A model of random distribution of objects into a matrix according to different scales of resolution.
Homeostasis - Tendency of a system to resist external perturbations and to maintain itself in a stable equilibrium.
Home Range - The area in which a species normally live.
Hyperspace - The space created by «Euclidean» vectors, each describing a niche function.
Edaphic (factor) - The physical and chemical conditions of soil.
Hedgerow - A strip of shrubs or trees planted in a rural landscape to delineate properties or to protect crops from windstorms.
Hotspot - A special place in which some processes are particularly evident.
Human Landscape - The spatial arrangement of natural and artificial objects, maintained and arranged by human activity.
Inbreeding - Genetic exchange between related individuals.
Incorporation - The process by which a system reduces the effects of a disturbance.
Intactness - The attribute of an ecological system unaffected by human intrusion
Interior species - Species living far from forest edges.
Key stone species - Species that shape the habitat in which they live allowing the presence of other species.
Landscape - A spatial configuration of patches of dimensions relevant for the phenomenon under consideration.
Landscape patchiness - A land mosaic composed of many patches.
Landslide - The movement of soil due to gravitational effects.

Leaching - The removal of soil components by water solution.
Layer - A map component of a GIS system.
Litter - Vegetation material recently fallen on the ground and only partially decomposed.
Local extinction - The disappearance of a species from a patch.
Long term studies - Studies planned in particular sites, regions and areas in order to track ecological processes for a long period of time.
Macrochore - A region composed from an aggregation of mesochores.
Macro-scale - The level between meso and mega-scale.
Management debt - Accumulation of management deficit into a landscape, previously managed.
Matrix - The dominant component of a landscape mosaic.
Megascale - The upper level of scaling.
Mesochore - An aggregation of microchore.
Mesoscale - An intermediate level between micro and macro-scale.
Metacommunity - The union of different neighboring communities.
Metapopulation - Sub-populations that are connected by movement (immigration-emigration of individuals).
Microchore - An aggregation of ecotopes.
Micromorphology - The soil morphology at a few cm wide.
Mobbing - The attacking behavior of a species against a predator.
Naturalness - Attribute of land meaning the intactness or integrity of ecosystems.
Nutrients (in the soil) - Elements necessary for plant nutrition.
Organismic approach - The vision of landscape according to the perception of an organism.
Outbreaks - Organism demographic explosion, generally applied to pests (mice, insect, weeds).
Parish - A religious division of a landscape common to all western Europe. One or more villages pertain to a parish.
Patterned landscape - The structural component of a landscape.
Pathogen - Any microorganism that produces diseases.
Percolation - The property of fluid by which it occupies a porous medium.
Percolation thresholds - The value of 0.5928, calculated on large theoretical lattices , by which a fluid percolates, moving from one side of a matrix to another.
Perturbation - A discrete event that modifies the status of a system without catastrophic consequences. Syn. of Disturbance.
Physiotope - A land unit with an homogeneous soil character. see Ecotope.
Phytoplankton - Plankton plants which are free living in waters.
Pixel - Abbreviation of Picture Element, the smallest unit of information of a map or a raster image.
Population patches - When the distribution of populations is represented in a spatial dimension.
Process approach - The study of complexity using processes as indicators.
Porosity - The capacity of an ecotone to be crossed by an organism.
Raster - A representation by grid cells of an object in a computer memory.
Regional biodiversity - The collection of organisms along specified areas of relevant interest for humans.
Resilience - A process by which a system incorporates a disturbance by small

changes in internal structure and function.

Shifting-mosaic steady state - The condition in which a landscape changes in the distribution of patches due to different causes but at the end maintains the same character.

Sink - A population that becomes extinct without external immigration. May also be applied to habitats.

Site uniqueness - The presence of unique characters linked to a particular site.

Sky-scape - The spatial arrangement of atmospheric humidity.

Source - A population that has a positive balance between births and deaths. May also be applied to habitats.

Spacing - The reaction of an organism to it's perception of the neighboring environment.

Spatial heterogeneity - The variation across space of vegetation type or land cover

Stewardship debt - Refers to the abandonment of cultural landscapes.

Stop-over migrants - Migratory birds that spend a short time in selected habitats along the migratory route to replenish energies.

Structural patch - A patch characterized by a recognizable pattern (for instance a soil type associated with a particular plant association).

Strip Corridor - A narrow belt of shrubs or woodland in an agricultural or urban area.

Termites - Group of Isoptera insects with a prevailing fossorial habit.

Topological component - The functional component of landscape complexity

Topology - Pattern of linkage between geographical elements.

Total human ecosystem - Conceptualization of modifications and effect of human life on the Earth (Naveh & Lieberman 1984).

Trampling - Soil compactation due to animal passage.

Traps (ecological) - Habitats that attract species with favorable conditions but in which some functions, such as breeding are not possible or are suddenly interrupted by human disturbance or predation.

Vector - An object that has magnitude and direction.

Vertical Landscape - The spatial arrangement along vertical surfaces like cliffs.

Water-scape - The spatial arrangement of sea/fresh-water masses and organisms.

Zooplankton - The animal component of minute organisms spread into water with scarce or no locomotive organs.

References

Allaby, M. (1985) *The Oxford dictionary of natural history* . Oxford University Press.

Bates, R.L. & Jackson, J.A. (1987) *Glossary of geology* (Third edition). American Geological Institute, Alexandria, VA.

Calow, P. (1998) *The encyclopedia of ecology & environmental management.* Blackwell Science Ltd, Oxford.

Lincoln, R.J. , Boxshall, G.A., Clark, P.F. (1982) *A dictionary of ecology, evolution and systematics.* Cambridge University Press.

Species Index

A

Ailanthus spp., 251
Alauda arvensis, 135, 136, 230
Alces alces, 209, 215
Alnus glutinosa, 201
Anthus spinoletta, 230
Anthus trivialis, 230
Apus apus, 16
Arctocephalus gazella, 96
Austrian pine, 234

B

Beaver, 131
Bison, 131, 199, 210
Bison bison, 131, 210
Blue tit, 95
Bouteloua gracilis, 228
Brachypodium, 145
Bufalo cafrus, 215
Bufo bufo, 202
Bufo viridis, 202

C

Canis lupus, 53, 209
Capra hircus, 135
Carduelis carduelis, 230
Carduelis chloris, 230
Caribou, 78
Castor canadensis, 105, 131, 215
Cervus elaphus, 210
Cettia cetti, 53
Choloepus didactylus, 41
Coccinella septempunctata, 96
Colinus virginianus, 94
Connochaetes sp, 55
Cornus mas, 102
Crag martin, 40
Crocodylus niloticus, 215
Crocus sp., 213
Crocus vernus, 214

D

Dicosmoecus gilvipes, 224
Dogwood, 102

E

Elaphe longissima, 202
Elianthus tuberosum, 202
Erica carnea, 145
Erinaceous europaeus, 165
Erithacus rubecula, 32, 89
European hare, 55
European wolf, 53

F

Fagus grandiflora, 142
Felis concolor, 54
Felis concolor coryi, 106
Florida panther, 106
Florida scrub lizard, 211
Fraxinus ornus, 79
Fringilla coelebs, 230

G-H

Garrulus glandarius, 79
Grass snake, 65
Hieraceum sp., 20
Hippopotamus, 215
Hippopotamus amphibius, 215
Hirundo rupestris, 40
Hydromantes italicus, 32

I-J

Italian stone pine, 226
Jays, 79

L

Lacerta sicula campestris, 202
Larrea, 228
Lepus timidus, 55

Leuciscus cabeda, 202
Loxodonta africana, 215
Lullula arborea, 135, 136
Lynx, 115
Lynx canadensis, 115

M-N

Martes foina, 79
Meles meles, 18, 93
Mountain lion, 54
Mule deer, 54
Muscardinus avellanarius, 18
Mustela nivalis, 59
Natrix natrix, 65, 202

O-P

Odocoileus, 135, 154
Odocoileus hemionus, 54
Odocoileus virginianus, 135
Parus caeruleus, 95
Parus major, 47, 95
Peromyscus leucopus, 154
Peromyscus maniculatus, 154
Phoca vitulina, 64
Pine, 101
Pinus nigra, 234
Ponderosa pine, 153, 235
Primula apennina, 40
Prosopis, 228

Q-R

Quercus cerris, 151
Quercus sp., 230
Rana dalmatina, 32, 202
Rangifer tarandus, 78
Ring ouzel, 51
Ringed plover, 249
Robinia pseudoacacia, 128, 251
Rubus spp., 65, 201

S

Salamandra salamandra, 32
Salix erbacea, 102, 103, 126
Salix spp., 201
Sceloporus woodi, 211
Sempervivum arachnoideum, 126
Sempervivum montanus, 20
Solidago canadensis, 96
Spanish goat, 135
Sus scrofa, 202, 222
Sylvia atricapilla, 65, 102

T

Taraxacum officinale, 25, 26
Tree lark, 136
Turdus iliacus, 230
Turdus philomelos, 230
Turdus torquatus, 51
Turkey oak, 151

U-V-W

Uroleucon nigrotuberculatum, 96
Ursus arctos horribilis, 221
Vaccinium myrtillus, 70, 127
Vibrio cholerae, 155
Vibrio spp., 155
Vipera aspis, 202
Vulpes vulpes, 54, 165, 202
White-footed mouse, 154
Wild boar, 202
Wormwood, 77, 78

Author Index

A

Acuna J.A. 184
Agger P. 8, 241
Ahern J. 262
Alados C.L, 184
Allen M.F.19
Allen R.B. 101
Allen T.F.H. 9
Anderson A.N. 184
Anderson J.E. 159
Andreassen H.P. 53
Andren H. 212
Andrews J. 140
Angelstam P. 197
Angermeier P.L. 119, 120
Antrop M. 241
Arthur W.B. 5
Ashton P.S. 207
Ashworth P.J. 15
Austad I. 8
Avissar R. 44
Awimbo J.A. 114

B

Bailey R.G. 149
Baker B.D. 94
Baker W.L, 10, 168, 231
Bakker J.P. 226
Bamkin K.L. 212
Bancroft G.T. 149
Barak P. 184
Barker W. 131
Baudry J. 61, 90
Beerling D.J. 102
Beier P. 53, 58
Bennett B.C. 255, 256, 257
Bennett K.D. 38
Benvenuti S. 78
Berthold P. 19
Bevers M. 10
Bierregaard R.O. Jr. 98, 149
Blackburn N. 49, 50
Blake J.G. 44
Blevins R.L. 184
Blondel J. 41, 75
Bogaert J. 168, 173, 174, 187
Box E.O. 29, 187
Boyd I.L. 96, 97
Brandt J. 8, 241
Brittingham M.C. 66
Brothers T.S. 98
Brown J.H. 133
Buchmann S.L. 19
Burel F. 61
Burger L.D. 212
Burkey T.V. 98
Burrough P.A. 10, 161, 190, 191
Butler D.R. 105

C

Cadenasso M.L. 9
Cai Y. 10, 168
Campbell R.D. 184
Carlile D.W. 29
Chambers C.L. 218
Chapin III F.S. 73, 74
Chapman M.G. 184
Chen S.G. 184
Cheng K. 17
Clarks J.S. 141, 142
Claussen M, 44
Clements F.C. 61
Collinge S.K. 243
Colwell R.C. 154, 155
Corlett R.T. 213
Correll D.L. 44
Costanza R. 233, 236

Cutler A. 212
Cuvillier R. 41

D

Dale V. 118
Daubenmire R. 61
Dayton P.K. 29
De Leo G.A. 119
Decamps H. 61, 119
Delcourt H.R. 29, 30, 61
Delcourt P.A. 29, 30, 61
Desaigues B. 61
Dias P.C. 75
DiCastri F. 9
Diffendorfer J.E. 95
Diggle P.J. 10
Donovan T.M. 75
Dudley D.R. 119

E-F

Eghball B. 184
Etzenhouser M.J. 135, 184, 188
Faarborg J. 98
Fahrig L. 82, 260
Farina A. 9, 11, 12, 16, 17, 38, 40, 41, 61, 66, 85, 116, 118, 159, 168, 184, 201, 202, 215, 222, 230, 241, 248
Feder J. 10, 184, 188
Foran B. 137
Forman R.T.T. 9, 10, 11, 12, 29, 65, 243
Foster D.R. 137, 140, 141
Fourcassié V. 184
Frank D.A, 11, 228
Freemark K. 82
Frontier S.A. 184
Furness R.W. 146

G

Galen C. 19
Galli A.E. 98
Gardner R.H. 9, 10, 121, 165, 167, 168
Gates J.E. 66
Gibbs J.P. 98
Gilmour D.A. 234
Gilpin M. E. 9, 81
Givnish T.J. 218
Godron M. 9, 11, 12, 29
Goldenfeld N. 5
Golley F.B. 11, 25
Goossens R. 10
Gosz J.R. 61
Gotzmark F. 159
Green B.H. 12, 254
Greenwood J.J.D. 56, 146
Gregory S.V. 87
Grelsson G. 121
Groffman P.M. 15
Gross J.E. 44
Grover H.D. 10
Grumbine R.E. 236, 238
Guastalli G. 223
Gustafson E.J. 168
Gysel L.W. 66

H

Haber W. 11, 241
Haeuber R.A. 250, 252, 253
Haila Y. 95, 152, 153
Haines-Young R.H. 10
Hall F.G. 10
Halladay D. 234
Hansen A.J. 10, 219, 258, 259
Hanski I.A. 9, 81
Hansson L. 92, 156, 197
Hantush M. M. 44
Haralick R.M. 10
Hargis C.C. 117, 121
Harris L.D, 10, 153, 258
Harrison R.L. 53
Harrison S. 81
Hastings A. 9
Hastings H.M.184
Healy S. 17
Henein K. 92
Herkert J.R. 96
Hill G.E. 146
Hinsley S.A. 98
Hobbs R.J. 212
Hoekstra T.W. 9, 29, 31
Hof J. 10
Hokit D.G. 211
Holland L.D. 9, 61
Holland M.M. 63
Holling C.S. 29, 125, 162
Hopkins A.J.M. 212

Hoppes W.G. 44
Hornberg G. 219
Hubbell S.P. 103, 104
Huff D.E. 209
Huggett R.J. 44
Hughes T.P. 34, 35
Hulse D.W. 10, 169
Hunter M.L. 65
Hutchinson G.E. 12, 79

I-J

Iannaccone P.M. 184
Ichoku C. 184
Impens I. 168, 173
Isaaks E.H. 10
Ismagilov R.F. 5
Iturbe I. R. 184
IUCN, 234, 236, 262
Jeffrey D.W. 146
Joalé P. 78
Johnson A.R. 93, 184
Johnson G.D. 184
Johnston C.A. 105

K

Kadanoff L.P. 5
Kareiva P.M. 19, 96
Karr J.R. 44, 119, 120
Kattan G.H. 96
Kavanagh R.P. 212
Kay B.D. 184
Kenkel N.C. 187
Kesner B.T. 82
Khokha M. 184
Kief T. L. 143, 144, 145
King A.W. 28, 66,164
Klaassen W. 44
Klein B.C. 96
Knapp A.K. 198, 199
Knick S.T. 19
Knopf F.L. 227, 248
Koenig W.D. 45, 115
Kolasa J. 9, 48
Kosko B. 190
Kotliar N.B. 48
Kozak E. 184
Kozakiewicz M. 94
Krajick K. 41
Krebs J.R. 47
Krummel J.R. 168, 184, 187
Kurki S. 33

L

Larsen K. 10, 169
Laszlo E. 11, 21
Lauenroth W.K, 228
Le Maitre D.C. 234
Leach M.K. 218
Leduc A. 184
Leimgruber P. 212
Leopold A. 61
Levin S. 29, 119
Levins R. 9, 81
Li B.L. 184
Li H. 116, 168, 177
Lidicker W.Z. Jr, 11, 95
Lieberman A.S. 9, 11, 50
Likens G.E. 44
Lima S.L. 19
Lindemayer D.B. 53
Liu J. 207, 211
Loehle C. 184
Londo G. 226
Lord J.M. 96
Lucas O.W.R. 117

M

MacArthur J.W. 41
MacArthur R.H. 9, 41, 48, 156
Mack M.C. 128, 129
Madden B. 146
Magnuson J.J. 11
Maguire D.J. 10
Malanson G.P. 184
Mandelbrot B.B. 10, 183, 184, 187
Margules C.R, 98
Marino M.A. 44
Marks B.J. 135, 168, 187
May R.M. 184, 187
McAuliffe J. R. 44
McBratney A.B. 184
McCleery R.H. 47
McDonnell R.A.. 10, 161, 190, 191
McDowell W.H. 44

McGarigal K. 135, 168, 187
McNeely J.A. 236, 238
Meentemeyer V. 29, 82
Merriam G. 56, 90, 92, 260
Merrill S.B. 253
Messier F. 66
Meyers N. 74, 163, 245
Michener W.K. 250, 252
Miller D.E. 44
Miller K.R. 238
Mills L.S. 129, 130, 132
Milne B.T. 10, 11, 16, 184, 187
Mladenoff D.J. 209, 220
Moore P.N. 10, 65
Morreale S.J. 53
Musick H.B. 10

N

Naeem S. 118
Naiman R.J. 53, 61, 105, 119, 215, 216, 217, 248
Nardelli R. 32
Naveh Z. 8, 9, 11, 12, 50, 197, 219, 230, 241, 251
Nepstad D.C. 212
Newmark W.D. 98
Nilsson C. 121
Norton B.G. 243
Norton D.A. 96, 98
Noss R.F. 53, 58, 132, 134, 213

O-P

Odum E.P. 29
Odum W.E. 61
Olsen E.R. 187
Opdam P.F.M. 98, 212
Ostfeld R.S. 98
Pachepsky Y.A. 184
Paine R.T. 129
Parker G.R. 168
Parrish J.K. 5
Pasitschniak M. 66
Pearson S.M. 98, 165, 167
Perevolotsky A. 228
Perfect E. 184
Perrier E. 184
Peterjohn W.T. 44
Peterson D.L. 184,187
Peterson G. 125, 127
Peterson R.O. 209
Phillips A. 241
Pickett S.T.A, 9, 46, 48, 99
Pielke R.A. 44
Pimentel D. 256
Plachter H. 8
Plotnick R.E. 10
Plowright R.C. 19
Powell D.M. 15
Power J.F. 184
Power M.E. 130, 198
Probst J.R. 28
Pulliam H.R. 9, 75, 76, 77, 78, 80

R

Ranney J.W. 61
Ranta E. 115
Rapport D.J. 145, 147, 148, 153, 155, 235
Rasiah V. 184
Rebane M. 140
Redpath S.M. 19
Reid W.V. 141, 246
Rex K.D. 184
Reynolds J.F. 168, 177
Ricklefs R.E. 61
Riitters K.H. 116, 117
Rinaldo A. 184
Ripley B.D. 10
Risser P.G. 9, 15, 61, 65
Rogers K.H. 46, 53, 215, 216, 217
Roland J. 98
Rollo C.D. 9
Root R.B. 19
Rosenberg D.K, 46, 56, 57, 58
Rossler M. 8
Rotenberry J.T. 19
Rudis V.A. 98
Rusek J. 61
Russell R.W. 184

S

Saab V, 205, 206
Saltz D. 48
Samson F. 227
Saunders D.A. 98

Schieck J. 19
Schimel D. 44
Schippers P. 93
Schlesinger W.H. 44
Schlosser I.J. 87
Schmidt J.C. 249, 250
Schwarz W.L. 44
Scoones I. 19
Scott J.M. 74
Seligman N.G. 228
Shafer C.L. 260
Shannon C. 72, 168, 179, 180
Shea K. 202
Shelford V.E. 61
Shorrocks B. 19
Simberloff D. 53
Simpson J.E. 37, 179, 180, 181
Skagen S.K. 252
Skole D.L. 98, 138, 139
Smith H. 140
Smith T.G. Jr. 184
Soukup M. 208
Southwick E.E. 19
Sparks R.E. 11
Spetch M.L. 17
Spingarn A. 98
Srivastava R.M. 10
Stamp J.A.. 61
Stanford J.A. 221
Starr T.B. 9
Stauffer D. 9, 165
Stortelder A. 50, 144, 159
Sugihara G. 184, 187
Sutherland W.J. 77
Swanson F.J. 11, 258
Swingland I.R. 19, 56
Szacki J. 94

T

Tansley A.G. 25, 236
Tardiff S.E. 221
Taylor A.D. 81
Taylor P.D. 98
Tegner M.J. 29
Temple S.A.. 66
Thomas J.W. 63
Tilman D. 163, 251
Tomlin C.D. 10, 169
Trimble Navigation, 10
Troll, 9, 12
Tscharntke T. 212
Turner C.L. 15
Turner I.M. 213
Turner M.G. 9, 11, 31, 90, 115, 116, 137, 210, 232
Turner S.J. 115

U-V

Ulanowicz R.E. 72, 243
Underwood A.J. 184
UNEP, 236
Urban D.L. 30, 67, 201
van der Heijden M.G.A, 71
van Droste B. 254
van Hees W.W.S. 184, 185
van Wilgen B.W. 234, 235
Villard M.A. 98
Vos W. 50, 144, 159

W-Z

Wagener S.M. 86
Walker D.A. 87
Walker D.J. 187
WallisDeVries M.F. 225
Ward D. 48
Wardle K. 137
Watkinson A.R. 9, 77
Wear D.N. 232, 234, 239, 240
Weaver J.E. 61
Weaver W.168
Wegner J.F. 56
Weinrich J. 28
Weng G. 5
Werner B.T. 5
White P.S. 9, 99
Whitesides G. 5
Whitford W.G. 153
With K.A. 164
Whittaker R.H. 50, 85
Wiens J.A. 10, 11, 16, 48, 61, 65, 82, 85, 91, 184, 185, 213
Wilcove D.S. 98, 119
Willson M.E. 215
Willson M.F. 212
Wilson E.O. 9, 48

With K.A. 34, 76, 130, 143, 184, 185, 198, 219, 236, 263
Woodmansee R.G. 44
Wooton J.T. 224
Wright R.G. 209
WWF, 236
Ziff R. 165
Zollner P.A. 19
Zonneveld I.S. 8, 9, 11, 22, 50, 52

Subject Index

A

absorption, 15
acid deposition, 125, 240
acid rain, 163
acoustic, 17
active bed, 6, 87
active landscape, 42
adsorption, 15
aerial photographs, 133, 137
aerial-plankton, 16
aesthetic value, 153
African ungulates, 55
aggregation, 98
agricultural subsidies, 158, 163
agro-ecosystems, 65, 234
airplane navigation, 36, 37
Alfa diversity, 71
alfa-alfa, 72
algal bloom, 154, 160
alibi, 53
alien plants, 98
Alpine-type, 87
Amazon forest, 255
amenity-caring science, 268
American Northwest, 258, 259
amount of edges, 71, 164, 165
anadromous streams, 215
animal activity, 20, 217
animal behavior, 19, 44, 184
animal dispersion, 39, 44, 90, 215
animal energy balance, 13
animal guilds, 162
animal landscape, 13, 39
animal suitability, 112
animals, 7, 16, 17, 18, 19, 29, 31, 32, 36, 38, 39, 40, 41, 46, 47, 48, 49, 52, 53, 55, 57, 90, 91, 93, 95, 96, 102, 106, 128, 130, 135, 136, 156, 157, 161, 162, 165, 215, 216, 217, 221, 228, 230, 244, 251, 255, 260
antelope, 217
Appalachians, 240
arid grasslands, 30
Arizona, 15, 229, 252
Arizona shrubland, 229
arthropods, 36, 37, 184
artic, 38
Asia, 256
astronomic cues, 19
atmosphere, 24, 26, 36
atmospheric gases, 25
Australia, 34, 74, 106, 122, 231, 245
autocatalitic patches, 197, 198, 203, 204
autopoietic, 63, 203, 239, 262
avalanches, 125, 141, 254
average patch size, 116

B

bacteria, 7, 39, 49, 50, 55, 128, 134, 155, 156, 184, 213
Bangladesh, 155
bare soil, 14, 221, 223
barrier, 56, 60, 93
basking, 260
beaver dams, 63
beech forest, 51, 62, 231
behavioral ecology, 47, 48
Belgium, 262
Beta diversity, 71
biodiversity, 4, 41, 60, 61, 73, 74, 89, 90, 98, 104, 114, 118, 119, 132, 133, 149, 152, 153, 157, 160, 162, 163, 195, 196, 198, 200, 201, 202, 203, 213, 215, 218, 219, 220, 225, 226, 230, 234, 237, 239, 242, 243, 245, 248, 251, 252, 254, 258, 260, 261, 262, 268
biodiversity erosion, 89
biomass, 13, 31, 39, 71, 82, 86, 128, 148, 168, 199, 216, 219, 230, 231, 234, 243, 249, 254, 255, 256
biomes, 61, 87, 156, 214, 228, 231, 248

biotechnologies, 261
birds, 16, 17, 32, 36, 41, 45, 47, 55, 59, 78, 89, 95, 104, 106, 135, 142, 146, 156, 198, 202, 205, 206, 207, 212, 215, 230, 246, 248, 252, 253
Boltzman, 72
boolean (crisp) sets, 190
boreal forests, 157, 217
boreal swamp forests, 219
box (grid) dimension method, 187
Brazil, 74, 138, 139, 245
breeding, 16, 17, 27, 32, 41, 45, 55, 76, 78, 81, 89, 95, 116, 136, 187, 198, 202, 205, 207, 230, 246, 251, 260
breeding habitat, 27
Brownian Motion, 168
buffer, 9, 47, 61, 65, 140, 160, 181, 220, 261, 262
buffer functions, 9
buffer zone, 47, 65, 140, 160, 220, 261, 262
burn intensity, 31
burned stands, 32
butterfly, 52, 56, 156, 246

C

C/N ratio, 199
cadastrial maps, 259
calcium carbonate, 39, 143, 145
California, 64, 101, 105, 111, 115, 215, 224, 231
California vernal ponds, 215
caliper, 64, 91, 117, 184, 185, 186
Canada, 62, 239
canopy, 103, 104
carrot, 67, 68
catabolism, 85
catastrophe risk, 254
catchment area, 15, 153
cauliflowers, 183
cellulose, 39
Central New England, 140
Central Tuscany, 65
centripetal process, 83, 84
chaotic, 5
charcoal, 231
chemical pollutants, 148
chicks, 17
children, 17
chitinaceous zooplankton, 155
chorological dimension, 22, 23, 24, 26
ciliates, 50
Civiglia stream, 200
class resolution, 169, 171
clearing, 44, 99, 156, 157, 222
climate, 85, 261
climatic changes, 68, 96, 133, 137, 141
climatic stress, 32, 164
closed-canopy forest, 103
cloud, 24, 36
cluster (mass) dimension, 188, 189
clusters, 12, 49, 165
cnidarians, 39
CO_2, 118, 233
coarse level, 28
coastal water, 155
coastlines, 128
cognitive maps, 94
cold regions, 157
Colombia, 74, 245
colonial bats, 104
colony, 16, 28
Colorado, 229, 249, 250
community, 3, 9, 13, 19, 22, 25, 29, 30, 31, 42, 49, 71, 87, 95, 99, 100, 105, 118, 122, 123, 125, 130, 139, 140, 143, 152, 153, 154, 161, 168, 198, 200, 203, 211, 217, 220, 225, 226, 228, 230, 241, 251, 260
complexity, 3, 5, 7, 8, 11, 12, 25, 26, 28, 29, 34, 41, 42, 48, 61, 65, 66, 71, 72, 73, 82, 83, 84, 86, 87, 88, 99, 105, 109, 114, 119, 121, 122, 123, 125, 132, 133, 135, 137, 140, 149, 151, 152, 162, 168, 169, 183, 184, 185, 186, 187, 197, 199, 203, 210, 211, 217, 218, 221, 228, 236, 238, 244, 246, 248, 249, 260
computer networks, 44
connectedness, 90
connectivity, 4, 14, 53, 57, 60, 90, 91, 92, 93, 104, 105, 116, 122, 156, 164, 178, 212, 220, 252, 261
contagion, 45, 85, 100, 116, 118, 133, 161, 169, 177, 178, 239
coral, 34, 35, 39, 215

core area, 47, 60, 169, 181, 182, 206, 220, 244, 261, 262
Coriolis forces, 37
corn, 94, 124
corridor, 52, 53, 54, 55, 56, 57, 58, 59, 60, 84, 92, 117, 125, 215, 216, 217, 220, 244, 252, 262
Costarica, 74
cottonwood fores, 205, 206, 207
countryside, 163, 242, 254, 267, 268
countryside heritage, 254
county, 112, 148
cover type, 117, 118, 162
craters, 31
creativity, 157
creosotebush, 143, 144, 145
crisp patterned objects, 190
crop lands, 78, 95
cultivated fields, 117
cultural model, 111, 113, 238
culture, 13, 157, 183

D

damage, 101, 146, 147, 250
decision theory, 202
decision-makers, 237
decomposition, 4, 85
deep sea, 8
degree of deforestation, 148
Delta diversity, 71
delta region, 114
deltas, 218, 254
Denmark, 262
desert shrubland, 143
deserts, 38, 123
deterministic chaos, 183
diffusion process, 93
digital information, 14
disclimax, 144
disease, 44, 45, 102, 154, 155, 267
dispersion, 25, 26, 38, 56, 58, 60, 78, 82, 84, 95, 104, 141, 142, 164, 260
disturbance, 4, 7, 9, 14, 19, 29, 30, 31, 41, 48, 51, 59, 61, 68, 71, 72, 74, 78, 82, 85, 95, 98, 99, 100, 101, 102, 103, 104, 105, 109, 120, 122, 123, 125, 126, 128, 129, 139, 140, 142, 144, 146, 149, 150, 151, 152, 156, 157, 159, 162, 163, 164, 195, 199, 200, 204, 208, 215, 217, 218, 219, 220, 222, 223, 225, 228, 229, 230, 231, 235, 241, 242, 244, 248, 251, 254, 258, 262, 267
disturbance regime, 9, 30, 31, 48, 61, 71, 72, 85, 95, 102, 122, 123, 128, 129, 139, 140, 142, 144, 149, 150, 152, 156, 157, 200, 204, 215, 217, 218, 219, 220, 224, 225, 228, 229, 230, 235, 241, 242, 248, 258, 262, 267
disturbance risk, 59
diversity, 4, 71, 72, 74, 85, 110, 111, 179, 180, 244
divider (caliper) method, 184
Dolomites (Italy), 246
dominance, 71
Douglas-fir forest, 218
downstream, 86, 87, 249
dragonflies, 246
drought, 56, 63, 78, 143, 217, 224, 248
dung deposition, 199
dunghills, 32
dynamic equilibrium, 143

E

earthquakes, 101
East Divide, 87
Eastern Africa, 39
ECOLECON, 211
ecological complexity, 5, 8, 10, 11, 13
ecological debt, 110, 148, 162
ecological systems, 7, 84, 119, 120, 164, 225
ecological traps, 59, 66, 78, 79
economy, 13, 14, 38, 82, 138, 232
ecorigion, 75, 112, 156, 161
ecosphere, 20, 148, 233
ecosystem, 3, 13, 14, 22, 23, 25, 29, 72, 73, 74, 89, 90, 99, 118, 122, 130, 137, 145, 147, 148, 151, 153, 155, 157, 159, 164, 195, 198, 209, 210, 225, 231, 233, 234, 235, 236, 237, 239, 246, 250, 251, 254, 256, 260, 261
ecosystem distress syndrome, 153
ecosystem management, 234, 236, 251

ecosystem service, 148, 157, 195, 233, 234, 237, 260, 261
ecotones, 4, 9, 24, 61, 62, 63, 64, 65, 66, 117, 119, 133, 140, 143, 150, 195, 214, 218, 223, 240
ecotopes, 52, 116, 159, 192
edge, 47, 53, 61, 62, 91, 116, 156, 171, 174, 175, 176, 177, 181, 185, 214, 220, 244, 258
edge contrast, 176
El Nino, 38, 154, 155
electrical power, 149
electromagnetic field, 21
elevation gradient, 87
energy waves, 6
environmental discontinuity, 163
environmental service, 145, 154
environmental synchrony, 115
ephemeral, 27, 47, 49, 73, 115, 152, 197, 198, 200, 202, 213, 214, 216, 227, 238, 248, 251, 262
epigean biomass, 49
epyphit colony, 137
erosive process, 7, 91, 128, 248
Euclidean geometry, 90, 183
Eurasia, 78
Europe, 8, 9, 10, 32, 102, 103, 106, 111, 138, 141, 218, 220, 222, 225, 226, 232, 239, 254, 256, 259
evaluation procedure, 109, 111, 112, 113, 145, 152
evenness, 169, 179
exotic species invasion, 59
expanded niche, 79, 80
experiment protocol, 56
explicit model, 116, 143, 169
extinction, 43, 45, 48, 53, 76, 81, 82, 98, 116, 122, 123, 126, 137, 149, 150, 162, 163, 199, 212, 218, 219, 255, 256, 260

F

farm, 32, 66, 67, 68, 140, 158, 159, 267
farmers, 163, 232, 254, 255
ferns, 40, 183
fire hazard, 153
fire ignition, 123
fire suppression, 120, 143, 153
Flint Hills (USA), 198, 199
flock, 16, 28, 42, 47
flooding, 5, 7, 63, 68, 77, 84, 100, 104, 123, 125, 141, 151, 200, 219, 220, 223, 251
Florida peninsula, 104
Florida scrub, 211
focal forest, 207
food searches, 17
foraging habitat, 27, 252
forest area, 208, 240
forest floor, 86
forest fragmentation, 19, 122, 152
forest remnants, 39, 53, 58, 73, 100, 113, 153, 162, 198, 200, 218, 267
forestry, 73, 220, 240, 253
FORMOSAIC, 207
fractal dimension, 64, 116, 135, 136, 161, 174, 183, 187
fractal geometry, 10, 135, 183, 184
fractional Brownian motion, 168
fragility, 109, 121, 122, 123, 245
fresh-water scape, 37
frost, 38, 103
frugivorous birds, 95
fuel load, 128, 153
functional landscapes, 60
functional patch, 28, 43, 47, 100, 159
fundamental niche, 79, 80
fungi, 71, 86, 213
fuzziness, 161, 190
fuzzy k-mean, 192
fuzzy logic, 190
fuzzy sets, 190, 191, 192
Fuzzy theory, 110, 190
fymbos vegetation, 234

G

Gamma diversity, 71
GAP analysis, 74, 133
gaps, 59, 98, 102, 103, 198, 238
gardens, 32
gas, 42, 233
Gaussian distribution, 168
gene flow, 53
genotypes, 245
geobotanical landscape, 3, 38

Geographic Information Systems, 10, 112
geographical ecology, 9
geographical scale, 27, 85, 89, 95, 124, 147, 148, 150, 246
geometric module, 17
geophytes, 49, 213
Georgia Piedmont landscape, 204
GIS, 6, 112, 133, 211, 223
glaciers, 104
Glen Canyon dam (USA), 249
Global Positioning Systems, 10
goats, 135, 136, 228, 229, 230
grain, 23, 48, 112, 160, 188, 189
grain size, 112
Gran Sasso d'Italia, 88
grass, 94, 128, 199, 218, 227, 228, 232, 257
grass, 51, 143, 200, 220, 228
gravitational movements, 68
gravity gradients, 25
grazing area, 55
Great Divide, 87
Great Plain (USA), 228
green ways, 57
grid cell, 208
grizzly, 130, 215, 221
guano, 104
guilds, 123, 150, 161, 162, 203
habitat, 27, 28, 53, 55, 56, 57, 59, 60, 61, 71, 73, 76, 78, 80, 81, 87, 89, 93, 96, 98, 106, 116, 117, 120, 126, 135, 149, 165, 166, 167, 188, 195, 204, 210, 212, 215, 216, 218, 220, 241, 242, 245, 246, 248, 251, 252, 260
habitat patches, 56, 75, 78, 81, 89, 90, 91, 94, 117, 118, 133, 159, 162, 203, 260

H

habitat suitability, 115, 210
habitat vrarity, 244
Hausdorff Besicovitch dimension, 183
headwater, 87
health, 109, 145, 146, 147, 148, 150, 152, 153, 154, 159, 160, 162, 239, 242, 250
healthy conditions, 114, 244
heating, 84
hedgerows, 15, 159, 215
heterogeneity, 4, 9, 25, 26, 27, 31, 41, 42, 44, 48, 49, 61, 64, 74, 75, 85, 89, 90, 91, 95, 99, 100, 104, 105, 114, 123, 124, 135, 143, 149, 156, 199, 200, 204, 217, 219, 225, 228, 244, 246, 251, 267
hierarchical model, 9
hierarchy theory, 66, 68
high altitude mountains, 38
holistic approach, 8, 13, 236, 238, 268
Holocene migration, 142
home range, 54, 56, 59, 60, 66, 116, 161, 162, 188, 189, 204, 243
homeorethic capacity, 137
homing, 17, 18
homogeneity, 4, 89, 99
horizontal patchiness, 41
horses, 218, 226, 229
hostile environment, 17, 63, 90
hostile mosaic, 41
hotspots, 4, 74, 149, 195, 200, 203, 219, 245, 246, 254, 258
human barriers, 148, 149
human capital, 234
human dominated landscapes, 111, 122, 149, 162, 195, 231, 232, 240, 244
human landscape, 3, 12, 13, 20, 39, 46, 66, 90, 111, 118, 125, 232, 239, 240
human scale, 29, 31, 73, 137, 209, 232
human stewardship, 51, 64, 158, 222
humidity, 36, 41
hurricanes, 100, 101, 104
hydrological regime, 85
hyper-space, 22
hyper-volume, 12

I

image analysis procedures, 10
immobilization, 15, 86
indeterminacy, 72
Indonesia, 74, 245
information theory, 161
infrastructure maps, 111
insect outbreaks, 100, 220
insects, 19, 25, 41, 55, 94, 213, 214, 215, 224, 260
insular theory, 9

integrity, 109, 111, 114, 115, 118, 119, 120, 121, 159, 161, 162, 220, 236, 237, 238, 239, 260
interference field, 21
interior species, 258
inter-site relationship, 152
invertebrates, 19, 85, 198
Isle Royal National Park, 209
Italy, 6, 51, 65, 69, 111, 112, 126, 136, 148, 209, 218, 226, 229, 230, 232, 246, 247, 248, 259, 262
IUCN, 234, 236, 262

J

Jatrun Sacha Biological Station (Ecuador), 257

K

Kansas, 198, 199
keystone species, 39, 109, 129, 130, 131, 152, 198
Konza Prairie Natural Area, 131, 198, 199
krill, 38, 96, 97

L

lacunarity, 10
lagoons, 44, 125
lake, 7, 43, 112, 184, 254
lake district, 112
lake shores, 254
Land cover, 11, 161
land cover maps, 111, 116
land fragmentation, 44
land mosaic, 41, 42, 61, 65, 78, 82, 87, 90, 95, 110, 112, 120, 122, 130, 137, 140, 143, 148, 164, 195, 197, 199, 200, 203, 207, 208, 212, 213, 220, 221, 225, 232, 246, 261, 262, 267
land sickness, 153
land tenure, 33
Landmark, 18, 19, 32
landscape amenity, 112
landscape complexity, 116, 190, 234
landscape design, 57, 219
landscape homeostasis, 72
landscape hotspots, 246
landscape management, 157, 234
landscape memory, 85
landscape rarity, 244
landslide, 83, 123
landslide falls, 123
largest patch size, 116
larval, 89
lavic cover, 125
Lazio (Italy), 226
leaching, 85
level of fragmentation, 148, 149
lianas, 257
lichens, 40
life trait, 27, 99, 126, 213, 221
light, 25, 37, 38, 42, 68, 102
linear habitat, 4, 53, 57, 59, 195, 215, 248
litter comminution, 85
livestock trails, 55
living range, 8
Logan Pass (USA), 221
Logarghena (Italy), 51, 136, 247
logged stand, 60, 220
Los Angeles, 105
Lyme disease, 154, 155

M

macrochore, 52
macrophytes, 86
Madagascar, 74, 245
magnetic cues, 19
Magra river, 6, 200, 201
maize, 117
management, 25, 74, 112, 120, 126, 148, 150, 151, 153, 156, 157, 195, 196, 197, 198, 199, 202, 203, 204, 205, 208, 209, 210, 211, 215, 217, 218, 220, 225, 230, 232, 234, 236, 237, 238, 239, 249, 251, 252, 260, 262, 263
manure, 105, 128, 159, 216, 221, 228
Maremma area, 226
marshes, 44, 122, 204, 205, 249
mating, 17
matrix, 3, 24, 27, 30, 40, 45, 46, 52, 56, 57, 58, 60, 61, 63, 82, 84, 89, 90, 95, 98, 112,

matrix (cont)118, 119, 137, 152, 158, 160, 165, 167, 168, 177, 189, 195, 197, 198, 200, 203, 204, 205, 207, 209, 212, 213, 215, 230, 239, 242, 252, 253, 258, 260, 261, 262
matrix biodiversity, 195, 213
meadows, 51
meanders, 251
Mediterranean, 46, 49, 51, 78, 89, 111, 126, 128, 140, 198, 213, 215, 220, 226, 228, 230, 231, 234, 253, 262
Mediterranean Australia, 140
Mediterranean Europe, 128
mega landscape, 9, 156, 245, 267
megachore, 52
megadiversity, 74
mesochore, 52
meta-community, 95
metapopulation, 9, 31, 45, 81, 82, 92, 93, 163, 203, 212
meteorology, 36
metropolitan landscape, 39
Mexico, 111
microbial activity, 13, 86
microbial respiration, 86
microchore, 52
micromammals, 260
micro-organisms, 49, 156, 213
micro-patches, 28, 49
micro-world, 156
migration, 18
Mississippi, 148, 149
mitigation, 57, 60, 112, 251, 268
mobbing behavior, 41
Monterey, 64, 101
Moran effect, 115
morphology, 14, 85, 99, 125, 227, 232, 249
mosaic diversity, 85, 227
mosaic habitat, 32
mosaic matrix, 140
mosaic rarity, 244
mosses, 40
mound nest, 39
mountain catena, 112
mountain moraine gully, 68
mountain passes, 53
mountain prairies, 30, 145
Mountain Rainier National Park, 258
mulches, 257
multiethnic societies, 12
mushroom clone, 83
mutualism, 130, 149

N

National Park of Abruzzi (Italy), 229
natural capital, 234, 248
natural history, 203
naturalness, 159
NDVI, 87
neotropical forests, 104
nest competition, 105
nest parasitism, 105
nest predation, 105
neutral models, 10, 118, 121, 164, 166, 167
New Mexico, 228
New Zealand, 74, 101, 106, 114
niche theory, 12
nomadic predators, 114
North America, 9, 10, 78, 141, 142, 215, 220, 227, 239, 245, 248, 256
Northern Apennines, 40, 62, 68, 69, 126, 127, 136, 138, 222, 229, 230, 247
Northern Italy, 51, 62, 200, 201
null models, 164
number of patches, 116, 165, 172, 176
nutrient cycles, 12, 57, 162

O

oases, 196, 252
offspring, 76, 78
Ohio river basin, 204
old-growth forest, 41, 219, 220, 259
old-growth stands, 219
olfactory, 17
olive orchard, 51
olive trees, 51
opportunistic species, 154, 258
Oregon, 56, 218, 258
Oregon Cascades, 258
organic matter, 49, 86, 128, 257
organism-centered landscapes, 218

organismic-centered perception, 12
ornamental traits, 146
Ospedalaccio pass (Italy), 68, 69
outbreak, 115, 154
output patches, 15

P

Pacific (USA), 153
Pacific North America, 258
Padanian valley (Italy), 248
Palaeolithic culture, 157
palatable plants, 31
Panama, 103
pandemic diseases, 155, 164
parasites, 66
parish, 8, 66, 67, 68, 112, 148, 237
parks, 196, 204, 209, 234, 239, 258, 260, 268
particle, 21, 22, 86
passive filter, 61
passive landscape, 42
patch biodiversity, 195, 213
patch irregularity, 173
patch isolation, 19, 169
patch rarity, 244
patch resolution, 110, 169, 170, 171
patch type class, 171, 175
patchiness, 4, 37, 48, 87, 93, 96, 97, 109, 152, 199, 258
patchy mosaic, 30, 95, 160
path, 19, 41, 52, 55, 95, 135, 136, 158, 186
pathogen infestation, 101
pattern ecotones, 61
patterned landscapes, 60
pelagic organisms, 96
penguins, 38
people, 5, 28, 138, 140, 157, 190, 242, 251, 254, 255, 256, 263, 268
percolation theory, 9, 82, 162, 165
periphyton, 86
permanent habitat patches, 198
permeability of edges, 64
pest outbreaks, 153, 267
physiotope, 50
pixel, 52, 99, 116, 121, 165, 173, 189
planet mosaic, 137
plant dispersal, 44
plant photosynthesis, 13
plant productivity, 15
plant recovery, 123
plants, 7, 19, 29, 31, 36, 38, 40, 49, 53, 58, 66, 68, 74, 77, 78, 91, 92, 93, 102, 103, 127, 128, 130, 136, 139, 142, 143, 157, 164, 213, 218, 225, 226, 230, 244, 246, 251, 255, 258, 260
Pleistocene diversity, 157
plough, 157
pluvial forests, 39
Po, 36, 148
policy, 13, 74, 119, 153, 198, 203, 209, 219, 238, 239, 250, 251, 252, 253, 261
pollens, 36
pollution, 36, 119, 125, 190
ponds, 49, 55, 204, 217, 251
population, 3, 22, 25, 29, 31, 35, 39, 42, 44, 48, 56, 60, 75, 76, 77, 78, 79, 80, 81, 85, 87, 89, 93, 95, 96, 98, 99, 105, 111, 114, 115, 119, 122, 125, 126, 128, 139, 141, 142, 153, 156, 161, 163, 164, 195, 197, 203, 209, 210, 211, 212, 215, 232, 236, 238, 251, 255, 256, 257, 260, 268
post-fire succession, 220
potential energy, 82, 127
poverty, 71, 155, 268
predation rates, 59
predators, 17, 59, 66, 78, 79, 96, 106, 116, 122, 130, 215, 224
predatory risk, 47, 66
prehistoric overkill, 157
Preston Park (USA), 221
prey over-exploitation, 53
prey-predator relationship, 96, 98
primary producers, 86
pristine old-growth forest, 219
pristine region, 52
problem-solving science, 268
Process approach, 7
process ecotones, 61
productivity, 25, 64, 118, 122, 123, 148, 153, 157, 216, 231, 234, 250, 251
protected areas, 114, 196, 204, 227, 236, 239, 244, 246, 253, 258, 260
province, 62, 112, 148, 201

prune, 157
pruned, 51
pseudo-sink, 77

Q

quanta, 21
Quebec, 62
Quijos Quichua (Ecuador), 257

R

random walking, 55, 93
real landscape, 111, 118, 167
realized niche, 79, 80
recreation, 201, 232, 255
redundancy, 71, 126, 168
refugia, 195, 200
regional (gamma) diversity, 85
remnant patches, 51, 114, 156, 163, 213
remnants, 39, 53, 58, 73, 100, 113, 153, 162, 198, 200, 218, 267
remote sensing, 5, 8, 10, 31, 33, 112, 133, 137, 154
remote upland areas, 114
representativeness, 109, 113, 114
reproductive surplus, 76
resilience, 64, 71, 109, 111, 114, 125, 126, 127, 148, 156, 162, 217, 225, 228, 231, 254, 256
resistance, 45, 109, 114, 125, 149
restoration, 53, 57, 112, 195, 197, 198, 248, 249, 250, 268
retention, 42, 73, 128, 160, 161, 205, 216, 219, 220, 248
retention rate, 205
retina, 18
return journey, 19
riparian vegetation, 15, 161, 200, 205, 215, 217, 248, 249, 252
river, 4, 5, 7, 8, 14, 15, 24, 44, 77, 78, 84, 85, 86, 87, 111, 117, 122, 125, 149, 150, 156, 160, 161, 164, 195, 197, 200, 215, 216, 223, 224, 248, 249, 250, 251, 254
river buffers, 117
river confluence, 254
river meanders, 44
river transportation, 7, 15
road, 19, 53, 54, 111, 120, 232
road maps, 111
rocky fissures, 20
rocky intertidal zone, 130
Rocky Mountains, 54, 87
roosting, 17, 41, 55, 64, 100, 260
root system, 25, 128, 230
roughness, 14, 185

S

Sahara, 225
saline depression, 38
salinity, 37, 155
salt water, 155
San Francisco, 111
sand deposit, 5, 6, 44
sand islands, 251
satellite imagines, 137
scale-invariant, 183
scattered trees, 89, 159
scenic beauty, 114, 254
sea, 7, 24, 26, 38, 41, 60, 104, 154, 155, 156, 183, 218
sea currents, 37, 155
sea-breeze fronts, 37
sea-scapes, 37, 43, 64
seasonal, 27, 49, 72, 94, 95, 136, 200, 202, 225, 251, 262
seed banks, 213
self-defensive, 5
self-similarity, 26, 85, 183, 187
semi-arid regions, 157
sensitive species, 157
sere, 220, 258, 259
Serengeti Park, 55, 228
sessile organisms, 27
Shannon, 72, 168, 179, 180
shape, 14, 23, 38, 39, 43, 44, 48, 53, 61, 64, 91, 118, 121, 135, 136, 137, 150, 158, 159, 164, 165, 168, 169, 173, 183, 184, 210, 240, 260
sheep, 126, 145, 229
shellfish, 155
shelterbelts, 53, 63
shifting mosaic, 48, 241, 244, 256

shoreline, 184
short life span, 31
shrub, 51, 89, 136, 220, 222, 230
sink, 59, 75, 76, 77, 78, 80, 87
sink habitat, 76, 78, 80
sky turbulence, 37
sliding window, 116, 188
small reserves, 260
Snake river (Idaho), 205
snow, 31, 32, 87, 90, 102, 103, 126, 145
snowfall, 58
social poverty, 163
socio-economic context, 61, 148
soil genesis, 44
soil implanting, 26
solar radiation, 13, 41
soliflux, 102
source habitat, 59, 76, 78, 80, 203
source-sink, 9, 27, 45, 75, 76, 77, 78, 79, 81
South Africa, 234, 235, 246
South America, 232, 256
South Fork (Idaho), 205
South Island (NZ), 114
Spain, 111, 209
Spanish dehesa, 157
spatial autocorrelation, 44, 168
spatial map, 17
spatial statistics, 112
spatially explicit models, 31, 123, 210
spatio-temporal scale, 7, 29, 99, 119, 137, 197, 203
species turnover, 122, 126, 244
spores, 36, 39, 92
spray, 78
stability, 65, 71, 72, 74, 82, 109, 114, 121, 122, 125, 127, 145, 154, 157, 203, 209, 215, 228, 244
stakeholders, 237, 261
stand density, 208
starvation, 38, 209
stone wall, 59
storms, 82
stream, 7, 14, 84, 87, 159, 160, 201, 224, 237, 240, 257
stressors, 150
strip corridors, 53
structural patch, 47
sub-atomic particle, 22
sublitter region, 86
sub-montane prairies, 247
suburban areas, 57
succession, 14, 29, 63, 82, 83, 84, 125, 139, 140, 158, 162, 163, 198, 200, 214, 222, 251
sun light, 25
super-indicator, 148
super-organism, 267
symbiosis, 29

T

Taverone stream, 200, 201
taxa, 29, 75
temperature, 32, 36, 37, 42, 48, 117, 126, 155, 161, 248, 249
temporal scale, 5, 7, 11, 13, 19, 24, 25, 26, 30, 31, 37, 38, 42, 114, 115, 122, 152, 164, 208, 233
temporary landscape, 16, 17
termite, 39
territory, 17, 41, 48, 54, 55, 66, 140, 159, 204, 232, 246
theoretical landscape, 118
thermal currents, 40
thermal spring, 30
thickness, 128, 230, 231
timber volume, 208
topological dimension, 22, 23, 24, 26, 52, 183
topsoils, 87
topstory layer, 231
toxic spreading, 82
traditional agro-ecosystem, 234
trait, 7, 52, 90, 131, 132, 146, 205
trampling, 63, 83, 105, 128, 200, 217, 220, 225, 228, 262
trans-disciplinary approach, 8, 148
trans-Saharan migrants, 230
tree fall, 100, 103
tree pruning, 243
tree recruitment, 208
tropical, 30, 38, 39, 41, 103, 149, 157, 207, 208, 213, 225, 232, 255
tropical forests, 30, 41, 103, 208, 225, 255

tropical sea, 39
Tucson, 15
tundra, 78, 123
turbulence, 5
Tuscany, 157, 159

U-V

umbrella species, 123
ungulates, 31, 130, 225, 228, 231
unhealthy, 152, 153, 154, 157
unsuitable habitats, 27
upland prairies, 51
urban landscape, 16, 44
urine, 199
vacuoles, 49
vagile animals, 42, 90, 92
vegetation defoliation, 199
vegetation diversity, 112
vegetation dominance, 63
vegetation dynamism, 139
vertebrates, 18
vertical landscape, 39, 40
vessel, 38
vigor, 148, 163, 231
virus, 7, 128, 156
visual, 17, 117
volcanoes, 104
voles, 115

W-Y-Z

wallowing, 131, 199, 200, 227
warm deserts, 39
water, 5, 7, 8, 14, 15, 19, 20, 24, 36, 37, 42, 44, 48, 65, 68, 74, 78, 82, 84, 87, 88, 91, 92, 101, 104, 111, 117, 122, 123, 131, 140, 143, 145, 148, 152, 155, 157, 159, 160, 161, 184, 196, 197, 198, 200, 217, 227, 234, 237, 239, 240, 243, 248, 249, 252, 256, 268
water deficit, 20, 252
water drainage, 8, 111
water flow, 14, 157, 237
water vapor, 36
waterflood, 58
watershed, 111, 112, 117, 148, 160, 161, 234, 236, 237, 251
weed, 59, 72, 77, 78, 220
weed invasion, 59, 72, 220
West Divide, 87
West Paleartic, 40
Westerholt (NL), 226
Western Cape (South Africa), 235
Western plains, 54
white bear, 38
willow shrublands, 249
wind exposure, 20
wind storms, 58, 104
windbreaks, 257
wintering birds, 51, 218
Wisconsin, 218, 220
wood, 71, 95, 101, 222, 231
wood charcoal, 231
woodland, 32, 65, 85, 94, 95, 99, 222, 247
WWF, 236
Yellowstone National Park, 31, 90, 209, 210, 228, 260
Zeri commune, 138